Lecture Notes in Statistics 124

Springer
New York
Berlin
Heidelberg
Barcelona
Budapest
Hong Kong
London
Milan
Paris
Santa Clara
Singapore
Tokyo

Christine H. Müller

Robust Planning and Analysis of Experiments

Christine H. Müller
Freie Universität Berlin
Fachbereich Mathematik und Informatik, WE1
Arnimallee 2-6
14195 Berlin
Germany

CIP data available.
Printed on acid-free paper.

Camera-ready copy provided by the author.
Printed and bound by Braun-Brumfield, Ann Arbor, MI.
Printed in the United States of America.

9 8 7 6 5 4 3 2 1

ISBN 0-387-98223-X Springer-Verlag New York Berlin Heidelberg SPIN 10573403

To my mother

Preface

Up to now the two different areas of "Optimal Design of Experiments" and "Robust Statistics", in particular of "Outlier Robust Statistics", are very separate areas of Statistics. There exist several books on optimum experimental design like those of Fedorov (1972), Silvey (1980), Pázman (1986), Shah and Sinha (1989), Atkinson and Donev (1992), Pukelsheim (1993) and Schwabe (1996a). There exist also several books on robust statistics like those of Huber (1981), Hampel et al. (1986), Tiku et al. (1986), Rousseeuw and Leroy (1987), Kariya and Sinha (1988), Staudte and Sheather (1990), Büning (1991), Rieder (1994) and Jurečková and Sen (1996). But there is almost no overlapping between the books on optimum experimental design and on robust statistics. Now the presented book will give a first link between these two areas. It will show that a robust inference will profit from an optimal design and that an optimal design is more reasonable if it allows also a robust analysis of the data.

The first part of the presented book gives an overview on the foundations of optimum experimental design. In this classical approach a design is optimal if the efficiency of the least squares estimator (or the classical F-test) is maximized within all possible designs. But the least squares estimators and the F-tests are not robust against outliers. In the presence of outliers they can be biased very much.

Outlier robust estimators and tests are derived in the second part of the book which provides the foundations of outlier robust statistics especially for planned experiments. This already differs very much from the approaches in the books and the majority of the papers on robust statistics because they usually do not regard planned experiments. The majority of the approaches in robust statistics assumes that the independent variables in regression experiments are random so that outliers can appear also in these independent variables. For planned experiments and for models with qualitative factors these assumptions make no sense so that the existing main concepts of outlier robust inference have to be specified for planned experiments. This is done in the second part of the book. As main qualitative concepts of outlier robustness the continuity and Fréchet differentiability of statistical functionals are regarded. From these qualitative concepts the quantitative concepts of outlier robustness are derived. These quantitative concepts are the breakdown point and the bias in shrinking contamination neighbourhoods, which is closely related to Hampel's concept based on influence functions. These concepts are mainly explained for general linear models but also their meaning for nonlinear problems is discussed.

In the third part the efficiency and robustness of estimators and tests are linked and the influence of the designs on robustness and efficiency is studied. For the designs this leads to new nontrivial optimality problems. These new optimality problems are solved for several situations in the third part which consists of Section 7, 8 and 9.

In Section 7 and 8 it is shown that the classical A- and D-optimal designs provide highest robustness and highest efficiency under robustness constraints if the robustness measure is based on the bias in shrinking contamination neighbourhoods. Thereby it turns out that the most robust estimators and tests and the most efficient robust estimators and tests have a very simple form at the classical optimal designs. Moreover, most robust tests and most efficient robust test can be only characterized at the D-optimal designs. Similar results also hold if a nonlinear aspect should be estimated or if the model is nonlinear. All these results show that the robust statistical analysis profits very much from an optimal choice of the design.

While in the robustness concept based on shrinking contamination the classical optimal designs are also optimal for the robust inference, the opposite is true for the robustness measure based on the breakdown point. This is demonstrated in Section 9. There it is shown that the designs which are optimal with respect to a high breakdown point are in general very different from the classical optimal designs. Because of this difference it is also discussed how to find a design which combines high breakdown point and high efficiency of the estimators.

The problem of combining high breakdown point and high efficiency provides for the designs plenty of new optimality problems which are not solved up to now. Besides these open problems there are also many other open problems for further research which are discussed in the outlook.

Finally, I would like to express my thanks to all who supported me in finishing the present work. Above all I am indebted to Prof. Dr. V. Kurotschka for his fruitful discussions and criticism. Thanks also to all my colleagues, in particular to Dr. W. Wierich, who teached me a lot of statistics, and to Dr. R. Schwabe, who was always very co-operative. I also thank Prof. Dr. H. Rieder for providing some preprints which initiated the present work. Moreover, I am particularly thankful to the corresponding editor J. Kimmel of the *Lecture Notes in Statistics* and to the referees for very valuable comments.

At last I would like to thank my family and in particular my husband and my mother for their support. Especially, I am very grateful to my mother because without her engaged care for my children this work would not have been possible.

Christine H. Müller, *Berlin, March 1997*

Contents

Part I

Efficient Inference for Planned Experiments

1

Planned Experiments

In this chapter the main parts of a planned experiment are explained. In particular, it is explained what deterministic and random designs are (Section 1.1), what linear and nonlinear models are (Section 1.2) and at which designs a linear or nonlinear aspect of a linear or nonlinear model is identifiable (Section 1.3).

1.1 Deterministic and Random Designs

In many experimental situations the outcome of the experiment depends on some factors of influence, say k factors. Usually the particular value $t = (t^1, ..., t^k)$ of these k factors can vary within some *experimental domain* $\mathcal{T}$ and the experiment is realized at different *experimental conditions* $t_{1N}, ..., t_{NN} \in \mathcal{T}$. In planned experiments the experimental conditions $t_{1N}, ..., t_{NN}$ are chosen by the experimenter from the experimental domain $\mathcal{T}$, prior to running the experiment. The experimental conditions can be chosen by a *deterministic design* $d_N = (t_{1N}, ..., t_{NN})' \in \mathcal{T}^N$ or by a *random design measure* δ. If the experimental conditions are chosen by a random design measure, then the design $d_N = (t_{1N}, ..., t_{NN})'$ is a realization of a random design $D_N = (T_{1N}, ..., T_{NN})'$ where $T_{1N}, ..., T_{NN}$ are independent identically distributed random variables each with distribution δ and the probability measure δ is chosen by the experimenter. While deterministic designs also can be used for small sample sizes N random designs are only useful for large sample sizes.

Many robust methods can only be judged by their asymptotic behaviour so that it is important to know the asymptotic behaviour of the designs. For random designs D_N we have with the strong law of large numbers

$$\frac{1}{N} \sum\nolimits_{n=1}^{N} 1_{\{t\}}(T_{nN}) \overset{N \to \infty}{\longrightarrow} \delta(\{t\})$$

with probability 1 for all $t \in \text{supp}(\delta)$, where $1_{\mathcal{A}}$ is the indicator function on the set $\mathcal{A}$ and $\text{supp}(\delta)$ denotes the support of the random design measure δ, i.e. the smallest closed set $\mathcal{D} \subset \mathcal{T}$ with $\delta(\mathcal{D}) = 1$. A similar property also should be satisfied by deterministic design sequences $(d_N)_{N\in\mathbb{N}}$. I.e. we should have for some design measure δ

$$\frac{1}{N}\sum_{n=1}^{N} 1_{\{t\}}(t_{nN}) \overset{N\to\infty}{\longrightarrow} \delta(\{t\}) \tag{1.1}$$

for all $t \in \text{supp}(\delta)$. In this context the design measure δ is called the *asymptotic design measure* of the sequence $(d_N)_{N\in\mathbb{N}}$. Instead of the deterministic designs $d_N = (t_{1N}, ..., t_{NN})'$ we can also regard their corresponding designs measures $\delta_N := \frac{1}{N}\sum_{n=1}^{N} e_{t_{nN}}$, where e_t denotes the one-point measure (Dirac-measure) on t. If the supports of δ and all δ_N are finite, then condition (1.1) is equivalent with

$$\delta_N \overset{N\to\infty}{\longrightarrow} \delta \qquad \text{weakly.} \tag{1.2}$$

Therefore we define generally the convergence of deterministic designs as follows.

Definition 1.1
A sequence of deterministic designs $(d_N = (t_{1N}, ..., t_{NN})')_{N\in\mathbb{N}}$ is converging to an asymptotic design measure δ if the corresponding design measures $\delta_N := \frac{1}{N}\sum_{n=1}^{N} e_{t_{nN}}$ are weakly converging to δ for $N \to \infty$.

For planning experiments mainly *discrete designs*, i.e. design measures δ with a finite support, make sense. But for deriving optimal designs (see Section 2.2) it is often very helpful to regard also more general designs, namely all probability measures on $\mathcal{T}$. Therefore any probability measure δ on $\mathcal{T}$ is called *generalized design* or briefly *design*.

Definition 1.2
a) Any probability measure δ on $\mathcal{T}$ is called a design (design measure).

b) The set of all designs on $\mathcal{T}$ is denoted by Δ_0.

1.2 Linear and Nonlinear Models

Often the *observations* $Y_{1N}, ..., Y_{NN}$ of the experiment at the different experimental conditions $t_{1N}, ..., t_{NN}$ depend on the experimental conditions in a functional relationship of the form

$$Y_{nN} = \mu(t_{nN}, \beta) + Z_{nN}, \qquad n = 1, ..., N,$$

where $\mu : \mathcal{T} \times \mathcal{B} \to \mathbb{R}$ is a known response function, $\beta \in \mathcal{B} \subset \mathbb{R}^r$ is an unknown parameter and $Z_{1N}, ..., Z_{NN}$ are independent random errors.

In *linear models* the response function μ is a linear function of β, i.e. $\mu(t, \beta) = a(t)'\beta$ for all $t \in \mathcal{T}$ and $\beta \in \mathcal{B}$.

Definition 1.3 (Linear model)
The response function $\mu : \mathcal{T} \times \mathcal{B} \to I\!R$ *is given by a linear model if* μ *has a linear parametrization, i.e. if a function* $a : \mathcal{T} \to I\!R^r$ *exists so that* $\mu(t, \beta) = a(t)'\beta$ *for all* $t \in \mathcal{T}$ *and* $\beta \in \mathcal{B}$.

Hence in a linear model the observations are given by

$$Y_{nN} = a(t_{nN})'\beta + Z_{nN}, \qquad n = 1, ..., N,$$

where the "regression" function $a : \mathcal{T} \to I\!R^r$ is known. These relations also can be collected in matrix notation according to

$$Y_N = A_{d_N}\beta + Z_N, \tag{1.3}$$

where $Y_N = (Y_{1N}, ..., Y_{NN})'$, $Z_N = (Z_{1N}, ..., Z_{NN})'$ and $A_{d_N} = (a(t_{1N}), ..., a(t_{NN}))'$. The realizations of Y_N and Z_N will be denoted by $y_N = (y_{1N}, ..., y_{NN})'$ and $z_N = (z_{1N}, ..., z_{NN})'$, respectively.

If no function $a : \mathcal{T} \to I\!R^r$ with $\mu(t, \beta) = a(t)'\beta$ for all $t \in \mathcal{T}$ exists, then the model is called a *nonlinear model*.

Definition 1.4 (Nonlinear model)
The response function $\mu : \mathcal{T} \times \mathcal{B} \to I\!R$ *is given by a nonlinear model if* μ *has no linear parametrization, i.e. for every* $a : \mathcal{T} \to I\!R^r$ *there exists* $t \in \mathcal{T}$ *and* $\beta \in \mathcal{B}$ *with* $\mu(t, \beta) \neq a(t)'\beta$.

The aim of statistical inference in linear and nonlinear models is to estimate β or an *aspect* $\varphi(\beta)$ of β or to test a hypothesis about β or $\varphi(\beta)$ by using the observations $Y_{1N}, ..., Y_{NN}$. The interesting aspect $\varphi(\beta)$ can be a *linear aspect*, where $\varphi : \mathcal{B} \to I\!R^s$ is a linear function, i.e. $\varphi(\beta) = L\beta$ with $L \in I\!R^{s \times r}$, or it can be a *nonlinear aspect*, where $\varphi : \mathcal{B} \to I\!R^s$ is not a linear function. Because β is a special linear aspect $\varphi(\beta)$, namely $\varphi(\beta) = L\beta$ with $L = E_r$, where E_r denotes the $r \times r$ identity matrix, in the following general presentation we regard only aspects $\varphi(\beta)$. Moreover, for linear aspects in a linear model we assume $\mathcal{B} = I\!R^r$.

1.3 Identifiability of Aspects

The statistical inference about $\varphi(\beta)$ only will be successful if it is possible to derive $\varphi(\beta)$ uniquely from observations $Y_{1N}, ..., Y_{NN}$ without errors. This is only the case if for all $\beta_1, \beta_2 \in \mathcal{B}$ the equality $\mu(t_{nN}, \beta_1) = \mu(t_{nN}, \beta_2)$ for all $n = 1, ..., N$ implies $\varphi(\beta_1) = \varphi(\beta_2)$. This property is called the *identifiability* of $\varphi(\beta)$. In estimation problems it is also called *estimability* and in testing problems it is called *testability*. The identifiability depends on the model, given by μ, on the linear aspect φ and on the design d_N. The existence of a design d_N at which φ is identifiable depends on the question whether φ is identifiable at the model which is given by the response function μ and the experimental region $\mathcal{T}$. So at first we define the identifiability of φ at a given set $\mathcal{D} \subset \mathcal{T}$ which provides the most general definition of identifiability. Because usually the response function μ is given and fixed we suppress in the definition the dependence on the response function. Only in situations in which different response functions are regarded the response functions are mentioned.

Definition 1.5
An aspect $\varphi : \mathcal{B} \to \mathbb{R}^s$ is identifiable at $\mathcal{D}$ if for all $\beta_1,\ \beta_2 \in \mathcal{B}$ we have the implication

$$\mu(t, \beta_1) = \mu(t, \beta_2) \text{ for all } t \in \mathcal{D} \Longrightarrow \varphi(\beta_1) = \varphi(\beta_2).$$

For deriving breakdown points (see Section 4.1 and Section 4.2) the sets $\mathcal{D}$ at which φ is not identifiable are important. These sets will be called *nonidentifying sets.*

Definition 1.6 (Nonidentifying set)
A set $\mathcal{D} \subset \mathcal{T}$ is nonidentifying for the aspect φ if φ is not identifiable at $\mathcal{D}$.

For deriving breakdown points the following obvious lemma will be important.

Lemma 1.1 *If in a linear model the set $\mathcal{D}$ is nonidentifying for a linear aspect φ with $\varphi(\beta) = L\beta$, then there exists $\beta_0 \in \mathbb{R}^r$ such that $a(t)'\beta_0 = 0$ for all $t \in \mathcal{D}$ and $L\,\beta_0 \neq 0$.*

Via the Definition 1.5 also the identifiability of the parametrization of the model, of φ at the model, of φ at a deterministic design and of φ at a design measure can be defined.

Definition 1.7
a) The parametrization of the model is identifiable if β is identifiable at $\mathcal{T}$.

b) An aspect $\varphi : \mathcal{B} \to \mathbb{R}^s$ is identifiable in the model if φ is identifiable at $\mathcal{T}$.

c) An aspect $\varphi : \mathcal{B} \to \mathbb{R}^s$ is identifiable at the deterministic design $d_N = (t_{1N}, ..., t_{NN})'$ if φ is identifiable at $\{t_{1N}, ..., t_{NN}\}$.

d) An aspect $\varphi : \mathcal{B} \to \mathbb{R}^s$ is identifiable at the design measure δ if φ is identifiable at $supp(\delta)$.

While a deterministic design d_N should be chosen so that φ is identifiable at d_N, at a realized random design d_N the identifiability may be violated. But for large sample sizes N, at least for $N \to \infty$, the identifiability should be ensured, so that φ should be identifiable at the random design measure δ.

Definition 1.8
The set of all design measures on $\mathcal{T}$ at which the aspect φ is identifiable is denoted by $\Delta(\varphi)$.

If we have a linear aspect φ in a linear model, then the following characterization of the identifiability of φ - which is also called *linear identifiability* in this case - is very useful (see for example Rao (1973), p. 223, or Christensen (1987), p. 15).

Lemma 1.2 *If φ is a linear aspect of a linear model with $\varphi(\beta) = L\,\beta$, then the following assertions are equivalent:*

a) φ is identifiable at d_N.

b) There exists a matrix $K \in I\!R^{s \times n}$ such that $L = K\,A_{d_N}$.

c) There exists a matrix $K \in I\!R^{s \times r}$ such that $L = K\,A'_{d_N}\,A_{d_N}$.

For the identifiability at a design measure δ a similar characterization holds. If δ is a random design measure (see Section 1.2), then we have that the expectation of $A'_{D_N}\,A_{D_N}$ is

$$\mathcal{I}(\delta) := \int a(t)\,a(t)'\delta(dt)$$

and that $\frac{1}{N}A'_{D_N}\,A_{D_N}$ is converging to $\mathcal{I}(\delta)$ for $N \to \infty$ with probability 1. But δ also can be the asymptotic design measure of a converging sequence of deterministic designs $(d_N)_{N\in I\!N}$ in the sense of (1.2). If the regression function a is continuous and bounded on the supports of δ and all δ_N, then the weak convergence also provides that

$$\frac{1}{N}A'_{d_N}\,A_{d_N} \overset{N\to\infty}{\longrightarrow} \mathcal{I}(\delta). \tag{1.4}$$

In particular this convergence holds for all regression functions a if the supports of δ and all δ_N are contained in a finite subset of $\mathcal{T}$. Hence there is a close relation between $\frac{1}{N}A'_{d_N}\,A_{d_N}$ and $\mathcal{I}(\delta)$. The matrix $\mathcal{I}(\delta)$ is called *information matrix* and plays also in the covariance matrix of the optimal estimators an important role (see Section 2.2 and Chapter 7).

Lemma 1.3 *If φ is a linear aspect of a linear model with $\varphi(\beta) = L\,\beta$, then the following assertions are equivalent:*

a) φ is identifiable at δ.

b) There exists a matrix $K \in I\!R^{s \times r}$ such that $L = K\,\mathcal{I}(\delta)$.

Proof. For δ with finite support the assertion follows from Lemma 1.2. For δ with infinite support the assertion follows from the theorem of Carathéodory (see Silvey (1980), p. 16, 72) which provides that for every δ a design $\tilde{\delta}$ with finite support exists with $\mathcal{I}(\delta) = \mathcal{I}(\tilde{\delta})$ and $\text{supp}(\tilde{\delta}) \subset \text{supp}(\delta)$.□

If φ is a linear or a nonlinear aspect in a linear model, then Lemma 1.2 can be extended as follows. For that define $A_{\mathcal{D}} = (a(\tau_1), ..., a(\tau_I))'$ for a finite set $\mathcal{D} = \{\tau_1, ..., \tau_I\} \subset \mathcal{T}$.

Lemma 1.4 *The aspect φ is identifiable at a finite set $\mathcal{D}$ if and only if there exists $\varphi^* : \mathbb{R}^I \to \mathbb{R}^s$ with $\varphi(\beta) = \varphi^*(A_{\mathcal{D}}\beta)$ for all $\beta \in \mathcal{B}$.*

If φ is a nonlinear aspect in a linear model, then its identifiability can be attributed to linear identifiability. For that we need the partial derivatives with respect to β of φ which depend on β because of the nonlinearity of φ. Then we have the following matrix of derivatives

$$\dot{\varphi}_\beta := \frac{\partial}{\partial \tilde{\beta}} \varphi(\tilde{\beta}) /_{\tilde{\beta}=\beta} \in \mathbb{R}^{s \times r}.$$

With these derivatives the identifiability of φ at a set $\mathcal{D} \subset \mathcal{T}$ can be characterized.

Lemma 1.5 *Let $\mathcal{B}$ be a convex and open subset of $\mathbb{R}^r$ and φ continuously differentiable on $\mathcal{B}$. Then the aspect φ is identifiable at $\mathcal{D}$ if and only if for all $\beta \in \mathcal{B}$ the linear aspect φ_β with $\varphi_\beta(\tilde{\beta}) = \dot{\varphi}_\beta \tilde{\beta}$ is identifiable at $\mathcal{D}$.*

Proof. Without loss of generality, we can assume that φ is a one-dimensional function, i.e. $s = 1$. Let $\beta_1, \beta_2 \in \mathcal{B}$ be any parameters so that $a(t)'\beta_1 = a(t)'\beta_2$ for all $t \in \mathcal{D}$. The mean value theorem and the convexity of $\mathcal{B}$ provide at once that the identifiability of all linear aspects φ_β implies the identifiability of φ. To show the converse implication assume that $\varphi(\beta_1) = \varphi(\beta_2)$ and $\dot{\varphi}_\beta\beta_1 \neq \dot{\varphi}_\beta\beta_2$ for some $\beta \in \mathcal{B}$. Then for all $\lambda > 0$ we have $\dot{\varphi}_\beta\lambda(\beta_1 - \beta_2) \neq 0$. Because $\mathcal{B}$ is open there exists λ_0 with $\beta + \lambda(\beta_1 - \beta_2) \in \mathcal{B}$ for $\lambda \in [0, \lambda_0]$. The identifiability of φ and $a(t)'(\beta + \lambda(\beta_1 - \beta_2)) = a(t)'\beta$ implies $\varphi(\beta + \lambda(\beta_1 - \beta_2)) = \varphi(\beta)$ for all $\lambda \in [0, \lambda_0]$. Hence the derivative of φ at β in direction of $\beta_1 - \beta_2$ must be equal to 0 which is a contradiction. □

In a nonlinear model there is no simple characterization of the identifiability of φ. We only can give sufficient conditions for local identifiability of the whole parameter β.

Definition 1.9
The parameter β is locally identifiable at $\mathcal{D}$ if for all $\beta \in \mathcal{B}$ a neighbourhood $U(\beta)$ exists so that for all $\beta_1, \beta_2 \in U(\beta)$ the equality $\mu(t, \beta_1) = \mu(t, \beta_2)$ for all $t \in \mathcal{D}$ implies $\beta_1 = \beta_2$.

A sufficient condition for the local identifiability is based on the derivatives of the response function μ with respect to β, i.e. on

$$\dot{\mu}(t,\beta) := \left(\frac{\partial}{\partial\tilde{\beta}}\mu(t,\tilde{\beta})/_{\tilde{\beta}=\beta}\right)' \in I\!R^r.$$

Lemma 1.6 *If for all $\tilde{\beta} \in \mathcal{B}$ in the linear model given by $\tilde{\mu}(t,\beta) = \dot{\mu}(t,\tilde{\beta})'\beta$ the aspect β is identifiable at $\mathcal{D}$, then β is locally identifiable at $\mathcal{D}$.*

Proof. If β is identifiable at δ in the model given by $\tilde{\mu}(t,\beta) = \tilde{a}(t)'\beta$ where $\tilde{a}(t) = \dot{\mu}(t,\tilde{\beta})$, then a set $\mathcal{D}_0 = \{\tau_1,...,\tau_r\} \subset \mathcal{D}$ with r elements exists so that in the model given by $\tilde{\mu}(t,\beta) = \tilde{a}(t)'\beta$ the parameter β also is identifiable at $\mathcal{D}_0$ or $d = (\tau_1,...,\tau_r)'$, respectively. Then Lemma 1.2 provides that the matrix

$$\tilde{A}_d = (\tilde{a}(\tau_1),...,\tilde{a}(\tau_r))' = (\dot{\mu}(\tau_1,\tilde{\beta}),...,\dot{\mu}(\tau_r,\tilde{\beta}))' \in I\!R^{r\times r}$$

is regular. With the inverse function theorem this implies that for a neighbourhood $U(\tilde{\beta}) \subset \mathcal{B} \subset I\!R^r$ of $\tilde{\beta}$ the function

$$\mu_{\mathcal{D}} : U(\tilde{\beta}) \ni \beta \to \mu_{\mathcal{D}}(\beta) := (\mu(\tau_1,\beta),...,\mu(\tau_r,\beta))' \in I\!R^r$$

is one-to-one which means that we have $(\mu(\tau_1,\beta_1),...,\mu(\tau_r,\beta_1))' \neq (\mu(\tau_1,\beta_2),...,\mu(\tau_r,\beta_2))'$ for all $\beta_1,\beta_2 \in U(\tilde{\beta})$ with $\beta_1 \neq \beta_2$. Hence there exists no $\beta_1,\beta_2 \in U(\tilde{\beta})$ with $\beta_1 \neq \beta_2$ and $\mu(t,\beta_1) = \mu(t,\beta_2)$ for all $t \in \mathcal{D}$. □

Because of Lemma 1.3 the identifiability of β at δ in the model given by $\tilde{\mu}(t,\beta) = \dot{\mu}(t,\tilde{\beta})'\beta$ is equivalent to the regularity of

$$\int \dot{\mu}(t,\tilde{\beta})\,\dot{\mu}(t,\tilde{\beta})'\delta(dt)$$

so that we have the following corollary.

Corollary 1.1 *If for all $\beta \in \mathcal{B}$ the matrix $\int \dot{\mu}(t,\beta)\,\dot{\mu}(t,\beta)'\delta(dt)$ is regular, then β is locally identifiable at δ.*

The following example shows that the regularity of $\int \dot{\mu}(t,\beta)\,\dot{\mu}(t,\beta)'\delta(dt)$ for all $\beta \in \mathcal{B}$ is not necessary for local identifiability.

Example 1.1
Assume $\mu(t,\beta) = \beta^{3t}$ for $\beta \in I\!R$, $t \in (0,\infty)$ and $\text{supp}(\delta) = \{1\}$. Then β is identifiable at δ and therefore also locally identifiable because $\beta_1^3 = \mu(1,\beta_1) = \mu(1,\beta_2) = \beta_2^3$ implies $\beta_1 = \beta_2$. But $\dot{\mu}(1,0) = 0$ so that $\int \dot{\mu}(t,0)\,\dot{\mu}(t,0)'\delta(dt) = 0$ is not regular. □

Moreover the regularity of $\int \dot{\mu}(t,\beta)\,\dot{\mu}(t,\beta)'\delta(dt)$ for all $\beta \in \mathcal{B}$ or, equivalently, the identifiability of β at δ in the model given by $\tilde{\mu}(t,\beta) = \dot{\mu}(t,\tilde{\beta})'\beta$ for all $\tilde{\beta} \in \mathcal{B}$ implies local identifiability but not identifiability in the sense of Definition 1.5 and 1.7. This shows the following example.

Example 1.2
Assume $\mu(t,\beta) = \mu(t,(\alpha,\beta)) = e^{\alpha}\sin(\beta+t)$ and $\text{supp}(\delta) = \{0, \frac{\pi}{2}\}$. Then for all $\beta = (\alpha,\beta)' \in I\!R^2$ the matrix

$$\begin{pmatrix} \dot{\mu}(0,(\alpha\,\beta))' \\ \dot{\mu}(\frac{\pi}{2},(\alpha\,\beta))' \end{pmatrix} = \begin{pmatrix} e^{\alpha}\sin(\beta) & e^{\alpha}\cos(\beta) \\ e^{\alpha}\sin(\beta+\frac{\pi}{2}) & e^{\alpha}\cos(\beta+\frac{\pi}{2}) \end{pmatrix}$$
$$= \begin{pmatrix} e^{\alpha}\sin(\beta) & e^{\alpha}\cos(\beta) \\ e^{\alpha}\cos(\beta) & -e^{\alpha}\sin(\beta) \end{pmatrix}$$

is regular so that also $\int \dot{\mu}(t,\beta)\,\dot{\mu}(t,\beta)'\delta(dt)$ is regular for all $\beta \in I\!R^2$. But for example for $\beta_1 = (0,0) \neq (0,2\pi) = \beta_2$ we have $\mu(0,\beta_1) = \sin(0) = 0 = \sin(2\pi) = \mu(0,\beta_2)$ and $\mu(\frac{\pi}{2},\beta_1) = \sin(\frac{\pi}{2}) = 1 = \sin(2\pi+\frac{\pi}{2}) = \mu(\frac{\pi}{2},\beta_2)$. □

2

Efficiency Concepts for Outlier-Free Observations

In this chapter the main definitions and results about an efficient inference in planned experiments is given for situations where the error distribution is ideal, i.e. where no outliers or other deviations appear. At first in Section 2.1 the ideal distribution of the errors is given. Then in Section 2.2 the efficiency concepts are given for estimating or testing a linear aspect in a linear model. Efficiency concepts for estimating a nonlinear aspect in a linear model or for estimation in a nonlinear model are presented in Section 2.3.

2.1 Assumptions on the Error Distribution

In classical approaches it is usually assumed that in a model of the form $Y_{nN} = \mu(t_{nN}, \beta) + Z_{nN}$, $n = 1, ..., N$, the errors $Z_{1N}, ..., Z_{NN}$ are independent, the expectation of $Z_N = (Z_{1N}, ..., Z_{NN})'$ is $E(Z_N) = 0$ and the covariance matrix of Z_N is $Cov(Z_N) = \sigma^2 E_N$, where $\sigma \in I\!R^+ \setminus \{0\}$ and E_N is the $N \times N$ identity matrix. In particular the classical assumptions do not include the possibility of outlying observation because outliers would provide $E(Z_N) \neq 0$ or $Cov(Z_N) \neq \sigma^2 E_N$.

If the distribution of Z_N satisfies the classical assumptions, then the distribution of the observation vector Y_N depends in particular on β and on the form of the design. If Y_N is the observation vector at a deterministic design d_N, then the distribution of Y_N will be denoted by P_β^N while the distribution of Z_N is denoted by $P^N := \bigotimes_{n=1}^N P$. If we have observations at a random design $D_N = (T_{1N}, ..., T_{NN})'$, which is distributed according to $\bigotimes_{n=1}^N \delta$, then with $(Z_{1N}, T_{1N}), ..., (Z_{NN}, T_{NN})$ also $(Y_{1N}, T_{1N}), ..., (Y_{NN}, T_{NN})$ are independent and identically distributed. The distribution of each Y_{nN} given $T_{nN} = t_{nN}$ is a Markov kernel $P_\beta(\cdot, t_{nN})$ so that the distribution of each (Y_{nN}, T_{nN}) can be denoted by $P_\beta \otimes \delta$. Then for the distributions of (Z_N, D_N) and (Y_N, D_N) we write $P^N := \bigotimes_{n=1}^N P \otimes \delta$ and $P_\beta^N := \bigotimes_{n=1}^N P_\beta \otimes \delta$, respectively.

2.2 Optimal Inference for Linear Problems

In this section we present optimal statistical procedures and characterizations of optimal designs for estimating or testing a linear aspect in a linear model with classical assumptions at the random errors.

If the aspect φ with $\varphi(\beta) = L\beta$ is identifiable at d_N, then under the assumptions $E(Z_N) = 0$ and $Cov(Z_N) = \sigma^2 E_N$ the best linear unbiased estimator is the Gauss-Markov estimator which is based on the least squares estimator $\widehat{\beta}_N^{LS}$ (see for example Christensen (1987), p. 18, Rao (1973), p. 223).

Definition 2.1 (Least squares estimator for a linear model)
An estimator $\widehat{\beta}_N : \mathbb{R}^N \times \mathcal{T}^N \to \mathbb{R}^r$ is a least squares estimator for β and denoted by $\widehat{\beta}_N^{LS}$ if

$$\widehat{\beta}_N(y_N, d_N) \in \arg\min\{\sum_{n=1}^{N} (y_{nN} - a(t_{nN})'\beta)^2;\ \beta \in \mathcal{B}\}$$

for all $y_N \in \mathbb{R}^N$ and $d_N \in \mathcal{T}^N$.

Definition 2.2 (Gauss-Markov estimator for a linear aspect)
An estimator $\widehat{\varphi}_N : \mathbb{R}^N \times \mathcal{T}^N \to \mathbb{R}^s$ is a Gauss-Markov estimator for the linear aspect $\varphi : \mathcal{B} \to \mathbb{R}^s$ and denoted by $\widehat{\varphi}_N^{LS}$ if $\widehat{\varphi}_N(y_N, d_N) = \varphi(\widehat{\beta}_N^{LS}(y_N, d_N))$ for all $y_N \in \mathbb{R}^N$ and $d_N \in \mathcal{T}^N$ where $\widehat{\beta}_N^{LS}$ is a least squares estimator for β.

Any least squares estimator can be explicitly represented by

$$\widehat{\beta}_N^{LS}(y_N, d_N) = (A'_{d_N} A_{d_N})^- A'_{d_N} y_N.$$

Thereby $A^- \in \mathbb{R}^{n \times m}$ denotes a *generalized inverse* (briefly *g-inverse*) of $A \in \mathbb{R}^{m \times n}$, i.e. a matrix satisfying $A\,A^- A = A$. Then the Gauss-Markov estimator for φ at d_N is given by

$$\widehat{\varphi}_N^{LS}(y_N, d_N) := \varphi(\widehat{\beta}_N^{LS}(y_N, d_N)) = L\,(A'_{d_N} A_{d_N})^- A'_{d_N} y_N$$

for all $y_N \in \mathbb{R}_N$ (see for example Christensen (1987), p.17). If β itself is identifiable at d_N, then the matrix $A'_{d_N} A_{d_N}$ is regular and therefore its generalized inverse and the least squares estimator is unique. If β itself is not identifiable at d_N, then the generalized inverse of $A'_{d_N} A_{d_N}$ is not unique but the identifiability of φ implies that $L\,(A'_{d_N} A_{d_N})^- A'_{d_N}$ is unique. For these properties and other properties of generalized inverses see for example Rao (1973), pp. 24, or Christensen (1987), pp. 336.

Under the assumption of $Cov(Z_N) = \sigma^2 E_N$ the covariance matrix of the Gauss-Markov estimator is

$$Cov(\widehat{\varphi}_N^{LS}(Y_N, d_N)) = L\,(A'_{d_N} A_{d_N})^- L'$$

(see for example Christensen (1987)). If we have a sequence of deterministic designs d_N converging to a design δ as in (1.2), then (1.4) implies

$$N\,Cov(\widehat{\varphi}_N^{LS}(Y_N, d_N)) = L\,\mathcal{I}(\delta_N)^- L' \overset{N\to\infty}{\longrightarrow} L\,\mathcal{I}(\delta)^- L'$$

if the regression function a is bounded and continuous on the supports of δ and all δ_N. Moreover for random designs we have with the strong law of large numbers $L\,(A'_{D_N} A_{D_N})^- L' \overset{N\to\infty}{\longrightarrow} L\,\mathcal{I}(\delta)^- L'$ with probability 1. For deterministic design sequences as well as for random designs $\widehat{\varphi}(Y_N, d_N)$ is also asymptotically normally distributed (see Eicker (1963, 1966), Schmidt (1975), Malinvaud (1970), Huber (1973), Bunke and Bunke (1986), pp. 89, Staudte and Sheather (1990), pp. 238).

Theorem 2.1 *If the linear aspect φ is identifiable at δ and $(d_N)_{N\in\mathbb{N}}$ are deterministic designs converging to δ or $(D_N)_{N\in\mathbb{N}}$ are random designs given by δ, then the Gauss-Markov estimator is asymptotically normally distributed, i.e.*

$$\mathcal{L}(\sqrt{N}(\widehat{\varphi}_N^{LS} - \varphi(\beta))|P_\beta^N) \overset{N\to\infty}{\longrightarrow} \mathcal{N}(0,\, \sigma^2\, L\,\mathcal{I}(\delta)^- L')$$

for all $\beta \in \mathcal{B}$.

Now consider the problem of testing a hypothesis of the form $H_0 : L\,\beta = l$ against the alternative $H_1 : L\,\beta \neq l$, where without loss of generality the rank of $L \in \mathbb{R}^{s\times r}$ is s. In situations where φ with $\varphi(\beta) = L\,\beta$ is identifiable at d_N and the error vector Z_N is normally distributed with expectation 0 and covariance matrix $\sigma^2 E_N$, i.e. $Z_N \sim \mathcal{N}(0, \sigma^2 E_N)$, the uniformly most powerful invariant test is the F-test based on the test statistic

$$T_N^{LS}(y_N, d_N) := \frac{(\widehat{\varphi}_N^{LS}(y_N, d_N) - l)'\,[L\,(A'_{d_N} A_{d_N})^- L']^{-1}\,(\widehat{\varphi}_N^{LS}(y_N, d_N) - l)/\text{rk}(L)}{\widehat{\sigma}_N^{LS}(y_N, d_N)^2},$$

with

$$\widehat{\sigma}_N^{LS}(y_N, d_N)^2 := \frac{y'_N(E_N - A_{d_N}(A'_{d_N} A_{d_N})^- A'_{d_N})y_N}{\text{rk}(E_N - A_{d_N}(A'_{d_N} A_{d_N})^- A'_{d_N})}$$

where rk(A) is the rank of a matrix A. Under the assumption of $Z_N \sim \mathcal{N}(0, \sigma^2 E_N)$ the test statistic $T_N^{LS}(Y_N, d_N)$ has a F-distribution with degrees of freedom of rk(L) and rk$(E_N - A_{d_N}(A'_{d_N} A_{d_N})^- A'_{d_N})$ and a noncentrality parameter of $(L\,\beta - l)'[L\,(A'_{d_N} A_{d_N})^- L']^{-1}(L\,\beta - l)$. See Lehmann (1959) p. 268, Christensen (1987), pp. 40, Rao (1973) pp. 236.

Theorem 2.1 and the convergence of $\widehat{\sigma}_N^{LS}(y_N, d_N)^2$ to σ^2 and of $N\,L\,(A'_{d_N} A_{d_N})^- L'$ and $N\,L\,(A'_{D_N} A_{D_N})^- L'$ to $L\,\mathcal{I}(\delta)^- L'$ provide that for $L\beta = l$ the test statistic $T_N^{LS}(Y_N, d_N)$ has an asymptotic central chi-squared distribution with s degrees of freedom. Moreover for contiguous alternatives $L\beta_N = l + N^{-1/2}\gamma$, $\gamma \neq 0$, (for contiguity see for example Hájek and Šidák (1967), pp. 201) the test statistic $T_N^{LS}(Y_N, d_N)$ has an asymptotic chi-squared distribution with s degrees of freedom and a noncentrality parameter of $\gamma'[\sigma^2 \, L\mathcal{I}(\delta)^- L']^{-1}\gamma$. This property follows from the general Theorem 6.1 proved in Section 6.2.

Theorem 2.2 *If the linear aspect φ is identifiable at δ and $(d_N)_{N\in\mathbb{N}}$ are deterministic designs converging to δ or $(D_N)_{N\in\mathbb{N}}$ are random designs given by δ, then the test statistic T_N^{LS} has an asymptotic chi-squared distribution, i.e.*

$$\mathcal{L}(s\,T_N^{LS}|P_{\beta_N}^N) \overset{N\to\infty}{\longrightarrow} \chi^2(s, \gamma'[\sigma^2 \, L\mathcal{I}(\delta)^- L']^{-1}\gamma)$$

for all $\beta_N = \beta + N^{-1/2}\overline{\beta} \in \mathcal{B}$ with $L\beta_N = l + N^{-1/2}\gamma$.

Hence the matrix $L\mathcal{I}(\delta)^- L'$ plays an important role for estimation and testing problems. For random designs as well as for deterministic designs it is the asymptotic covariance of the Gauss-Markov estimator and the asymptotic power of the F-test depends on it. Therefore for any design $\delta \in \Delta$ it will be interpreted as a covariance matrix. If φ is identifiable at δ, then Lemma 1.3 provides that the matrix $L\mathcal{I}(\delta)^- L'$ is unique.

If $\varphi(\beta) = L\,\beta$ is a one-dimensional aspect, i.e. $L \in \mathbb{R}^{1\times r}$, then usually there exists a design δ^* which minimizes the scalar $L\mathcal{I}(\delta)^- L'$ within all designs at which φ is identifiable, i.e.

$$\delta^* \in \arg\min\{L\mathcal{I}(\delta)^- L';\ \delta \in \Delta(\varphi)\}.$$

If $\varphi(\beta) = L\,\beta$ is not a one-dimensional aspect, i.e. $L \in \mathbb{R}^{s\times r}$ with $s > 1$, then a design $\delta_U \in \Delta(\varphi)$ is called an *universally optimal design* or briefly an *U-optimal design* in Δ if and only if

$$L\mathcal{I}(\delta_U)^- L' \leq L\mathcal{I}(\delta)^- L' \qquad \text{for all } \delta \in \Delta \cap \Delta(\varphi),$$

where the relation $\leq$ between positive-semidefinite matrices is meant in the positive-semidefinite sense, i.e. $C_1 \leq C_2$ if and only if $C_2 - C_1$ is positive-semidefinite. Thereby, let Δ be any subset of Δ_0, the set of all probability measures on $\mathcal{T}$.

Definition 2.3 (U-optimality)
δ_U is U-optimal for φ in Δ if $L\mathcal{I}(\delta_U)^- L' \leq L\mathcal{I}(\delta)^- L'$ for all $\delta \in \Delta\cap\Delta(\varphi)$.

Except in trivial cases an U-optimal design does not exist. Therefore in the literature several other optimality criteria were investigated (see for example Fedorov (1972), Bandemer (1977), Krafft (1978), Bandemer and Näther (1980), Silvey (1980), Pázman (1986), Atkinson and Donev (1992), Pukelsheim (1993)). One famuous criterion is the *determinant optimality criterion* or briefly the *D-optimality criterion*, which is based on the determinant of the covariance matrix (or generalized variance), i.e. on $\det(L\mathcal{I}(\delta)^- L')$.

Definition 2.4 (D-optimality)
δ_D is D-optimal for φ in Δ if $\delta_D \in \arg\min\{\det(L\mathcal{I}(\delta)^- L');\ \delta \in \Delta \cap \Delta(\varphi)\}$.

This D-optimality criterion can be derived from the power of the associated F-tests and the size of confidence ellipsoids. Moreover it is invariant with respect to one-to-one linear transformations of the linear aspect. I.e. if instead of the linear aspect $\varphi(\beta) = L\beta$ a linear aspect $\tilde{\varphi}(\beta) = \tilde{L}\beta$ is regarded with $\tilde{L} = K\,L$, where $K \in I\!R^{s\times s}$ is a regular matrix, then the D-optimal design is independent of the special choice of the aspect because of

$$\det(\tilde{L}\mathcal{I}(\delta)^- \tilde{L}) = \det(K\,L\mathcal{I}(\delta)^- L'\,K') = (\det(K)^2)\ \det(L\mathcal{I}(\delta)^- L').$$

In particular for testing problems this invariance property of the D-optimality criterion is very important because for different but equivalent formulations of the hypotheses H_0 the optimal designs should not be different.

Another very suggestive criterion is the *average optimality criterion* or briefly the *A-optimality criterion*, which is based on the trace of the covariance matrix, i.e. on $\mathrm{tr}(L\mathcal{I}(\delta)^- L')$. This criterion is called A-optimality criterion because it is equivalent with the average of the variances of the estimators for the single components of $\varphi(\beta)$.

Definition 2.5 (A-optimality)
δ_A is A-optimal for φ in Δ if $\delta_A \in \arg\min\{\mathrm{tr}(L\mathcal{I}(\delta)^- L');\ \delta \in \Delta \cap \Delta(\varphi)\}$.

This A-optimality criterion is very sensitive to transformations of the interesting aspect. Only orthogonal transformations will leave it unchanged because of

$$\mathrm{tr}(K\,L\mathcal{I}(\delta)^- L'\,K') = \mathrm{tr}(L\mathcal{I}(\delta)^- L'\,K'\,K) = \mathrm{tr}(L\mathcal{I}(\delta)^- L')$$

for orthogonal matrices $K \in I\!R^{s\times s}$. Because the A-optimality criterion is not invariant to nonorthogonal transformations it makes no sense for testing hypotheses. But for estimation problems it could be very useful because it is directly attached to the parameters of interest and attempts to minimize the expected mean squared distance of the estimates.

D- and A-optimal designs can be characterized by equivalence theorems which both can be based on the general equivalence theorem due to Whittle (1973) (see for example Bandemer et al. (1977), Silvey (1980), Pukelsheim (1993)). For the special linear aspect $\varphi(\beta) = \beta$ originally the equivalence theorem for D-optimality was derived by Kiefer and Wolfowitz (1960) and the equivalence theorem for A-optimality is due to Fedorov (1971). These equivalence theorems can be easily transferred to general linear aspect in situations where β itself is identifiable at the optimal design δ^*, i.e. $\delta^* \in$

$\Delta(\beta)$. See Kiefer (1974), Pázman (1986), p.117, Schwabe (1996a), p. 12. For the D-optimality we only additionally have to assume that φ is a s-dimensional minimal linear aspect which means that the rank $\text{rk}(L)$ of L is equal to s if $\varphi(\beta) = L\beta \in I\!R^s$. Moreover, we always assume that the experimental region $\mathcal{T}$ is a compact set and that $a : \mathcal{T} \to I\!R^r$ is a continuous function.

Theorem 2.3 *Let $\delta_D \in \Delta(\beta)$ and let φ be an s-dimensional minimal linear aspect. Then:*

a) δ_D is D-optimal for φ in $\Delta(\varphi)$ if and only if

$$a(t)'\mathcal{I}(\delta_D)^{-1}L'\,[L\,\mathcal{I}(\delta_D)^{-1}L']^{-1}\,L\,\mathcal{I}(\delta_D)^{-1}a(t) \leq s \text{ for all } t \in \mathcal{T}.$$

b) If δ_D is D-optimal for φ in $\Delta(\varphi)$, then

$$a(t)'\mathcal{I}(\delta_D)^{-1}L'\,[L\,\mathcal{I}(\delta_D)^{-1}L']^{-1}\,L\,\mathcal{I}(\delta_D)^{-1}a(t) = s$$
$$\text{for all } t \in \text{supp}(\delta_D).$$

Theorem 2.4 *Let $\delta_A \in \Delta(\beta)$. Then:*

a) δ_A is A-optimal for φ in $\Delta(\varphi)$ if and only if

$$|L\,\mathcal{I}(\delta_A)^{-1}a(t)|^2 \leq \text{tr}(L\,\mathcal{I}(\delta_A)^{-1}L') \text{ for all } t \in \mathcal{T}.$$

b) If δ_A is A-optimal for φ in $\Delta(\varphi)$, then

$$|L\,\mathcal{I}(\delta_A)^{-1}a(t)|^2 = \text{tr}(L\,\mathcal{I}(\delta_A)^{-1}L') \text{ for all } t \in \text{supp}(\delta_A).$$

For optimal designs δ at which β itself is not identifiable, i.e. $\delta \notin \Delta(\beta)$, there exist some equivalence theorems (see Silvey (1978), Pukelsheim and Titterington (1983), Pázman (1980), Näther and Reinsch (1981), Gaffke (1987)). These theorems are rather complicated but also very generell. Here equivalence theorems for more restricted classes of designs are more useful. For that define for any $\mathcal{D}$ at which φ is identifiable

$$\Delta_{\mathcal{D}} := \{\delta \in \Delta_0;\ \text{supp}(\delta) = \mathcal{D}\}$$

and regard this restricted class of designs. Note that because of Definition 1.7 at every $\delta \in \Delta_{\mathcal{D}}$ the aspect φ is identifiable.

Theorem 2.5 *Let φ be an s-dimensional minimal linear aspect which is identifiable at $\mathcal{D}$. Then δ_D is D-optimal for φ in $\Delta_{\mathcal{D}}$ if and only if*

$$a(t)'\mathcal{I}(\delta_D)^{-}L'\,[L\,\mathcal{I}(\delta_D)^{-}L']^{-1}\,L\,\mathcal{I}(\delta_D)^{-}a(t) = s \text{ for all } t \in \mathcal{D}.$$

Proof. There exists a reparametrization with $a(t)'\beta = \tilde{a}(t)'\tilde{\beta}$ for all $t \in \mathcal{D}$ such that $\tilde{\beta}$ is identifiable at $\mathcal{D}$ in the reparametrized model. Then the assertion follows by applying Theorem 2.3 on the reparametrized model. □

Theorem 2.6 *Let φ be identifiable at $\mathcal{D}$. Then δ_A is A-optimal for φ in $\Delta_\mathcal{D}$ if and only if*

$$|L\,\mathcal{I}(\delta_A)^- a(t)|^2 = \mathrm{tr}(L\,\mathcal{I}(\delta_A)^- L') \text{ for all } t \in \mathcal{D}.$$

Proof. With the same reparametrization as in the proof of Theorem 2.5 the assertion follows by applying Theorem 2.4 on the reparametrized model. □

If the regressors $a(\tau_1), ..., a(\tau_I)$ on $\mathcal{D} = \{\tau_1, ..., \tau_I\}$ are linearly independent, then the A-optimal design and the D-optimal design in

$$\Delta_\mathcal{D}^* := \{\delta \in \Delta_0;\ \mathrm{supp}(\delta) \subset \mathcal{D}\}$$

have a very simple form (see Pukelsheim and Torsney (1991)). To show this the following two lemmata are useful.

Lemma 2.1 *If $A \in \mathbb{R}^{I\times r}$ is of rank I and $D \in \mathbb{R}^{I\times I}$ is a regular diagonal matrix, then $A\,(A'D\,A)^- A' = D^{-1}$.*

Proof. Set $B = D^{1/2}A$. Then $B(B'B)^- B'$ is the perpendicular projection operator onto the space spanned by the columns of B. But this space is the whole space $\mathbb{R}^I$ because with A also B is of full rank I. Hence we have

$$D^{1/2}A\,(A'D\,A)^- A'D^{1/2} = B(B'B)^- B' = E_I.$$

See for example Christensen (1987), p. 335-338. □

Lemma 2.2 *If $\varphi(\beta) = L\beta$ is identifiable at supp$(\delta) = \mathcal{D}_0 = \{\overline{\tau}_1, ..., \overline{\tau}_J\} \subset \mathcal{D} = \{\tau_1, ..., \tau_I\}$ and $A_\mathcal{D} = (a(\tau_1), ..., a(\tau_I))'$ is of full rank I, then*

$$L\,\mathcal{I}(\delta)^- a(t)\,\delta(\{t\}) = L\,(A_\mathcal{D}'A_\mathcal{D})^- a(t)$$

for all $t \in \mathcal{D}$, and in particular for $L = A_{\mathcal{D}_0} = (a(\overline{\tau}_1), ..., a(\overline{\tau}_J))'$ and $t \in \mathcal{D}_0$,

$$|A_{\mathcal{D}_0}\mathcal{I}(\delta)^- a(t)\,\delta(\{t\})| = |A_{\mathcal{D}_0}(A_{\mathcal{D}_0}'A_{\mathcal{D}_0})^- a(t)| = 1.$$

Proof. Without loss of generality assume $\overline{\tau}_i = \tau_i$ for $i = 1, ..., J$. Let be D the diagonal matrix with diagonal elements $\delta(\{\tau_1\}), ..., \delta(\{\tau_I\})$ and D_0 the diagonal matrix with diagonal elements $\delta(\{\tau_1\}), ..., \delta(\{\tau_J\})$. Because φ is identifiable at δ there exists $K \in \mathbb{R}^{s\times J}$ so that $L = K\,A_{\mathcal{D}_0}$ (see

Lemma 1.2). Then we also have $L = (K|0_{s\times(I-J)})\, A_{\mathcal{D}}$. This implies with Lemma 2.1

$$L\,(A'_{\mathcal{D}}A_{\mathcal{D}})^{-}A'_{\mathcal{D}} = (K|0_{s\times(I-J)}).$$

On the other hand, because $A_{\mathcal{D}_0}$ is also of full rank J, Lemma 2.1 provides

$$L\,\mathcal{I}(\delta)^{-}A'_{\mathcal{D}}D = K\,A_{\mathcal{D}_0}(A'_{\mathcal{D}_0}D_0\,A_{\mathcal{D}_0})^{-}(A'_{\mathcal{D}_0}D_0|0_{s\times(I-J)}) = (K|0_{s\times(I-J)}).$$

See also Müller (1987), Lemma 6.3. □

Theorem 2.7 *If $\varphi(\beta) = L\beta$ is identifiable at $\mathcal{D} = \{\tau_1, ..., \tau_I\}$ and $A_{\mathcal{D}} = (a(\tau_1), ..., a(\tau_I))'$ is of full rank I, then δ_A is A-optimal for φ in $\Delta^*_{\mathcal{D}}$ if and only if*

$$\delta_A(\{t\}) = \frac{|L\,(A'_{\mathcal{D}}A_{\mathcal{D}})^{-}a(t)|}{\sum_{\tau\in\mathcal{D}}|L\,(A'_{\mathcal{D}}A_{\mathcal{D}})^{-}a(\tau)|} \tag{2.1}$$

for all $t \in \mathcal{D}$, where

$$\mathrm{tr}(L\,\mathcal{I}(\delta_A)^{-}L') = \left(\sum\nolimits_{\tau\in\mathcal{D}}|L\,(A'_{\mathcal{D}}A_{\mathcal{D}})^{-}a(\tau)|\right)^2.$$

Proof. Let $\mathcal{D}_0 = \{t \in \mathcal{D};\ L(A'_{\mathcal{D}}A_{\mathcal{D}})^{-}a(t) \neq 0\}$. For any design $\delta \in \Delta^*_{\mathcal{D}} \cap \Delta(\varphi)$ Lemma 2.2 provides

$$\begin{aligned}
\mathrm{tr}(L\,\mathcal{I}(\delta)^{-}L') &= \int |L\,\mathcal{I}(\delta)^{-}a(t)|^2\delta(dt)\\
&= \sum\nolimits_{t\in\mathcal{D}_0}|L\,(A'_{\mathcal{D}}A_{\mathcal{D}})^{-}a(t)|^2\,\frac{1}{\delta(\{t\})}\\
&= \sum\nolimits_{t\in\mathcal{D}_0}|L\,(A'_{\mathcal{D}}A_{\mathcal{D}})^{-}a(t)|^2\,\frac{1}{\overline{\delta}(\{t\})}\,\frac{1}{\sum_{\tau\in\mathcal{D}_0}\delta(\{\tau\})}\\
&= \mathrm{tr}(L\,\mathcal{I}(\overline{\delta})^{-}L')\,\frac{1}{\sum_{\tau\in\mathcal{D}_0}\delta(\{\tau\})}\\
&\geq \mathrm{tr}(L\,\mathcal{I}(\overline{\delta})^{-}L')
\end{aligned}$$

with equality if and only if $\mathrm{supp}(\delta) = \mathcal{D}_0$, where $\overline{\delta}$ is given by

$$\overline{\delta}(\{t\}) = \frac{1}{\sum_{\tau\in\mathcal{D}_0}\delta(\{\tau\})}\,1_{\mathcal{D}_0}(t)\;\delta(\{t\}).$$

According to Theorem 2.6 and Lemma 2.2 a design δ is A-optimal in $\Delta_{\mathcal{D}_0}$ if and only if δ is given by 2.1. See also Pukelsheim and Torsney (1991). □

Theorem 2.8 *If $\varphi(\beta) = L\beta \in \mathbb{R}^s$ is identifiable at $\mathcal{D} = \{\tau_1, ..., \tau_s\}$ and $A_{\mathcal{D}} = (a(\tau_1), ..., a(\tau_s))'$ is of full rank s, then δ_D is D-optimal for φ in $\Delta_{\mathcal{D}}$ if and only if*

$$\delta_D(\{t\}) = \frac{1}{s}$$

for all $t \in \mathcal{D}$.

Proof. Because φ is identifiable at $\mathcal{D}$ there exist a regular matrix $K \in \mathbb{R}^{s\times s}$ with $L = K A_\mathcal{D}$, where $A_\mathcal{D} = (a(\tau_1), ..., a(\tau_s))'$. Then Lemma 2.1 provides for any $\delta \in \Delta_\mathcal{D}$

$$\begin{aligned}
&\det(L\mathcal{I}(\delta)^- L') \\
&= \det(K A_\mathcal{D} (A'_\mathcal{D}\mathrm{diag}(\delta(\{\tau_1\}), ..., \delta(\{\tau_s\})) A_\mathcal{D})^- A'_\mathcal{D} K') \\
&= \det(K \,\mathrm{diag}(\delta(\{\tau_1\}), ..., \delta(\{\tau_s\}))^{-1} K') \\
&= \det(K)^2 \frac{1}{\delta(\{\tau_1\})} \cdot \ldots \cdot \frac{1}{\delta(\{\tau_s\})}
\end{aligned}$$

which is minimized only by $\delta_D = \frac{1}{s}\sum_{\tau\in\mathcal{D}} e_\tau$. □

Example 2.1 (Linear regression)
In a linear regression model the observation at t_{nN} is given by

$$Y_{nN} = \beta_0 + \beta_1 t_{nN} + Z_{nN} = a(t_{nN})'\beta + Z_{nN}$$

for $n = 1, ..., N$, where $\beta = (\beta_0, \beta_1)' \in \mathbb{R}^2$ and $a(t) = (1, t)' \in \mathbb{R}^2$. If $\mathcal{T} = [-1, 1]$ is the experimental region, then the equivalence theorems (Theorem 2.3 and Theorem 2.4) provide that the A-optimal and D-optimal design for β is $\delta_A = \delta_D = \frac{1}{2}(e_{-1} + e_1)$. If $\mathcal{T} = [0, 1]$, then the D-optimal design is still $\delta_D = \frac{1}{2}(e_0 + e_1)$. But the A-optimal design is now $\delta_A = \frac{1}{\sqrt{2}+1}(\sqrt{2}e_0 + e_1)$. Note that in this case the A-optimal design puts more observations at $t = 0$ than the D-optimal design. This is due to the fact that at $t = 0$ the component β_0 can be estimated very precisely so that also the estimation of β_1, which is confounded with β_0 at $t = 1$, profits by a precise estimation of β_0. For testing, a design prefering $t = 0$ has no advantage because any hypothesis of the form $H_0 : \beta = (l_1, l_2)'$ is equivalent with $H_0 : \varphi(\beta) = (\beta_0, \beta_0 + \beta_1)' = (l_1, l_1 + l_2)'$ and even the A-optimal design for φ has the form $\frac{1}{2}(e_0 + e_1)$. □

Example 2.2 (One-way lay-out)
In a one-way lay-out model with I levels the observation at $t_{nN} = i$ is given by

$$Y_{nN} = \beta_i + Z_{nN} = a(t_{nN})'\beta + Z_{nN}$$

for $n = 1, ..., N$, where $\beta = (\beta_1, ..., \beta_I)' \in \mathbb{R}^I$, $a(t) = (1_{\{1\}}(t), ..., 1_{\{I\}}(t))' \in \mathbb{R}^I$ and $\mathcal{T} = \{1, ..., I\}$. If the level 1 is a control level, then an interesting aspect is $\varphi(\beta) = (\beta_2 - \beta_1, ..., \beta_I - \beta_1)' \in \mathbb{R}^{I-1}$. Then Theorem 2.5 provides that $\delta_D = \frac{1}{I}\sum_{t=1}^I e_t$ is the D-optimal design for φ in $\Delta(\varphi)$. But according to Theorem 2.7 the A-optimal design for φ in $\Delta(\varphi)$ is $\delta_A = \frac{1}{\sqrt{I-1}+I-1}(\sqrt{I-1}e_1 + \sum_{t=2}^I e_t)$. This example shows in particular that for testing $H_0 : \beta_1 = \beta_2 = ... = \beta_I$, which is equivalent with $H_0 : \varphi(\beta) = 0$, we should use a D-optimal design which does not depend on the choice of the control level. In opposite to testing the optimal design for estimation should depend on the choice of the control level because the effect of the control level appears in each component of $\varphi(\beta)$. This means that the estimation of each component of $\varphi(\beta)$ profits by a precise estimation of the effect of the control level. Therefore for estimation we should use an A-optimal design which puts more observations at the control level. Compare also with Kurotschka (1972, 1978). □

2.3 Efficient Inference for Nonlinear Problems

In this section we regard the problem of estimating a nonlinear aspect of a linear model and of estimating the whole parameter β of a nonlinear model. We will see that while asymptotically optimal estimators can be derived the situation for optimal designs is much more complicated. Only locally optimal designs can be derived und used for efficiency comparisons. In some situations designs can be derived which maximize the minimum efficiency. We start with the problem of estimating a nonlinear aspect in a linear model and continue with the problem of estimating the whole parameter β of a nonlinear model.

As the Gauss-Markov estimator for a linear aspect is based on the least squares estimator by $\widehat{\varphi}_N^{LS}(y_N, d_N) = \varphi(\widehat{\beta}_N^{LS}(y_N, d_N))$ we similarly can base an estimator for a nonlinear aspect on the least squares estimator.

Definition 2.6 (Gauss-Markov estimator for a nonlinear aspect) *An estimator $\widehat{\varphi}_N : \mathbb{R}^N \times \mathcal{T}^N \longrightarrow \mathbb{R}^s$ is a Gauss-Markov estimator for the nonlinear aspect $\varphi : \mathcal{B} \longrightarrow \mathbb{R}^s$ and denoted by $\widehat{\varphi}_N^{LS}$ if $\widehat{\varphi}_N(y_N, d_N) = \varphi(\widehat{\beta}_N^{LS}(y_N, d_N))$ for all $y_N \in \mathbb{R}^N$ and $d_N \in \mathcal{T}^N$, where $\widehat{\beta}_N^{LS}$ is the least squares estimator for β.*

If $(d_N)_{N\in\mathbb{N}}$ is converging to a design measure δ in the sense of (1.2) or d_N is the realization of a random design D_N with distribution $\bigotimes_{n=1}^{N} \delta$, then under the classical assumptions for the error distribution the Gauss-Markov estimator is asymptotically normally distributed.

Theorem 2.9 *If $\mathcal{B}$ is a convex and open subset of $\mathbb{R}^r$, φ is continuously differentiable on $\mathcal{B}$ and identifiable at δ with finite support and $(d_N)_{N\in\mathbb{N}}$ are deterministic designs converging to δ or $(D_N)_{N\in\mathbb{N}}$ are random designs given by δ, then*

$$\mathcal{L}(\sqrt{N}(\widehat{\varphi}_N^{LS} - \varphi(\beta))|P_\beta^N) \xrightarrow{N\to\infty} \mathcal{N}(0,\, \sigma^2\, \dot{\varphi}_\beta \mathcal{I}(\delta)^- \dot{\varphi}_\beta')$$

for all $\beta \in \mathcal{B}$.

Proof. The identifiability of φ at the finite set $\mathcal{D} = \text{supp}(\delta)$ provides according to Lemma 1.4 that $\varphi(\beta) = \varphi^*(A_\mathcal{D}\beta)$ for all $\beta \in \mathcal{B}$. Because the

Gauss-Markov estimator $A_{\mathcal{D}}\widehat{\beta}_N^{LS}$ is a consistent estimator of $A_{\mathcal{D}}\beta$ and $\mathcal{B}$ is open, for every $\beta \in \mathcal{B}$ and $\epsilon > 0$ there exists N_ϵ such that a version of $\widehat{\beta}_N^{LS}$ is lying in $\mathcal{B}$ with probability greater than $1-\epsilon$ for $N \geq N_\epsilon$. Note that if β is not identifiable at δ, then $\widehat{\beta}_N^{LS}$ is not unique and the different versions of $\widehat{\beta}_N^{LS}$ are given by the different versions of the generalized inverse of $A'_{d_N}A_{d_N}$. If $\widehat{\beta}_N^{LS} \in \mathcal{B}$, then according to Lemma 1.4 we have $\widehat{\varphi}_N^{LS} = \varphi^*(A_{\mathcal{D}}\widehat{\beta}_N^{LS})$. Moreover, the differentiability of φ provides also the differentiability of φ^* with $\dot{\varphi}_\beta = \dot{\varphi}^*_\beta A_{\mathcal{D}}$, where $\dot{\varphi}^*_\beta := \frac{\partial}{\partial\eta}\varphi^*(\eta)/_{\eta=A_{\mathcal{D}}\beta}$. Then the differentiability of φ^* and the consistency and asymptotic normality of the Gauss-Markov estimator $A_{\mathcal{D}}\widehat{\beta}_N^{LS}$ provides

$$\begin{aligned}
&\sqrt{N}(\widehat{\varphi}_N^{LS} - \varphi(\beta)) - \sqrt{N}(\dot{\varphi}_\beta\widehat{\beta}_N^{LS} - \dot{\varphi}_\beta\beta) \\
&= \sqrt{N}(\varphi^*(A_{\mathcal{D}}\widehat{\beta}_N^{LS}) - \varphi^*(A_{\mathcal{D}}\beta)) - \sqrt{N}(\dot{\varphi}_\beta\widehat{\beta}_N^{LS} - \dot{\varphi}_\beta\beta) \\
&= \sqrt{N}|A_{\mathcal{D}}\widehat{\beta}_N^{LS} - A_{\mathcal{D}}\beta|\frac{\varphi^*(A_{\mathcal{D}}\widehat{\beta}_N^{LS}) - \varphi^*(A_{\mathcal{D}}\beta) - \dot{\varphi}^*_\beta(A_{\mathcal{D}}\widehat{\beta}_N^{LS} - A_{\mathcal{D}}\beta)}{|A_{\mathcal{D}}\widehat{\beta}_N^{LS} - A_{\mathcal{D}}\beta|} \\
&\to 0
\end{aligned}$$

in probability for $(P_\beta^N)_{N\in\mathbb{N}}$ so that the assertion follows from the asymptotic normality of the Gauss-Markov estimators $\dot{\varphi}_\beta\widehat{\beta}_N^{LS}$ (see Theorem 2.1). □

The estimator based on the least squares estimator is the estimator which minimize uniformly the asymptotic covariance matrix for all $\beta \in \mathcal{B}$ within all estimators of the form $\varphi(\widehat{\beta}_N)$ where $\widehat{\beta}_N$ is some estimator for β. This is shown in Section 7.3 in a more general context.

For deriving optimal designs it is as for linear aspect in general not possible to find a design which minimizes the asymptotic covariance matrix uniformly. But additionally, because the covariance matrix depends on β, it is in general also not possible to find a design which minimizes some one-dimensional function of the covariance matrix simultaneously for all $\beta \in \mathcal{B}$. Therefore only *locally optimal* designs can be derived. Because the asymptotic covariance matrix of the Gauss-Markov estimator for a nonlinear aspect at β is equal to the covariance matrix of the Gauss-Markov estimator for φ_β given by $\varphi_\beta(\tilde{\beta}) = \dot{\varphi}_\beta\tilde{\beta}$, the locally optimal designs are optimal designs for estimation of φ_β.

Definition 2.7 (Local D-optimality)
$\delta_{\beta,D}$ is locally D-optimal for φ in Δ at β if

$$\delta_{\beta,D} \in \arg\min\{\det(\dot{\varphi}_\beta\,\mathcal{I}(\delta)^-\dot{\varphi}'_\beta);\ \delta \in \Delta \cap \Delta(\varphi_\beta)\}.$$

Definition 2.8 (Local A-optimality)
$\delta_{\beta,A}$ is locally A-optimal for φ in Δ at β if

$$\delta_{\beta,A} \in \arg\min\{\mathrm{tr}(\dot{\varphi}_\beta\,\mathcal{I}(\delta)^-\dot{\varphi}'_\beta);\ \delta \in \Delta \cap \Delta(\varphi_\beta)\}.$$

Several strategies were involved to overcome the β-dependence of the designs, see for example Silvey (1980), Buonaccorsi (1986a,b), Buonaccorsi and Iyer (1986), Ford et al. (1989) amd Mandal and Heiligers (1992), Kitsos (1986, 1992). One strategy is to use the locally optimal designs for efficiency comparisons as Silvey (1980), Kitsos (1992) and Müller (1995a) have done. Here only the locally A-optimal designs will be of interest so that we define *relative efficiency* of a design δ at β as

$$E(\delta, \beta) := \frac{\mathrm{tr}(\dot{\varphi}_\beta\, \mathcal{I}(\delta_{\beta,A})^- \dot{\varphi}_\beta')}{\mathrm{tr}(\dot{\varphi}_\beta\, \mathcal{I}(\delta)^- \dot{\varphi}_\beta')} \in [0, 1].$$

A design δ_M which maximizes the minimum relative efficiency is called a *maximin efficient design*.

Definition 2.9 (Maximin efficient design)
δ_M is maximin efficient for φ in Δ if

$$\delta_M \in \arg\max\{\min_{\beta\in\mathcal{B}} \frac{\mathrm{tr}(\dot{\varphi}_\beta\, \mathcal{I}(\delta_{\beta,A})^- \dot{\varphi}_\beta')}{\mathrm{tr}(\dot{\varphi}_\beta\, \mathcal{I}(\delta)^- \dot{\varphi}_\beta')};\ \delta \in \Delta \cap \Delta(\varphi)\}.$$

By straightforward calculation Silvey (1980) derived the maximin efficient design for estimating the maximum point of a one-dimensional quadratic function and Kitsos (1992) derived the maximin efficient design for linear calibration. In Müller (1995a) a general approach for deriving maximin efficient designs is presented. In this approach we restrict ourselves to designs with a support included in a finite set $\mathcal{D}$, i.e. to

$$\Delta_{\mathcal{D}}^* := \{\delta \in \Delta_0;\ \mathrm{supp}(\delta) \subset \mathcal{D}\},$$

because in many situations all locally A-optimal designs have a support which is included in a set $\mathcal{D} = \{\tau_1, ..., \tau_I\}$. Usually in this situations the regressors $a(\tau_1), ..., a(\tau_I)$ are also linearly independent. Then according to Theorem 2.7 the locally A-optimal designs $\delta_{\beta,A}$ in $\Delta_{\mathcal{D}}^* \cap \Delta(\varphi_\beta)$ are given by

$$\delta_{\beta,A}(\{t\}) := \frac{|\dot{\varphi}_\beta(A_{\mathcal{D}}' A_{\mathcal{D}})^- a(t)|}{\sum_{\tau\in\mathcal{D}} |\dot{\varphi}_\beta(A_{\mathcal{D}}' A_{\mathcal{D}})^- a(\tau)|} \qquad \text{for } t \in \mathcal{D},$$

where $A_{\mathcal{D}} = (a(\tau_1), ..., a(\tau_I))'$. Müller (1995a) presented besides other results the following theorem which also can be generalized for robust estimation (see Section 7.3).

Theorem 2.10 *Let φ be identifiable at $\mathcal{D} = \{\tau_1, ..., \tau_I\}$, $a(\tau_1), ..., a(\tau_I)$ are linearly independent and for all $t \in \mathcal{D}$*

$$\max\{\delta_{\beta,A}(\{t\});\ \beta \in \mathcal{B}\} = 1.$$

Then δ_M is maximin efficient for φ in $\Delta_{\mathcal{D}}^*$ if and only if $\delta_M = \frac{1}{I}\sum_{t\in\mathcal{D}} e_t$, and the maximin efficiency is equal to $\frac{1}{I}$.

Example 2.3 (Linear calibration)
In the linear calibration problem a linear regression model is assumed, where the observation at t_{nN} is given by

$$Y_{nN} = \beta_0 + \beta_1 t_{nN} + Z_{nN} = a(t_{nN})'\beta + Z_{nN}$$

for $n = 1, ..., N$, where $\beta = (\beta_0, \beta_1)' \in \mathcal{B} = I\!R \times (I\!R \setminus \{0\})$ and $a(t) = (1, t)' \in I\!R^2$. Assume that the experimental region is $\mathcal{T} = [-1, 1]$. Then the interesting aspect of β is that experimental condition t_y which would provide some given value y if there are no errors Z_{nN}, i.e. $\beta_0 + \beta_1 t_y = y$. But this means that the interesting aspect is $\varphi(\beta) = t_y = \frac{y-\beta_0}{\beta_1}$ which is a nonlinear aspect of β. Then we have $\dot{\varphi}_\beta = \frac{-1}{\beta_1^2}(\beta_1, y - \beta_0)' \in I\!R^2$ so that Elfving's theorem (see Elfving (1952), Pukelsheim (1981)) provides that every locally A- and D-optimal design $\delta_{\beta,A} = \delta_{\beta,D}$ in Δ_0 has a support which is included in $\mathcal{D} = \{-1, 1\}$. The locally A-optimal designs, which are also locally D-optimal, are given by Theorem 2.7 and satisfy $\max\{\delta_{\beta,A}(\{t\});\ \beta \in \mathcal{B}\} = 1$ (see Müller (1995a)). Hence according to Theorem 2.10 the maximin efficient design in $\Delta_{\mathcal{D}}^*$ is $\delta_M = \frac{1}{2}(e_{-1} + e_1)$, and the maximin efficiency is $\frac{1}{2}$.
□

For estimating β in a nonlinear model given by the nonlinear response function $\mu(t, \beta)$ we also can use a least squares estimator.

Definition 2.10 (Least squares estimator for a nonlinear model)
An estimator $\widehat{\beta}_N : I\!R^N \times \mathcal{T}^N \longrightarrow I\!R^r$ is a least squares estimator for β and denoted by $\widehat{\beta}_N^{LS}$ if

$$\widehat{\beta}_N(y_N, d_N) \in \arg\min\{\sum\nolimits_{n=1}^N (y_{nN} - \mu(t_{nN}, \beta))^2;\ \beta \in \mathcal{B}\}$$

for all $y_N \in I\!R^N$ and $d_N \in \mathcal{T}^N$.

Under some regularity conditions this least squares estimator is consistent and in particular asymptotically normally distributed and asymptotically optimal (see Jennrich (1969), Seber and Wild (1989) pp. 563, Gallant (1987) pp. 255, Bunke and Bunke (1989) pp. 30, Läuter (1989), Wu (1981), for the asymptotic optimality see also Section 7.4). For example under the following regularity conditions given by Seber and Wild (1989) the least squares estimator is asymptotically normally distributed under P_β^N. For that the first derivative

$$\dot{\mu}(t, \beta) := \left(\frac{\partial}{\partial \tilde{\beta}} \mu(t, \tilde{\beta})/_{\tilde{\beta}=\beta}\right)' \in I\!R^r$$

and the second derivative

$$\ddot{\mu}(t, \beta) := \frac{\partial^2}{\partial^2 \tilde{\beta}} \mu(t, \tilde{\beta})/_{\tilde{\beta}=\beta} = \frac{\partial}{\partial \tilde{\beta}} \dot{\mu}(t, \tilde{\beta})/_{\tilde{\beta}=\beta} \in I\!R^{r \times r}$$

of the nonlinear response function μ are important:

- β is identifiable at δ. (2.2)
- $\mathcal{I}_\beta(\delta) := \int \dot\mu(t,\beta)\,\dot\mu(t,\beta)'\delta(dt)$ is nonsingular. (2.3)
- $\mathcal{B}$ and $\mathcal{T}$ are closed and bounded subsets of $\mathbb{R}^r$ and $\mathbb{R}^k$, respectively. (2.4)
- β is an interior point of $\mathcal{B}$. (2.5)
- μ is continuous on $\mathcal{T} \times \mathcal{B}$. (2.6)
- $\dot\mu$ and $\ddot\mu$ are continuous on $\mathcal{T} \times \mathcal{B}$. (2.7)
- $(d_N)_{N\in\mathbb{N}}$ are deterministic designs converging to δ. (2.8)

Note that condition (2.2) does not imply condition (2.3) (see Example 1.1).

Theorem 2.11 *Under the assumptions (2.2) - (2.8) the least squares estimator is asymptotically normally distributed at β, i.e.*

$$\mathcal{L}(\sqrt{N}(\widehat{\beta}_N^{LS} - \beta)|P_\beta^N) \overset{N\to\infty}{\longrightarrow} \mathcal{N}(0,\ \sigma^2\, \mathcal{I}_\beta(\delta)^{-1}).$$

Because the asymptotic covariance matrix of a least squares estimator in a nonlinear model depend on β only locally optimal designs as for estimating nonlinear aspects can be defined.

Definition 2.11 (Local D-optimality for a nonlinear model) *$\delta_{\beta,D}$ is locally D-optimal for φ in Δ at β if*

$$\delta_{\beta,D} \in \arg\min\{\det(\mathcal{I}_\beta(\delta)^{-1});\ \delta \in \Delta \cap \Delta(\beta)\}.$$

Definition 2.12 (Local A-optimality for a nonlinear model) *$\delta_{\beta,A}$ is locally A-optimal for φ in Δ at β if*

$$\delta_{\beta,A} \in \arg\min\{\mathrm{tr}(\mathcal{I}_\beta(\delta)^{-1});\ \delta \in \Delta \cap \Delta(\beta)\}.$$

If we have only designs with a support included in some $\mathcal{D} = \{\tau_1, ..., \tau_I\}$, where $a(\tau_1), ..., a(\tau_I)$ are linearly independent, then Theorem 2.7 provides the locally A-optimal designs. Therefor set

$$A_{\mathcal{D},\beta} := (\dot\mu(\tau_1,\beta), ..., \dot\mu(\tau_I,\beta))'.$$

Then the locally A-optimal design $\delta_{\beta,A}$ in $\Delta_{\mathcal{D}}^*$ is given by

$$\delta_{\beta,A}(\{t\}) = \frac{|(A_{\mathcal{D},\beta}'A_{\mathcal{D},\beta})^{-1}\,\dot\mu(t,\beta)|}{\sum_{\tau\in\mathcal{D}} |(A_{\mathcal{D},\beta}'A_{\mathcal{D},\beta})^{-1}\,\dot\mu(\tau,\beta)|}$$

for $t \in \mathcal{D}$.

Example 2.4 (Generalized linear model)
We regard a generalized linear model, where the observation at t_{nN} is given by

$$Y_{nN} = e^{\beta_0+\beta_1 t_{nN}} + Z_{nN} = \mu(t_{nN}, \beta) + Z_{nN},$$

for $n = 1, ..., N$, where $\beta = (\beta_0, \beta_1) \in I\!R^2$, $\mu(t, \beta) = e^{\beta_0+\beta_1 t}$ and $t \in \mathcal{T} = [0, 1]$. If we restrict ourselves to designs with a support included in $\mathcal{D} = \{0, 1\}$, then we have

$$A_{\mathcal{D},\beta} = \begin{pmatrix} e^{\beta_0} & 0 \\ e^{\beta_0+\beta_1} & e^{\beta_0+\beta_1} \end{pmatrix},$$

where

$$\dot{\mu}(t, \beta) = e^{\beta_0+\beta_1 t} \begin{pmatrix} 1 \\ t \end{pmatrix}.$$

Hence according to Theorem 2.7 the locally A-optimal design in $\Delta_{\mathcal{D}}^*$ is given by

$$\delta_{\beta,A} = \frac{1}{\sqrt{2} + e^{-\beta_1}} (\sqrt{2}\, e_0 + e^{-\beta_1}\, e_1).\square$$

Part II

Robust Inference for Planned Experiments

3

Smoothness Concepts of Outlier Robustness

In this chapter concepts of outlier robustness are presented which base on some smoothness properties of the estimators and corresponding functionals. In Section 3.1 error distributions are given, which take the presence of outliers into account, and neighbourhoods around the ideal error distribution are introduced which include the outlier modelling distributions. Basing on these neighbourhoods in Section 3.2 the main smoothness concepts of outlier robustness are given. In Section 3.3 the strongest smoothness concept, the Fréchet differentiability, is derived for M-functionals which provide M-estimators.

3.1 Distributions Modelling Outliers

If we have observations without outliers, then for a model as given in Section 1.2 a common assumption for the distribution of the error vector $Z_N = (Z_{1N}, ..., Z_{NN})'$ is $E(Z_N) = 0$ and $Cov(Z_N) = \sigma^2 E_N$ (see Section 2.1). Additionally, often it is assumed that except for σ^2 the distribution of Z_{nN} is known, i.e. that the distribution of Z_{nN} is given by some parametric family $\{P_\sigma;\ \sigma \in \mathbb{R}^+\}$. For $\sigma = 1$ we set $P = P_\sigma$ so that P is the distribution of $\frac{1}{\sigma} Z_{nN}$. Then the distributions of the observations $Y_{1N}, ..., Y_{NN}$ are given by the parametric family $\{P_\theta;\ \theta \in \Theta\}$, where $\Theta \subset \mathbb{R}^r \times \mathbb{R}^+$ and $\theta = (\beta', \sigma)'$.

In models with different experimental conditions (independent variables) $t_{1N}, ..., t_{NN}$ the members of the parametric family will be Markov kernels, where $P_{\theta,t} := P_\theta(\cdot, t)$ is the distribution of Y_{nN} at the experimental condition t. Hence the probability measure $P_{\theta,t_{nN}}$ provides for deterministic designs $d_N = (t_{1N}, ..., t_{NN})'$ the distribution of Y_{nN} and for random designs $D_N = (T_{1N}, ..., T_{NN})'$ the distribution of Y_{nN} given $T_{nN} = t_{nN}$. Hence for outlier-free situations we have for deterministic designs $Cov_\theta(Y_N) = Cov_\theta(Z_N) = \sigma^2 E_N$, $E_\theta(Z_{nN}) = 0$ and $E_\theta(Y_{nN}) = \mu(t_{nN}, \beta)$ for $n = 1, ..., N$, where $P_\theta^N := \bigotimes_{n=1}^N P_{\theta,t_{nN}}$ denotes the distribution of $Y_N = (Y_{1N}, \ldots, Y_{NN})'$. For random designs we have $Cov_\theta(Y_N|D_N = d_N) = Cov_\theta(Z_N) = \sigma^2 E_N$, $E_\theta(Z_{nN}) = 0$ and $E_\theta(Y_{nN}|T_{nN} = t_{nN}) = \mu(t_{nN}, \beta)$

for $n = 1, ..., N$, where $P_\theta^N := \bigotimes_{n=1}^N P_{\theta,\delta}$ with $P_{\theta,\delta} := P_\theta \otimes \delta$ denotes the distribution of (Y_N, D_N). In particular we always have that $\int y\, P_{\theta,t}(dy) = \mu(t,\beta)$ holds and that $P^N := \bigotimes_{n=1}^N P$ is the distribution of $\frac{1}{\sigma} Z_N$.

But if the observations include some outliers or gross errors, then the errors $Z_{1N}, ..., Z_{NN}$ may remain independent but they will be not anymore identically distributed with expectation $E_\theta(Z_{nN}) = 0$. Thereby the amount and the form of outliers may be different for different experimental conditions t so that the distribution of a single error Z_{nN} or a single observation Y_{nN} may depend on the experimental condition t_{nN} at which the observation Y_{nN} is made.

For random designs this means that not any more the distribution of Y_{nN} will be given by the Markov kernel P_θ but will be given by another Markov kernel $Q_{N,\theta}$, where eventually $\int y\, Q_{N,\theta,t}(dy) \neq \mu(t,\beta)$. Then (Y_{nN}, T_{nN}) is distributed according to $Q_{N,\theta,\delta} := Q_{N,\theta} \otimes \delta$ and because $(Y_{1N}, T_{1N}), ..., (Y_{NN}, T_{NN})$ remain being independent the distribution of (Y_N, D_N) is given by $Q_\theta^N := \bigotimes_{n=1}^N Q_{N,\theta,\delta}$.

For deterministic designs the distribution $Q_{nN,\theta}$ of Y_{nN} also may depend on the value of t_{nN} so that $Q_{nN,\theta}$ also may be given by a Markov kernel $Q_{N,\theta}$, where $Q_{nN,\theta} = Q_{N,\theta,t_{nN}} = Q_{N,\theta}(\cdot, t_{nN})$. Then the distribution of Y_N is $Q_\theta^N := \bigotimes_{n=1}^N Q_{nN,\theta}$.

Because usually $Q_{N,\theta,\delta} = Q_{N,\theta} \otimes \delta$ will be in some neighbourhood of $P_{\theta,\delta} = P_\theta \otimes \delta$ (see below) we often can suppress θ by writing $Q_{N,\delta}$ and Q^N instead of $Q_{N,\theta,\delta}$ and Q_θ^N. In particular this is done if $\theta = (0_r', 1)'$, i.e. if $\frac{1}{\sigma} Z_{nN}$ is regarded. Sometimes also N will be suppressed. Recall also from Section 1.1 that for deterministic design sequences $(d_N)_{N\in\mathbb{N}}$ the asymptotic design measure is denoted by δ so that it makes sense to regard $Q_{N,\theta,\delta}$ and $P_{\theta,\delta}$ also for deterministic designs.

With the Markov kernel $Q_{N,\theta}$ the form of the contamination by outliers can be described. Often it can be assumed that at a given experimental condition t only with some probability $\tilde{\epsilon}(t) \in [0,1]$ an outlier appears and if an outlier appears its distribution is given by some distribution $\tilde{P}(\cdot, t) = \tilde{P}_t$. Then we have $Q_{N,\theta,t} = (1-\tilde{\epsilon}(t))P_{\theta,t} + \tilde{\epsilon}(t)\tilde{P}_t$. For asymptotic purposes it will be useful to split $\tilde{\epsilon} : \mathcal{T} \to [0,1]$ into a constant $\epsilon > 0$ and a contamination function $c : \mathcal{T} \to [0,\infty)$ such that with $\tilde{\epsilon}(t) = \epsilon\, c(t)$ we have $Q_{N,\theta,t} = (1 - \epsilon\, c(t))P_{\theta,t} + \epsilon\, c(t)\tilde{P}_t$. All such distributions provide a neighbourhood around the distribution $P_{\theta,\delta}$ for outlier-free experimental conditions. This neighbourhood generalizes the ϵ-contamination neighbourhood, which is defined for situations where all errors Z_{nN} have the same distribution (see for example Huber (1981), p. 11, Huber (1983)), and is called by Rieder (1994), p. 262, the *conditional neighbourhood with contamination curve* $c : \mathcal{T} \to [0,\infty)$.

The union of all these neighbourhoods with respect to all contamination curves c satisfying $\int c(t)\, \delta(dt) \leq 1$ provides the *average conditional neighbourhood* (see Rieder (1994), p. 263, but also Bickel (1981), (1984),

Rieder (1985), (1987), Kurotschka and Müller (1992)). For simplicity here this neighbourhood also will be called *contamination neighbourhood* and in particular *full contamination neighbourhood* if all probability measures of $\mathcal{P}$ are allowed for $\tilde{P}_t$ for every $t \in \mathcal{T}$. Thereby $\mathcal{P}$ is the set of all probability measures $\tilde{P} = \alpha \tilde{P}_* + (1-\alpha)\tilde{P}_{**}$ on the Borel σ-field of $\mathbb{R}$, where $\alpha \in [0,1]$, $\tilde{P}_*$ is a discrete measure and $\tilde{P}_{**}$ has a density with respect to the Lebesgue measure λ. Moreover let denote $\mathcal{P}^*$ the set of all probability measures on $\mathbb{R} \times \mathcal{T}$. Note that if the support $\text{supp}(\delta)$ of δ is finite, which is usually the case in designed experiments, then for every choice $\tilde{P}_t \in \mathcal{P}$ for $t \in \text{supp}(\delta)$ there exists a probability measure $\tilde{P} \otimes \delta \in \mathcal{P}^*$ with a Markov kernel $\tilde{P}$ such that $\tilde{P}_t = \tilde{P}(\cdot, t)$ for $t \in \text{supp}(\delta)$. Usually the neighbourhoods around $P_{\theta,\delta}$ are regarded but sometimes it will be useful to have neighbourhoods also around other probability measures $P_\delta \in \mathcal{P}^*$ with conditional distributions P_t.

Definition 3.1 (Contamination neighbourhood)
Let be $\epsilon > 0$.
a) $\mathcal{U}_c(P_\delta, \epsilon, c)$ is the (full) conditional neighbourhood of P_δ with contamination curve $c : \mathcal{T} \to [0, \infty)$ if

$$\begin{aligned} \mathcal{U}_c(P_\delta, \epsilon, c) \quad = \quad & \{Q \otimes \delta \in \mathcal{P}^*;\ Q \text{ is Markox kernel, where} \\ & Q(\cdot, t) = (1 - \epsilon\, c(t))P_t + \epsilon\, c(t)\tilde{P}_t \\ & \text{with } \tilde{P}_t \in \mathcal{P} \text{ for all } t \in \mathcal{T}\}. \end{aligned}$$

b) $\mathcal{U}_c(P_\delta, \epsilon)$ is the (full) contamination neighbourhood of P_δ if

$$\mathcal{U}_c(P_\delta, \epsilon) = \bigcup \{\mathcal{U}_c(P_\delta, \epsilon, c);\ c : \mathcal{T} \to [0, \infty),\ \int c(t)\, \delta(dt) \le 1\}.$$

While simple contaminations of the form $(1-\epsilon)P_\delta + \epsilon\, \tilde{P}_\delta$ are included in $U_c(P_\delta, \epsilon)$ simple contaminations of the form $(1-\epsilon)P_\delta + \epsilon\, \tilde{P}_{\tilde{\delta}}$, where $\tilde{\delta} \ne \delta$, are not included in $U_c(P_\delta, \epsilon)$. But sometimes the neighbourhoods of all these simple contaminations with

$$\tilde{P}_{\tilde{\delta}} \in \mathcal{P}^*(\text{supp}(\delta)) := \{Q_\xi \in \mathcal{P}^*;\ \text{supp}(\xi) \subset \text{supp}(\delta)\}$$

are also useful (see Section 3.2 and Section 4.2). These neighbourhoods are called *simple contamination neighbourhoods.*

Definition 3.2 (Simple contamination neighbourhood)
Let be $\epsilon > 0$. $\mathcal{U}_c^s(P_\delta, \epsilon)$ is the simple contamination neighbourhood of P_δ if

$$\begin{aligned} \mathcal{U}_c^s(P_\delta, \epsilon) \quad = \quad & \{Q_\xi \in \mathcal{P}^*;\ \text{there exists } \tilde{P}_{\tilde{\delta}} \in \mathcal{P}^*(supp(\delta)) \text{ with} \\ & Q_\xi = (1-\epsilon)P_\delta + \epsilon\, \tilde{P}_{\tilde{\delta}}\}. \end{aligned}$$

Some asymptotic results for $N \to \infty$ holds only if the radius of the neighbourhoods around $P_{\theta,\delta}$ decreases with $\sqrt{N}$ so that we also will regard *full shrinking contamination neighbourhoods* (see in particular Section 5.1).

Definition 3.3 (Full shrinking contamination neighbourhood)
$\mathcal{U}_{c,\epsilon}(P_{\theta,\delta})$ is the full shrinking contamination neighbourhood if

$$\begin{aligned}\mathcal{U}_{c,\epsilon}(P_{\theta,\delta}) \;=\; & \{(Q_{N,\delta})_{N\in I\!N};\ \textit{there exists } c:\mathcal{T}\to[0,\infty) \\ & \textit{with } \int c(t)\,\delta(dt)\le 1 \textit{ and } \tilde{P}_t\in\mathcal{P} \textit{ for } t\in\mathcal{T} \\ & \textit{such that for all } t\in\mathcal{T} \textit{ and } N\in I\!N \\ & Q_{N,t}=(1-N^{-1/2}\epsilon\, c(t))P_{\theta,t}+N^{-1/2}\epsilon\, c(t)\tilde{P}_t\}.\end{aligned}$$

In Section 5.1 it is shown that for some classes of estimators the full shrinking contamination neighbourhood is too large so that it is useful to restrict the shrinking neighbourhood to sequences $(Q_N)_{N\in I\!N}$ which are contiguous to the central distribution P_θ. This is achieved by allowing only probability measures $\tilde{P}_t$ which have a density $g(\cdot,t)$ with respect to the central probability measure $P_{\theta,\delta}$, where

$$\begin{aligned}g\in\mathcal{G}_\theta := \{g: I\!R\times\mathcal{T}\to[0,\infty);\ & \int g(y,t)^2\,P_{\theta,\delta}(dy,dt)<\infty, \\ & \|g\|_\infty<\infty,\ \int g(y,t)\,P_{\theta,t}(dy)=1 \text{ for all } t\in\mathcal{T}\}.\end{aligned}$$

Thereby $\|g\|_\infty$ denotes the essential supremum of g with respect to $P_{\theta,\delta}$. Here these shrinking neighbourhoods are called *restricted shrinking contamination neighbourhoods.*

Definition 3.4 (Restricted shrinking contamination neighbourhood)
Let be $\epsilon>0$. $\mathcal{U}^0_{c,\epsilon}(P_{\theta,\delta})$ is the restricted shrinking contamination neighbourhood if

$$\begin{aligned}\mathcal{U}^0_{c,\epsilon}(P_{\theta,\delta}) \;=\; & \{(Q_{N,\delta})_{N\in I\!N};\ \textit{there exists } c:\mathcal{T}\to[0,\infty) \textit{ with} \\ & \int c(t)\,\delta(dt)\le 1 \textit{ and } g\in\mathcal{G}_\theta \textit{ such that} \\ & Q_{N,t}=(1-N^{-1/2}\epsilon\, c(t))P_{\theta,t}+N^{-1/2}\epsilon\, c(t)g(\cdot,t)P_{\theta,t} \\ & \textit{for all } t\in\mathcal{T} \textit{ and } N\in I\!N\}.\end{aligned}$$

Note that besides the simple contamination neighbourhood in all other contamination neighbourhoods the distribution δ for the experimental conditions is fixed. This is due to the fact that in planned experiments the experimenter determines the distribution δ which provides the random design D_N or the asymptotic design for deterministic design sequences $(d_N)_{N\in I\!N}$.

To allow not only a small amount of gross errors and large outliers but also a large amount of small deviations from the central model P_θ we can regard neighbourhoods which are given by a metric. There are many possible metrics (see for example Huber (1981), pp. 25). Here the most useful metric is the Kolmogorov metric (see in Section 3.2 and Section 3.3). For designed experiments this metric can be generalized to distances between probability measures $P_\delta := P \otimes \delta$ and $Q_\xi := Q \otimes \xi$ on $\mathbb{R} \times \mathcal{T}$ by denoting by $F_t(\cdot)$ and $G_t(\cdot)$ the distribution functions of the Markov kernels $P_t = P(\cdot, t)$ and $Q_t = Q(\cdot, t)$, respectively.

Definition 3.5 (Generalized Kolmogorov metric)
If δ and ξ are probability measures on $\mathcal{T}$ with support included in a finite set $\mathcal{T}_0$, then

$$d_K(P_\delta, Q_\xi) := \max_{(y,t) \in \mathbb{R} \times \mathcal{T}_0} |F_t(y)\delta(\{t\}) - G_t(y)\xi(\{t\})|$$

is the generalized Kolmogorov metric on $\mathcal{P}^*$.

The following lemma shows the relation to the traditional Kolmogorov metric on $\mathbb{R}$.

Lemma 3.1 *Let δ and ξ be probability measures on $\mathcal{T}$ with support included in a finite set $\mathcal{T}_0$. Then we have:*

$$\begin{aligned}
&a)\quad \max_{t \in \mathcal{T}} |\delta(\{t\}) - \xi(\{t\})| \leq d_K(P_\delta, Q_\xi),\\
&b)\quad \max_{y \in \mathbb{R}} |F_t(y) - G_t(y)| \leq \frac{2\, d_K(P_\delta, Q_\xi)}{\min\{\delta(\{t\});\ t \in \operatorname{supp}(\delta)\}} \quad \textit{and}\\
&c)\quad d_K(P_\delta, Q_\xi) \leq \max_{t \in \mathcal{T}} |\delta(\{t\}) - \xi(\{t\})|\\
&\qquad\qquad + \max_{(y,t) \in \mathbb{R} \times \mathcal{T}} \delta(\{t\})\, |F_t(y) - G_t(y)|.
\end{aligned}$$

Proof. The assertion a) follows by estimating

$$\begin{aligned}
&d_K(P_\delta, Q_\xi)\\
&\quad\geq \max_{(y,t) \in \mathbb{R} \times \mathcal{T}} G_t(y)|\delta(\{t\}) - \xi(\{t\})|\\
&\qquad - \max_{(y,t) \in \mathbb{R} \times \mathcal{T}} \delta(\{t\})|F_t(y) - G_t(y)|
\end{aligned}$$

and letting $y \to \infty$. The assertion b) follows by using a) and estimating

$$\begin{aligned}
&d_K(P_\delta, Q_\xi)\\
&\quad\geq \max_{(y,t) \in \mathbb{R} \times \mathcal{T}} \delta(\{t\})|F_t(y) - G_t(y)|\\
&\qquad - \max_{(y,t) \in \mathbb{R} \times \mathcal{T}} G_t(y)|\delta(\{t\}) - \xi(\{t\})|\\
&\quad\geq \left(\min_{t \in \operatorname{supp}(\delta)} \delta(\{t\})\right) \max_{y \in \mathbb{R}} |F_t(y) - G_t(y)| - d_K(P_\delta, Q_\xi).
\end{aligned}$$

The assertion c) follows by estimating

$$\begin{aligned}
&d_K(P_\delta, Q_\xi)\\
&\quad\leq \max_{(y,t) \in \mathbb{R} \times \mathcal{T}} \delta(\{t\})|F_t(y) - G_t(y)|\\
&\qquad + \max_{(y,t) \in \mathbb{R} \times \mathcal{T}} G_t(y)|\delta(\{t\}) - \xi(\{t\})|. \square
\end{aligned}$$

Other generalizations of the one-dimensional Kolmogorov metric are

$$\overline{d}_K(P_\delta, Q_\xi) := \sum_{t \in \mathcal{T}_0} \max_{y \in I\!R} |F_t(y)\delta(\{t\}) - G_t(y)\xi(\{t\})|$$

and

$$d_K^*(P_\delta, Q_\xi) := \max_{(y,t) \in I\!R \times \mathcal{T}_0} |F_\delta(y,t) - G_\xi(y,t)|,$$

where F_δ and G_ξ are the distribution functions of P_δ and Q_ξ, respectively, i.e. $F_\delta(y,t) := \sum_{\bar{t} \leq t} F_{\bar{t}}(y)\delta(\{\bar{t}\})$. But these metrics are equivalent to d_K because $d_K(P_\delta, Q_\xi) \leq \overline{d}_K(P_\delta, Q_\xi) \leq I\, d_K(P_\delta, Q_\xi)$ and $\frac{1}{2} d_K(P_\delta, Q_\xi) \leq d_K^*(P_\delta, Q_\xi) \leq I\, d_K(P_\delta, Q_\xi)$ if $\mathcal{T}_0$ has I elements.

With the generalized Kolmogorov metric also a neighbourhood, the *Kolmogorov neighbourhood*, around the central model $P_{\theta,\delta}$ can be defined. If the neighbourhood should only contain distributions modelling outliers and other contaminations of the observations, then again the design δ should be fixed, i.e. we restrict ourselves to neighbourhoods containing only

$$Q_\xi \in \mathcal{P}_\delta^* := \{Q_\xi \in \mathcal{P}^*;\ \xi = \delta\}.$$

This neighbourhood will be called *restricted Kolmogorov neighbourhood.* For analytical purposes it also will be useful to regard *full Kolmogorov neighbourhoods* which contain $Q_\xi \in \mathcal{P}^*(\text{supp}(\delta))$. In particular these neighbourhoods also contain the empirical distributions of (y_N, d_N) which will be important in Section 3.2 and Section 5.1.

Definition 3.6 (Kolmogorov neighbourhood)
Let be $\epsilon > 0$.
a) $\mathcal{U}_K^0(P_\delta, \epsilon)$ *is the restricted Kolmogorov neighbourhood of* P_δ *if*

$$\mathcal{U}_K^0(P_\delta, \epsilon) = \{Q_\delta \in \mathcal{P}_\delta^*;\ d_K(P_\delta, Q_\delta) \leq \epsilon\}.$$

b) $\mathcal{U}_K(P_\delta, \epsilon)$ *is the full Kolmogorov neighbourhood of* P_δ *if*

$$\mathcal{U}_K(P_\delta, \epsilon) = \{Q_\xi \in \mathcal{P}^*(supp(\delta));\ d_K(P_\delta, Q_\xi) \leq \epsilon\}.$$

For sequences $(Q_{N,\delta})_{N \in I\!N}$ shrinking neighbourhoods can be defined as in Definition 3.3 and 3.4. But to expand the neighbourhood we will regard not only sequences approaching $P_{\theta,\delta}$ but also sequences approaching $P_{\theta_N,\delta}$, where $(\theta_N)_{N \in I\!N}$ is an element of

$$J_\theta := \{(\theta_N)_{N \in I\!N};\ \text{there exists } K > 0 \text{ such that } \sqrt{N}|\theta_N - \theta| \leq K \text{ for all } N \in I\!N\}.$$

Definition 3.7 (Shrinking Kolmogorov neighbourhood)
Let be $\epsilon > 0$. $\mathcal{U}_{K,\epsilon}(P_{\theta,\delta})$ *is the shrinking Kolmogorov neighbourhood of* $P_{\theta,\delta}$ *if*

$$\mathcal{U}_{K,\epsilon}(P_{\theta,\delta}) = \{(Q_{N,\xi_N})_{N \in I\!N};\ \textit{there exists } (\theta_N)_{N \in I\!N} \in J_\theta \textit{ with } Q_{N,\xi_N} \in \mathcal{U}_K(P_{\theta_N,\delta}, N^{-1/2}\epsilon) \textit{ for all } N \in I\!N\}.$$

Lemma 3.2 *We have*

$$\mathcal{U}_c^s(P_\delta,\epsilon) \subset \mathcal{U}_K(P_\delta,\epsilon) \text{ and } \mathcal{U}_c(P_\delta,\epsilon) \subset \mathcal{U}_K^0(P_\delta,\epsilon) \subset \mathcal{U}_K(P_\delta,\epsilon)$$

and in particular for the shrinking neighbourhoods

$$\mathcal{U}_{c,\epsilon}^0(P_{\theta,\delta}) \subset \mathcal{U}_{c,\epsilon}(P_{\theta,\delta}) \subset \mathcal{U}_{K,\epsilon}(P_{\theta,\delta}).$$

Proof. At first we show $\mathcal{U}_c^s(P_\delta,\epsilon) \subset \mathcal{U}_K(P_\delta,\epsilon)$. Let be $Q_\xi = (1-\epsilon)P_\delta + \epsilon\,\tilde{P}_{\tilde{\delta}}$ arbitrary. Then we have

$$\begin{aligned} Q_\xi &= \sum\nolimits_{t\in\text{supp}(\delta)} (1-\epsilon)P_t\,\delta(\{t\})e_t + \epsilon\tilde{P}_t\,\tilde{\delta}(\{t\})e_t \\ &= \sum\nolimits_{t\in\text{supp}(\delta)} \tilde{Q}_t\,\tilde{\xi}(\{t\})e_t \end{aligned}$$

with

$$\tilde{\xi}(\{t\}) = (1-\epsilon)\delta(\{t\}) + \epsilon\tilde{\delta}(\{t\})$$

and

$$\tilde{Q}_t = \frac{1}{(1-\epsilon)\delta(\{t\}) + \epsilon\tilde{\delta}(\{t\})}(1-\epsilon)\delta(\{t\})P_t + \epsilon\tilde{\delta}(\{t\})\tilde{P}_t,$$

where

$$\tilde{G}_t(\cdot) = \frac{1}{(1-\epsilon)\delta(\{t\}) + \epsilon\tilde{\delta}(\{t\})}(1-\epsilon)\delta(\{t\})F_t(\cdot) + \epsilon\tilde{\delta}(\{t\})\tilde{F}_t(\cdot)$$

is the distribution function of $\tilde{Q}_t$. This implies

$$\begin{aligned} d_K(P_\delta,Q_\xi) &= \max\nolimits_{(y,t)\in\mathbb{R}\times\text{supp}(\delta)} |F_t(y)\,\delta(\{t\}) - \tilde{G}_t(y)\,\tilde{\xi}(\{t\})| \\ &= \max\nolimits_{(y,t)\in\mathbb{R}\times\text{supp}(\delta)} \epsilon|\tilde{F}_t(y)\,\tilde{\delta}(\{t\}) - F_t(y)\,\delta(\{t\})| \\ &\leq \epsilon, \end{aligned}$$

i.e. $Q_\xi \in \mathcal{U}_K(P_\delta,\epsilon)$. Now we only have to show $\mathcal{U}_c(P_\delta,\epsilon) \subset \mathcal{U}_K^0(P_\delta,\epsilon)$. For that take any $Q_\delta \in \mathcal{U}_c(P_\delta,\epsilon)$. Then we have $Q_t = (1-\epsilon\,c(t))P_t + \epsilon\,c(t)\tilde{P}_t$ and its distribution function is $G_t(\cdot) = (1-\epsilon\,c(t))F_t(\cdot) + \epsilon\,c(t)\tilde{F}_t(\cdot)$ so that

$$d_K(P_\delta,Q_\delta) = \max\nolimits_{(y,t)\in\mathbb{R}\times\text{supp}(\delta)} \delta(\{t\})\,\epsilon\,c(t)|F_t(y) - \tilde{F}_t(y)| \leq \epsilon$$

because $\int c(t)\,\delta(dt) \leq 1$ implies $c(t) \leq \frac{1}{\delta(\{t\})}$ for all $t \in \text{supp}(\delta)$. □

3.2 Smoothness of Estimators and Functionals

An estimator will be robust against outliers if its behaviour does not change too much in the presence of outliers. Of course the change of the behaviour will depend on the amount of outliers. In the presence of only few outliers the change will be smaller than in the presence of many outliers. But at least in the presence of only few outliers the change of the estimator should not be too large. The presence of few outliers can be described by neighbourhoods around the ideal model $P_{\theta,\delta}$ as given in Section 3.1, and the behaviour of an estimator can be described by its distribution which is induced by the underlying distribution of Y_N.

Because many robust estimators for an aspect $\varphi(\beta)$ of β are based on an estimator for the unknown variance σ^2 we also here regard estimators of an aspect $\zeta(\theta)$ of the whole parameter θ. Thereby the parameter θ can be $\theta = (\beta', \sigma)' \in \Theta \subset I\!R^r \times I\!R^+$ but also $\theta = (\beta', \sigma_1, ..., \sigma_J)' \in \Theta \subset I\!R^r \times (I\!R^+)^J$ to include the heteroscedastic case, where we have different variances $\sigma_1^2, ..., \sigma_J^2$ for different disjoined subsets $\mathcal{T}_1, ..., \mathcal{T}_J$ of $\mathcal{T}$.

If we now have an estimator $\widehat{\zeta}_N : I\!R^N \times \mathcal{T}^N \to I\!R^q$ for an aspect $\zeta(\theta) \in I\!R^q$, then we can regard its distribution $(Q^N)^{\widehat{\zeta}_N}$ when Y_N or (Y_N, D_N), respectively, is distributed according to Q^N, where Q^N bases on $Q_{N,\delta}$ (see Section 3.1). Similarly if Y_N or (Y_N, D_N), respectively, is distributed according to P_θ^N, then the distribution of $\widehat{\zeta}_N$ is $(P_\theta^N)^{\widehat{\zeta}_N}$. Then the estimator will be robust against outliers when for all θ the distribution $(Q^N)^{\widehat{\zeta}_N}$ is close to $(P_\theta^N)^{\widehat{\zeta}_N}$ for all $Q_{N,\delta}$ in some neighbourhood $\mathcal{U}(P_{\theta,\delta}, \epsilon)$ of outlier modelling distributions.

If we have some metric d_1 providing the neighbourhood $\mathcal{U}(P_{\theta,\delta}, \epsilon)$ and some metric d_2 for the probability measures on $I\!R^q$, then this robustness means that for all θ the mapping $Q_{N,\delta} \to (Q^N)^{\widehat{\zeta}_N}$ is continuous at $P_{\theta,\delta}$ with respect to d_1 and d_2. This type of robustness we here call *general robustness*. Thereby the most useful choice of the metric d_1 will be the generalized Kolmogorov metric d_K which provides in particular $\mathcal{U}_K^0(P_{\theta,\delta}, \epsilon)$ for fixed designs (see Section 3.1). For the metric d_2 for probability measures on $I\!R^q$ with Borel σ-field $\mathcal{A}^q$ it will be useful to use the Prohorov metric d_P which is given by

$$\begin{aligned} d_P(P, Q) &:= \min\{\epsilon > 0;\ P(A) \leq Q(z \in I\!R^q; \\ &\qquad \min_{y \in A} |z - y| \leq \epsilon) + \epsilon \text{ for all } A \in \mathcal{A}^q\} \end{aligned}$$

(see for example Huber (1981), p. 27).

Definition 3.8 (General robustness)
An estimator $\widehat{\zeta}_N : I\!R^N \times \mathcal{T}^N \to I\!R^q$ for $\zeta(\theta)$ is generally robust if for all $\theta \in \Theta$ and all $\epsilon^* > 0$ there exists $\epsilon > 0$ such that $d_P((Q^N)^{\widehat{\zeta}_N}, (P_\theta^N)^{\widehat{\zeta}_N}) < \epsilon^*$ for all $Q_{N,\delta} \in \mathcal{U}_K^0(P_{\theta,\delta}, \epsilon)$.

Usually it is too difficult to verify the general robustness for an estimator $\widehat{\zeta}_N$ so that alternative robustness concepts were developed. One approach is to regard the asymptotic behaviour of the estimator. Another approach is to drop all distributional considerations and to compare the estimated value of the estimator $\widehat{\zeta}_N$ at a sample without outliers with the estimated value at a sample with some few outliers. This leads to the concepts of bias and breakdown point for finite samples (see Section 4.2). Here we start with the asymptotic approach because also in this approach asymptotic bias and asymptotic breakdown points play a role and appear in a more general context (see also Section 4.1).

At first we specify the general robustness to *large sample robustness* which often also is called *qualitative robustness* (see Huber (1981), p. 10) and which was introduced by Hampel (1971).

Definition 3.9 (Large sample robustness)

a) A sequence of estimators $\widehat{\zeta}_N : \mathbb{R}^N \times \mathcal{T}^N \longrightarrow \mathbb{R}^q$ for $\zeta(\theta)$, $N \in \mathbb{N}$, is large sample robust at $P_{\theta,\delta}$ if for all $\epsilon^* > 0$ there exists $\epsilon > 0$ and $N^* \in \mathbb{N}$ such that $d_P((Q^N)^{\widehat{\zeta}_N}, (P_\theta^N)^{\widehat{\zeta}_N}) < \epsilon^*$ for all $Q_{N,\delta} \in \mathcal{U}_K^0(P_{\theta,\delta}, \epsilon)$ and $N \geq N^*$.

b) A sequence of estimators $\widehat{\zeta}_N : \mathbb{R}^N \times \mathcal{T}^N \longrightarrow \mathbb{R}^q$ for $\zeta(\theta)$, $N \in \mathbb{N}$, is large sample robust if for all $\theta \in \Theta$ the estimators $\widehat{\zeta}_N$ are large sample robust at $P_{\theta,\delta}$.

Often for all observations y_N and corresponding designs d_N the estimated value $\widehat{\zeta}_N(y_N, d_N)$ of an estimator $\widehat{\zeta}_N$ can be expressed via the empirical distribution P_{y_N,d_N} as $\widehat{\zeta}_N(y_N, d_N) = \widehat{\zeta}(P_{y_N,d_N})$. Thereby $\widehat{\zeta} : \mathcal{P}^* \longrightarrow \mathbb{R}^q$ is a mapping (functional) from the space of all probability measures on $\mathbb{R} \times \mathcal{T}$ to $\mathbb{R}^q$ and the empirical distribution at the observation y_N and the design d_N is $P_{y_N,d_N} = \frac{1}{N}\sum_{n=1}^N e_{(y_{nN},t_{nN})}$, where $e_{(y,t)}$ is the one-point measure (Dirac measure) on $(y,t) \in \mathbb{R} \times \mathcal{T}$. If as in Section 1.1 $\delta_N = \frac{1}{N}\sum_{n=1}^N e_{t_{nN}}$ is the empirical distribution of $t_{1N}, ..., t_{NN}$, then we have a Markov kernel P_{y_N} such that $P_{y_N,d_N} = P_{y_N} \otimes \delta_N$. Again with $P_{y_N,t}$ we denote the probability measures $P_{y_N}(\cdot, t)$ on $\mathbb{R}$ and with $F_{y_N,t}$ its distribution function.

For functionals $\widehat{\zeta}$ providing consistent estimators by $\widehat{\zeta}_N(y_N, d_N) = \widehat{\zeta}(P_{y_N,d_N})$ the large sample robustness of $(\widehat{\zeta}_N)_{N \in \mathbb{N}}$ is equivalent with the continuity of the functional with respect to d_K. Thereby we should distinguish between continuity on $\mathcal{P}_\delta^* := \{Q_\xi \in \mathcal{P}^*;\ \xi = \delta\}$ and $\mathcal{P}^*(\mathrm{supp}(\delta)) := \{Q_\xi \in \mathcal{P}^*;\ \mathrm{supp}(\xi) \subset \mathrm{supp}(\delta)\}$. The second set has the advantage that it also includes the empirical distributions given by (y_N, d_N).

Definition 3.10

A functional $\widehat{\zeta} : \mathcal{P}^* \longrightarrow \mathbb{R}^q$ provides consistent estimators at Q_δ if

$$Q^N(|\widehat{\zeta}(P_{y_N,d_N}) - \widehat{\zeta}(P_\delta)| > \epsilon) \stackrel{N\to\infty}{\longrightarrow} 0$$

for all $\epsilon > 0$.

Theorem 3.1 *Let be* $supp(\delta_N) \subset supp(\delta) = \mathcal{T}_0 = \{t_1, ..., t_I\}$ *for all* $N \in \mathbb{N}$ *and for deterministic designs* $\lim_{N\to\infty} \delta_N(\{t\}) = \delta(\{t\})$ *for all* $t \in \mathcal{T}_0$.

a) If $\widehat{\zeta}$ *is continuous at* $P_{\theta,\delta}$ *with respect to* d_K *on* $\mathcal{P}^*(supp(\delta))$, *then* $\widehat{\zeta}$ *provides estimators which are large sample robust and consistent at* $P_{\theta,\delta}$.

b) If there exists $\epsilon > 0$ *such that* $\widehat{\zeta}$ *provides estimators which are consistent at all* $Q_\delta \in \mathcal{U}_K^0(P_{\theta,\delta}, \epsilon)$ *and large sample robust at* $P_{\theta,\delta}$, *then* $\widehat{\zeta}$ *is continuous at* $P_{\theta,\delta}$ *with respect to* d_K *on* $\mathcal{P}_\delta^*$.

Proof. The proof is the same as in Huber (1981), p. 41/42 and Hampel (1971) for the one-dimesional case. We only additionally have to show that for all $\epsilon_1 > 0$ and $\epsilon_2 > 0$ there exists N^* such that for all $N \in \mathbb{N}$ and $Q_{N,\delta} \in \mathcal{P}_\delta^*$ we have

$$Q_\delta^N(d_K(Q_{N,\delta}, P_{y_N,d_N}) \leq \epsilon_1) \geq 1 - \epsilon_2. \tag{3.1}$$

But this follows from Lemma 3.1 which provides

$$\begin{aligned} d_K(Q_{N,\delta}, P_{y_N,d_N}) \leq\ & \max_{t\in\mathcal{T}_0} |\delta(\{t\}) - \delta_N(\{t\})| \\ & + \max_{(y,t)\in\mathbb{R}\times\mathcal{T}_0} |Q_{N,t}(y) - P_{y_N,t}(y)|\, \delta(\{t\}) \end{aligned}$$

so that as in the one-dimesional case the uniform convergence in the Glivenko-Cantelli theorem (see for example Billingsley (1968), p. 103/104, Fisz (1963), p. 394, and Noether (1963)) provides (3.1). □

Theorem 3.1 shows that continuity of the functional $\widehat{\zeta}$ is closely related to the large sample robustness of the corresponding estimators. Because the functional $\widehat{\zeta}$ described mainly the asymptotic values of the estimators $\widehat{\zeta}_N$ given by $\widehat{\zeta}_N(y_N, d_N) = \widehat{\zeta}(P_{y_N,d_N})$ we will call the continuity property also *general asymptotic robustness*.

Definition 3.11 (General asymptotic robustness)
The estimators $\widehat{\zeta}_N$, $N \in \mathbb{N}$, *given by the functional* $\widehat{\zeta}$ *are generally asymptotically robust if for all* $\theta \in \Theta$ *the functional* $\widehat{\zeta}$ *is continuous at* $P_{\theta,\delta}$ *with respect to* d_K *on* $\mathcal{P}^*(supp(\delta))$, *i.e. for all* $\theta \in \Theta$ *and* $\epsilon^* > 0$ *there exists* $\epsilon > 0$ *such that* $|\widehat{\zeta}(Q_\xi) - \widehat{\zeta}(P_{\theta,\delta})| < \epsilon^*$ *for all* $Q_\xi \in \mathcal{U}_K(P_{\theta,\delta}, \epsilon)$.

General asymptotic robustness at $P_{\theta,\delta}$ describes only that the functional $\widehat{\zeta}$ is continuous at $P_{\theta,\delta}$. It does not describe how smooth the functional at $P_{\theta,\delta}$ is, what the maximal deviation from $\widehat{\zeta}(P_{\theta,\delta})$ in a given neighbourhood of $P_{\theta,\delta}$ is and how large a neighbourhood can be chosen so that the functional remains bounded. But these three properties are also important for robustness. The questions concerning the maximal deviation from $\widehat{\zeta}(P_{\theta,\delta})$ in a given neighbourhood and the smallest neighbourhood without singularity are regarded in Section 4.1 and Section 4.2. In this section we will regard further robustness concepts based on infinitesimal neighbourhoods of $P_{\theta,\delta}$.

A stronger robustness condition as continuity is *Lipschitz continuity*, for example for the generalized Kolmogorov metric d_K, i.e. for every $\theta \in \Theta$ we should have $|\widehat{\zeta}(Q_\xi) - \widehat{\zeta}(P_{\theta,\delta})| = O(d_K(P_{\theta,\delta}, Q_\xi))$ which means that there exists $K \in \mathbb{R}$ such that

$$\lim_{\epsilon \to 0} \max_{d_K(P_{\theta,\delta},Q_\xi) \leq \epsilon} \frac{|\widehat{\zeta}(Q_\xi) - \widehat{\zeta}(P_{\theta,\delta})|}{d_K(P_{\theta,\delta}, Q_\xi)} \leq K.$$

A still stronger robustness condition than Lipschitz continuity is *Fréchet differentiability*, which means that for every $\theta \in \Theta$ there exists a functional $\tilde{\zeta}_\theta : \mathcal{P}^\pm \to \mathbb{R}^q$ on the set of all signed measures on $\mathbb{R} \times \mathcal{T}$ which is linear in $Q_\xi - P_{\theta,\delta}$ such that $|\widehat{\zeta}(Q_\xi) - \widehat{\zeta}(P_{\theta,\delta}) - \tilde{\zeta}_\theta(Q_\xi - P_{\theta,\delta})| = o(d_K(P_{\theta,\delta}, Q_\xi))$, i.e.

$$\lim_{\epsilon \to 0} \max_{d_K(P_{\theta,\delta},Q_\xi) \leq \epsilon} \frac{|\widehat{\zeta}(Q_\xi) - \widehat{\zeta}(P_{\theta,\delta}) - \tilde{\zeta}_\theta(Q_\xi - P_{\theta,\delta})|}{d_K(P_{\theta,\delta}, Q_\xi)} = 0$$

(see Huber (1981), pp. 34). Again it is enough to regard only all $Q_\xi \in \mathcal{P}^*$ with $\text{supp}(\xi) \subset \text{supp}(\delta)$.

Definition 3.12 (Strong asymptotic robustness)
The estimators $\widehat{\zeta}_N$, $N \in \mathbb{N}$, given by the functional $\widehat{\zeta}$, are strongly asymptotically robust if for all $\theta \in \Theta$ the functional $\widehat{\zeta}$ is Fréchet differentiable at $P_{\theta,\delta}$ with respect to d_K on $\mathcal{P}^*(supp(\delta))$.

Often it is difficult to derive Fréchet differentiability of $\widehat{\zeta}$ so that Davies (1993) proposed Lipschitz continuity as important robustness criterion. But for designed experiments often Fréchet differentiability can be shown (see Section 3.3).

In Section 5.1 it is shown that some wellknown outlier robust estimators as the median are not strongly asymptotically robust. This means that Fréchet differentiability is not only connected with outlier robustness but also with inlier robustness (see Section 5.1).

Because Fréchet differentiability may be a too strong robustness condition we can regard weaker differentiability notions as for example *Gateaux differentiability*. While Fréchet differentiability can be interpreted as differentiability with respect to all bounded subsets of $\mathcal{P}^*(\text{supp}(\delta))$, Gateaux differentiability can be interpreted as differentiability with respect to all finite subsets of $\mathcal{P}^*(\text{supp}(\delta))$. Differentiability also can be regarded with respect to all compact subsets of $\mathcal{P}^*(\text{supp}(\delta))$ which provides *Hadamard differentiability*. In situations where Fréchet differentiability is too strong Hadamard differentiability can be also very helpful and important for robustness considerations (see van der Vaart (1991a,b), Maercker (1992), Rieder (1994), Ren (1994)). But because for designed experiments Fréchet differentiability often can be shown we here only will regard Gateaux differentiability besides Fréchet differentiability.

A functional $\widehat{\zeta}$ is Gateaux differentiable at $P_{\theta,\delta}$ on $\mathcal{P}^*(\text{supp}(\delta))$ if for all $\theta \in \Theta$ there exists a functional $\tilde{\zeta}_\theta : \mathcal{P}^\pm \to I\!R^q$ which is linear in $Q_\xi - P_{\theta,\delta}$ such that for all $\tilde{P}_{\tilde{\xi}} \in \mathcal{P}^*(\text{supp}(\delta))$

$$\lim_{\epsilon \to 0} \frac{|\widehat{\zeta}((1-\epsilon)P_{\theta,\delta} + \epsilon \tilde{P}_{\tilde{\xi}}) - \widehat{\zeta}(P_{\theta,\delta}) - \epsilon\, \tilde{\zeta}_\theta(\tilde{P}_{\tilde{\xi}} - P_{\theta,\delta})|}{\epsilon} = 0.$$

It is clear that if $\widehat{\zeta}$ is Fréchet differentiable, then it is also Gateaux differentiable and the Fréchet derivative $\tilde{\zeta}_\theta$ coincides with the Gateaux derivative.

Gateaux differentiability is such a weak concept that even functionals which provides nonrobust estimators are Gateaux differentiable. For example the mean $\widehat{\zeta}(P) = \int y\, P(dy)$, which provides the arithmetic mean, is Gateaux differentiable. But its Gateaux derivative $\tilde{\zeta}(Q - P) = \int y\,(Q - P)(dy)$ is unbounded in Q and functionals providing robust estimators should have a bounded Gateaux derivative.

At least as in approaches without different fixed experimental conditions the Gateaux derivatives in directions of one-point measures $\tilde{P}_{\tilde{\xi}} = e_{(y,t)}$ with $y \in I\!R$ and $t \in \text{supp}(\delta)$ should be bounded. Hampel (1968, 1974) interpreted these derivative as the influence of a single observation on the estimator and called the functions derived by these derivatives the influence function (see also Hampel et al. (1986)). Similarly, here the derivative in direction of $\tilde{P}_{\tilde{\xi}} = e_{(y,t)}$ expresses the influence of an observation y at the experimental condition t (see also Hampel (1978)).

Definition 3.13 (Influence function)
The influence function $IF(\widehat{\zeta}, P_{\theta,\delta}, \cdot) : I\!R \times supp(\delta) \to I\!R$ of $\widehat{\zeta}$ at $P_{\theta,\delta}$ is defined by

$$IF(\widehat{\zeta}, P_{\theta,\delta}, y, t) = \lim_{\epsilon \to 0} \frac{\widehat{\zeta}((1-\epsilon)P_{\theta,\delta} + \epsilon e_{(y,t)}) - \widehat{\zeta}(P_{\theta,\delta})}{\epsilon}.$$

If the influence function is bounded, then the estimators given by the functional $\widehat{\zeta}$ will be called *weakly asymptotically robust.*

Definition 3.14 (Weak asymptotic robustness)
The estimators $\widehat{\zeta}_N$ given by the functional $\widehat{\zeta}$ are weakly asymptotically robust if for all $\theta \in \Theta$

$$\max_{(y,t) \in I\!R \times \text{supp}(\delta)} |IF(\widehat{\zeta}, P_{\theta,\delta}, y, t)| < \infty.$$

Often the weak asymptotic robustness implies that all Gateaux derivatives in directions of $\tilde{P}_{\tilde{\xi}}$ with $\text{supp}(\tilde{\xi}) \subset \text{supp}(\delta)$ are bounded. In particular this holds for functionals $\widehat{\zeta}$ which are Fréchet differentiable at $P_{\theta,\delta}$ and continuous in a neighbourhood of $P_{\theta,\delta}$ because of the following lemma.

Lemma 3.3 *If the functional $\widehat{\zeta}$ is Fréchet differentiable at $P_{\theta,\delta}$ with respect to d_K on $\mathcal{P}^*(supp(\delta))$ and continuous in a neighbourhood $\mathcal{U}_K(P_{\theta,\delta},\epsilon)$ of $P_{\theta,\delta}$ with respect to d_K, then the Fréchet derivative has the form*

$$\tilde{\zeta}_\theta(Q_\xi - P_{\theta,\delta}) = \int \psi_\theta(y,t) Q_\xi(dy,dt), \tag{3.2}$$

where $\psi_\theta : \mathbb{R}\times\mathcal{T} \to \mathbb{R}^q$ is bounded and continuous and $\int \psi_\theta(y,t)P_{\theta,\delta}(dy,dt) = 0$.
If additionally $\widehat{\zeta}(P_{\theta,\xi}) = \zeta(\theta)$ for all ξ with $supp(\xi) = supp(\delta)$ is satisfied, then $\int \psi_\theta(y,t)P_{\theta,t}(dy) = 0$ for all $t \in supp(\delta)$.

Proof. In Huber (1981), p. 36/37, it is shown that the Fréchet derivative satisfies (3.2), where $\psi_\theta : \mathbb{R} \times \mathcal{T} \to \mathbb{R}^q$ is bounded and continuous and satisfies $\int \psi_\theta(y,t)P_{\theta,\delta}(dy,dt) = 0$. The additional assumption of $\widehat{\zeta}(P_{\theta,\xi}) = \zeta(\theta)$ for all ξ with $\text{supp}(\xi) = \text{supp}(\delta)$ provides in particular for $\delta(\epsilon) := (1-\epsilon)\,\delta + \epsilon\,\xi$

$$\begin{aligned}
0 &= \lim_{\epsilon\to 0} \frac{|\widehat{\zeta}(P_{\theta,\delta(\epsilon)}) - \widehat{\zeta}(P_{\theta,\delta}) - \tilde{\zeta}(P_{\theta,\delta(\epsilon)} - P_{\theta,\delta})|}{d_K(P_{\theta,\delta(\epsilon)}, P_{\theta,\delta})} \\
&= \lim_{\epsilon\to 0} \frac{1}{\max_{t\in\text{supp}(\delta)} |(1-\epsilon)\,\delta(\{t\}) + \epsilon\,\xi(\{t\}) - \delta(\{t\})|} \cdot \\
&\qquad \Big| \sum\nolimits_{t\in\text{supp}(\delta)} (1-\epsilon)\,\delta(\{t\}) \int \psi_\theta(y,t)\,P_{\theta,t}(dy) \\
&\qquad + \sum\nolimits_{t\in\text{supp}(\delta)} \epsilon\,\xi(\{t\}) \int \psi_\theta(y,t)\,P_{\theta,t}(dy) \Big| \\
&= \frac{|\sum_{t\in\text{supp}(\delta)} \xi(\{t\}) \int \psi_\theta(y,t)\,P_{\theta,t}(dy)|}{\max_{t\in\text{supp}(\delta)} |\delta(\{t\}) + \xi(\{t\})|}
\end{aligned}$$

for all ξ with $\text{supp}(\xi) = \text{supp}(\delta)$. This implies $\int \psi_\theta(y,t)P_{\theta,t}(dy) = 0$ for all $t \in \text{supp}(\delta)$. □

In particular under some regularity conditions M-functionals also have a derivative of the form (3.2) (see Section 3.3). For all functionals satisfying (3.2) we have $IF(\widehat{\zeta}, P_{\theta,\delta}, y, t) = \psi_\theta(y,t)$ and therefore

$$\begin{aligned}
&\frac{1}{\epsilon}\, |\tilde{\zeta}_\theta([(1-\epsilon)P_{\theta,\delta} + \epsilon\tilde{P}_{\tilde{\xi}}] - P_{\theta,\delta})| \\
&\quad = \Big|\int IF(\widehat{\zeta}, P_{\theta,\delta}, y, t)\tilde{P}_{\tilde{\xi}}(dy,dt)\Big| \\
&\quad \le \max_{(y,t)\in\mathbb{R}\times\text{supp}(\delta)} |IF(\widehat{\zeta}, P_{\theta,\delta}, y, t)|.
\end{aligned}$$

From the influence function at once a quantitative robustness measure can be derived, namely the *gross-error-sensitivity* which also was introduced by Hampel (1968, 1974) (see also Hampel et al. (1986)).

Definition 3.15 (Gross-error-sensitivity)
The gross-error-sensitivity $s^*(\widehat{\zeta}, P_{\theta,\delta})$ *of* $\widehat{\zeta}$ *at* $P_{\theta,\delta}$ *is defined as*

$$s^*(\widehat{\zeta}, P_{\theta,\delta}) := \max_{(y,t)\in \mathbb{R}\times \text{supp}(\delta)} |IF(\widehat{\zeta}, P_{\theta,\delta}, y, t)|.$$

In Section 5.1 we will see that often this robustness measure coincides with the bias in a shrinking contamination neighbourhood.

3.3 Fréchet Differentiability of M-Functionals

For linear models the least squares estimator $\widehat{\beta}_N^{LS}$ which is given by

$$\widehat{\beta}_N^{LS} \in \arg\min_{\beta\in\mathcal{B}} \sum_{n=1}^N (y_{nN} - a(t_{nN})'\beta)^2$$

can be generalized by substituting the squares by some other function $\phi : \mathbb{R}\times\mathcal{T} \to \mathbb{R}$ which may also depend on the experimental conditions t. For example ϕ may be given by $\phi(y,t) = |y|$ or $\phi(y,t) = |y|\min\{|y|, b(t)\}$ for some function $b : \mathcal{T} \to (0,\infty)$. This leads to estimators $\widehat{\beta}_N^{\phi}$ solving

$$\widehat{\beta}_N^{\phi} \in \arg\min_{\beta\in\mathcal{B}} \sum_{n=1}^N \phi(y_{nN} - a(t_{nN})'\beta, t_{nN}). \tag{3.3}$$

By taking the derivative with respect to β the solution $\widehat{\beta}_N^{\phi}$ of the minimization problem (3.3) is a root of the problem

$$\sum_{n=1}^N \psi(y_{nN} - a(t_{nN})'\beta, t_{nN}) = 0, \tag{3.4}$$

where $\psi : \mathbb{R}\times\mathcal{T} \to \mathbb{R}^r$ is defined by $\psi(y,t) = \frac{\partial}{\partial y}\phi(y,t)\, a(t)$. Any solution of problem (3.4) is called a M-estimate for β and any estimator satisfying (3.4) for all $(y_N, d_N) \in \mathbb{R}^N \times \mathcal{T}^N$ is called a *M-estimator for* β *with score function* ψ.

Problem (3.4) can be written also via the empirical distribution P_{y_N,d_N} so that we are regarding solutions of

$$\int \psi(y - a(t)'\beta, t)\, P_{y_N,d_N}(dy, dt) = 0.$$

This can be generalized by allowing arbitrary probability measures $Q_\xi \in \mathcal{P}^*$ instead of the empirical distributions. Then any functional $\widehat{\beta} : \mathcal{P}^* \to \mathbb{R}^r$ with

$$\int \psi(y - a(t)'\widehat{\beta}(Q_\xi), t)\, Q_\xi(dy, dt) = 0$$

for every $Q_\xi \in \mathcal{P}^*$ is called a *M-functional for* β with *score function* ψ.

M-functionals cannot only be defined for linear models. Generally for designed experiments they can be defined for arbitrary parameters $\theta \in \Theta \subset \mathbb{R}^{r+J}$ by using score functions $\tilde{\psi} : \mathbb{R} \times \mathcal{T} \times \Theta \to \mathbb{R}^{r+J}$. Then a *M-functional for* θ is a functional $\widehat{\theta} : \mathcal{P}^* \to \mathbb{R}^{r+J}$ which satisfies

$$\int \tilde{\psi}(y, t, \widehat{\theta}(Q_\xi))\, Q_\xi(dy, dt) = 0$$

for every $Q_\xi \in \mathcal{P}^*$.

Definition 3.16 (M-functional and M-estimator)
A functional $\widehat{\theta} : \mathcal{P}^* \to \mathbb{R}^{r+J}$ is called a M-functional for θ with score function $\tilde{\psi}$ if it satisfies

$$\int \tilde{\psi}(y, t, \widehat{\theta}(Q_\xi))\, Q_\xi(dy, dt) = 0$$

for every $Q_\xi \in \mathcal{P}^*$. An estimator $\widehat{\theta}_N : \mathbb{R}^N \times \mathcal{T}^N \to \mathbb{R}^{r+J}$ is called a M-estimator for θ if there exists a M-functional $\widehat{\theta}$ for θ such that $\widehat{\theta}_N(y_N, d_N) = \widehat{\theta}(P_{y_N, d_N})$ for all $(y_N, d_N) \in \mathbb{R}^N \times \mathcal{T}^N$.

Often the set

$$I(\tilde{\psi}, Q_\xi) := \{\theta \in \Theta;\ \int \tilde{\psi}(y, t, \theta)\, Q_\xi(dy, dt) = 0\}$$

has more than one element so that the M-functional is not unique. To get unique M-functionals we can use selection functions $\nu : \mathcal{P}^* \times \Theta \to \mathbb{R}$ so that for every $Q_\xi \in \mathcal{P}^*$ the solution

$$\widehat{\theta}_\nu(Q_\xi) \in \min\{\nu(Q_\xi, \theta);\ \theta \in I(\tilde{\psi}, Q_\xi)\}$$

is unique. Then $\widehat{\theta}_\nu$ is a unique M-functional. For example for the one-dimensional location problem we can use a selection function which is given by $\nu(Q, \theta) = \min_{\theta \in \Theta} |G^{-1}(\frac{1}{2}) - \theta|$, where $G^{-1}(\frac{1}{2})$ is the median of Q. For other selection functions see Clarke (1983) and Example 3.1 below.

The following theorem shows that under relatively general conditions M-functionals are Fréchet differentiable. This theorem was derived by Clarke (1986). Here we extend this theorem for the case that instead of the whole parameter θ only an aspect $\zeta(\theta) \in \mathbb{R}^q$ is of interest. For that let be for $Q_\xi \in \mathcal{P}^*$

$$K_{Q_\xi}(\theta) := \int \tilde{\psi}(y, t, \theta)\, Q_\xi(dy, dt)$$

and $\partial K_{Q_\xi}(\theta)$ the generalized Jacobian matrix, i.e. the convex hull of all $(r + J) \times (r + J)$ matrices obtained as a limit of a sequence of Jacobian matrices of $K_{Q_\xi}(\cdot)$ at $\overline{\theta}$ at which $K_{Q_\xi}(\cdot)$ is differentiable (see Clarke (1986), p. 198). Then the conditions are the following:

- $\widehat{\theta}_\nu(P_{\theta,\delta}) = \theta.$ (3.5)
- $\tilde{\psi}$ is continuous and bounded on $I\!R \times \mathcal{T} \times \overline{\Theta}$, where $\overline{\Theta} \subset \Theta$ is some nondegenerate compact subset containing θ in its interior. (3.6)
- $\tilde{\psi}(y,t,\cdot)$ is locally Lipschitz continuous at θ in the sense that there exists $\kappa \in I\!R$ such that $|\tilde{\psi}(y,t,\overline{\theta}) - \tilde{\psi}(y,t,\theta)| < \kappa|\overline{\theta} - \theta|$ for all $y \in I\!R$, $t \in \mathcal{T}$ and all $\overline{\theta}$ in a neighbourhood of θ. (3.7)
- $K_{P_{\theta,\delta}}$ is continuously differentiable at θ such that $M(\theta) \in \partial K_{P_{\theta,\delta}}(\theta)$ is nonsingular. (3.8)
- For every ϵ_1 there exists ϵ_2 such that for all $Q_\xi \in \mathcal{U}_K(P_{\theta,\delta}, \epsilon_2)$ (3.9)
  $\max_{\overline{\theta} \in \overline{\Theta}} |K_{Q_\xi}(\theta) - K_{P_{\theta,\delta}}(\theta)| < \epsilon_1$, and
  $\partial K_{Q_\xi}(\overline{\theta}) \subset \partial K_{P_{\theta,\delta}}(\overline{\theta}) + \epsilon_1 B_{(r+J)}$ for all $\overline{\theta} \in \overline{\Theta}$,
  where $B_{(r+J)} \subset I\!R^{(r+J)\times(r+J)}$ denotes the unit ball of matrices for which $B \in B_{(r+J)}$ implies $\max_{|y|\le 1} |By| \le 1$.
- $\displaystyle\int \tilde{\psi}(y,t,\theta)\,(Q_\xi - P_{\theta,\delta})(dy,dt) = O(d_K(Q_\xi, P_{\theta,\delta}))$ (3.10)
  for all $Q_\xi \in \mathcal{P}^*$.
- $\min_{\overline{\theta} \notin U}(\nu(P_{\theta,\delta}, \overline{\theta}) - \nu(P_{\theta,\delta}, \theta)) > 0$ for all neighbourhoods U of θ. (3.11)
- For every $\epsilon_1 > 0$ there exists $\epsilon_2 > 0$ such that $Q_\xi \in \mathcal{U}_K(P_{\theta,\delta}, \epsilon_2)$ (3.12) implies that $\nu(Q_\xi, \overline{\theta})$ is continuous in $\overline{\theta} \in \Theta$ and satisfies $\max_{\overline{\theta} \in \Theta} |\nu(Q_\xi, \overline{\theta}) - \nu(P_{\theta,\delta}, \overline{\theta})| < \epsilon_1$.
- ζ is differentiable at θ with $\dot{\zeta}(\theta) := \dfrac{\partial}{\partial \overline{\theta}} \zeta(\overline{\theta})/_{\overline{\theta}=\theta} \in I\!R^{q\times(r+J)}$. (3.13)

Theorem 3.2 *If the funtional $\widehat{\zeta} : \mathcal{P}^* \to I\!R^q$ is given by $\widehat{\zeta}(Q_\xi) = \zeta(\widehat{\theta}_\nu(Q_\xi))$, where $\widehat{\theta}_\nu : \mathcal{P}^* \to I\!R^{r+J}$ is a M-functional with score function $\tilde{\psi}$ and conditions (3.5) - (3.13) are satisfied, then $\widehat{\zeta}$ is Fréchet differentiable at $P_{\theta,\delta}$ on $\mathcal{P}^*(supp(\delta))$ with respect to d_K and has derivative*

$$\tilde{\zeta}_\theta(Q_\xi - P_{\theta,\delta}) = -\dot{\zeta}(\theta)\, M(\theta)^{-1} \int \tilde{\psi}(y,t,\theta)\,(Q_\xi - P_{\theta,\delta})(dy,dt).$$

Proof. For the selection function $\nu_\theta(Q_\xi, \overline{\theta}) := |\theta - \overline{\theta}|$ Theorem 4.1 in Clarke (1986) provides the Fréchet differentiability of the M-functional $\widehat{\theta}_{\nu_\theta}$ with score function $\tilde{\psi}$ at $P_{\theta,\delta}$ with respect to d_K. Moreover Theorem 3.1 in Clarke (1986) provides that there exists $\epsilon_1 > 0$ and $\epsilon_2 > 0$ such that for all $Q_\xi \in \mathcal{U}_K(P_{\theta,\delta}, \epsilon_1)$ we have

$$\{\overline{\theta} \in \Theta;\ K_{Q_\xi}(\overline{\theta}) = 0\} \cap \{\overline{\theta} \in \Theta;\ |\theta - \overline{\theta}| \le \epsilon_2\} = \{\widehat{\theta}_{\nu_\theta}(Q_\xi)\}.$$

Moreover Theorem 4.1 in Clarke (1983) (see in particular its proof) provides the existence of $\epsilon < \epsilon_1$ with $\widehat{\theta}_\nu(Q_\xi) = \widehat{\theta}_{\nu_\theta}(Q_\xi)$ for all $Q_\xi \in \mathcal{U}_K(P_{\theta,\delta}, \epsilon)$. Hence $\widehat{\theta}_\nu$ is also Fréchet differentiable at $P_{\theta,\delta}$ and the Fréchet differentiability of $\widehat{\zeta}$ follows by the chain rule. □

Example 3.1 (Fréchet differentiability of M-functionals for linear models)
By using the generalized Kolmogorov metric we can show the Fréchet differentiability of M-functionals for linear models without regarding product measures as Bednarski and Zontek (1994, 1996) have done to derive Fréchet differentiability in planned experiments. Let be $\mathcal{T} = \dot{\bigcup}_{j=1}^{J} \mathcal{T}_j$, $\text{supp}(\delta) = \{t_1, ..., t_I\} = \mathcal{T}_0$ and $\mathcal{T}_{0j} := \text{supp}(\delta) \cap \mathcal{T}_j \neq \emptyset$ for $j = 1, ..., J$. Assume that for all $\theta = (\beta', \sigma_1, ..., \sigma_J)' \in \Theta = \mathbb{R}^r \times (\mathbb{R}^+)^J$ the ideal distribution $P_{\theta,\delta}$ is given by

$$P_{\theta,t}(dy) = \sum_{j=1}^{J} 1_{\mathcal{T}_{0j}}(t) \frac{1}{\sigma_j} f\left(\frac{y - a(t)'\beta}{\sigma_j}\right) \lambda(dy), \tag{3.14}$$

where $f : \mathbb{R} \rightarrow [0, \infty)$ is the symmetric λ-density of a probability measure P on $\mathbb{R}$. Moreover let the score function $\tilde{\psi} : \mathbb{R} \times \mathcal{T} \times \Theta \rightarrow \mathbb{R}^{r+J}$ be given by $\tilde{\psi}(y, t, \theta) = \sum_{j=1}^{J} 1_{\mathcal{T}_{0j}}(t) \psi\left(\frac{y - a(t)'\beta}{\sigma_j}, t\right)$, where

$$\psi(y, t) = \begin{pmatrix} H\, a(t)\, \psi_0(y, t) \\ \psi_1(y, t) \\ \cdot \\ \cdot \\ \cdot \\ \psi_J(y, t) \end{pmatrix} \in \mathbb{R}^{r+J} \tag{3.15}$$

with nonsingular matrix $H \in \mathbb{R}^{r \times r}$, $\psi_0(y, t) = \text{sgn}(y) \min\{|y|, b(t)\}\, w(t)$ and $\psi_j(y, t) = [\psi_0(y, t)^2 - \int \psi_0(y, t)^2\, P(dy)]\, 1_{\mathcal{T}_{0j}}(t)$ for $j = 1, ..., J$. Regard also the selection function $\nu : \mathcal{P}^* \times \Theta \rightarrow \mathbb{R}$ which is given by

$$\begin{aligned} \nu(Q_\xi, \theta) = \sum_{t \in \mathcal{T}_0} & \xi(\{t\}) [|G_t^{-1}\left(\tfrac{1}{2}\right) - a(t)'\beta| \\ & + \sum_{j=1}^{J} 1_{\mathcal{T}_{0j}}(t) \left| \left(G_t^{-1}\left(\tfrac{3}{4}\right) - G_t^{-1}\left(\tfrac{1}{4}\right)\right) \left(F^{-1}\left(\tfrac{3}{4}\right) - F^{-1}\left(\tfrac{1}{4}\right)\right)^{-1} \right. \\ & \left. - \sigma_j \right|], \end{aligned} \tag{3.16}$$

where $G_t^{-1}(\alpha)$ is the α-quantile of Q_t and $F^{-1}(\alpha)$ is the α-quantile of P. If β is identifiable at δ, then for all $\theta \in \mathbb{R}^r \times (\mathbb{R}^+)^J$ the conditions (3.5) - (3.12) are satisfied so that for all $\theta \in \mathbb{R}^r \times (\mathbb{R}^+)^J$ the M-functional $\widehat{\theta}_\nu : \mathcal{P}^* \rightarrow \mathbb{R}^r \times (\mathbb{R}^+)^J$ for θ with score function $\tilde{\psi}$ is Fréchet differentiable at $P_{\theta,\delta}$.

We are now going to show that really the conditions (3.5) - (3.12) are satisfied. Thereby we mainly use the fact that in the one-dimensional location-scale case the Huber score function satisfies the conditions (3.5) - (3.12) (see Clarke (1986)) because the above score function $\tilde{\psi}$ is a generalization of the Huber score function. Thereby we use $\tilde{\psi}_0, ..., \tilde{\psi}_J$ for the single components of $\tilde{\psi}$. At first note that for every $\theta = (\beta', \sigma_1, ..., \sigma_J)'$ and $\alpha \in (0,1)$ we have $F_{\theta,t}^{-1}(\alpha) = \sigma_j F^{-1}(\alpha) + a(t)'\beta$ if $t \in \mathcal{T}_{0j}$. In particular we have

$$F_{\theta,t}^{-1}\left(\frac{1}{2}\right) = a(t)'\beta \tag{3.17}$$

and

$$\left(F_{\theta,t}^{-1}\left(\frac{3}{4}\right) - F_{\theta,t}^{-1}\left(\frac{1}{4}\right)\right)\left(F^{-1}\left(\frac{3}{4}\right) - F^{-1}\left(\frac{1}{4}\right)\right)^{-1} = \sigma_j \tag{3.18}$$

if $t \in \mathcal{T}_{0j}$.

Condition (3.5): Because f is symmetric and $\psi(\cdot, t)$ is antisymmetric for every $t \in \mathcal{T}$ we have

$$\int \sum\nolimits_{j=1}^{J} 1_{\mathcal{T}_{0j}}(t)\, \psi\left(\frac{y - a(t)'\beta}{\sigma_j}, t\right) P_{\theta,\delta}(dy, dt) = 0$$

for $\theta = (\beta', \sigma_1, ..., \sigma_J)'$. Because β is identifiable at δ, $\text{supp}(\delta) \cap \mathcal{T}_j \neq \emptyset$ for all $j = 1, ..., J$ and because of (3.17) and (3.18) the parameter θ is the only one with $\nu(P_{\theta,\delta}, \theta) = 0$. Hence $\widehat{\theta}_\nu(P_{\theta,\delta}) = \theta$.

Conditions (3.6) and (3.7) hold as in the location-scale case.

Condition (3.8): From the location-scale case we get (see Clarke (1986), p. 203):

$$\begin{aligned}
&\frac{\partial}{\partial \overline{\beta}} \int \tilde{\psi}_0(y, t, (\overline{\beta}', \sigma_1, ..., \sigma_J)')\, P_{\theta,t}(dy) /_{\overline{\beta}=\beta} \\
&\quad = \sum\nolimits_{j=1}^{J} 1_{\mathcal{T}_{0j}}(t)\, \tfrac{\partial}{\partial \overline{\mu}} \int \psi_0\left(\tfrac{y-\overline{\mu}}{\sigma_j}\right) \tfrac{1}{\sigma_j} f\left(\tfrac{y-\mu}{\sigma_j}\right) \lambda(dy) /_{\overline{\mu}=\mu}\, H\, a(t)\, a(t)', \\
&\frac{\partial}{\partial \overline{\beta}} \int \tilde{\psi}_j(y, t, (\overline{\beta}', \sigma_1, ..., \sigma_J)')\, P_{\theta,t}(dy) /_{\overline{\beta}=\beta} \\
&\quad = \sum\nolimits_{j=1}^{J} 1_{\mathcal{T}_{0j}}(t)\, \tfrac{\partial}{\partial \overline{\mu}} \int \psi_j\left(\tfrac{y-\overline{\mu}}{\sigma_j}\right) \tfrac{1}{\sigma_j} f\left(\tfrac{y-\mu}{\sigma_j}\right) \lambda(dy) /_{\overline{\mu}=\mu}\, a(t)' = 0, \\
&\frac{\partial}{\partial \overline{\sigma_j}} \int \tilde{\psi}_0(y, t, (\overline{\beta}', \sigma_1, ..., \overline{\sigma_j}, ..., \sigma_J)')\, P_{\theta,t}(dy) /_{\overline{\sigma_j}=\sigma_j} \\
&\quad = 1_{\mathcal{T}_{0j}}(t)\, \tfrac{\partial}{\partial \sigma} \int \psi_0\left(\tfrac{y-a(t)'\beta}{\sigma}\right) \tfrac{1}{\sigma_j} f\left(\tfrac{y-a(t)'\beta}{\sigma_j}\right) \lambda(dy) /_{\sigma=\sigma_j}\, H\, a(t) = 0,
\end{aligned}$$

$$\frac{\partial}{\partial \overline{\sigma_j}} \int \tilde{\psi}_j(y,t,(\overline{\beta}',\sigma_1,...,\overline{\sigma_j},...,\sigma_J)')\,P_{\theta,t}(dy)/_{\overline{\sigma_j}=\sigma_j}$$
$$= 1_{\mathcal{T}_{0j}}(t)\,\tfrac{\partial}{\partial\sigma}\int\psi_j\left(\tfrac{y-a(t)'\beta}{\sigma}\right)\tfrac{1}{\sigma_j}f\left(\tfrac{y-a(t)'\beta}{\sigma_j}\right)\lambda(dy)/_{\sigma=\sigma_j}$$
$$=: b(t) \neq 0,$$
$$\frac{\partial}{\partial \overline{\sigma_j}} \int \tilde{\psi}_i(y,t,(\overline{\beta}',\sigma_1,...,\overline{\sigma_j},...,\sigma_J)')\,P_{\theta,t}(dy)/_{\overline{\sigma_j}=\sigma_j} = 0 \text{ for } i\neq j,$$

for $i,j=1,...,J$. Hence with

$$\begin{aligned} D_b &:= \operatorname{diag}(\textstyle\sum_{t\in\mathcal{T}_{01}} b(t)\delta(\{t\}),...,\sum_{t\in\mathcal{T}_{0J}} b(t)\delta(\{t\})),\\ A_d &:= (a(t_1),...,a(t_I))' \text{ and}\\ D_h &:= \operatorname{diag}(h(t_1)\delta(\{t_1\}),...,h(t_I)\delta(\{t_I\})),\end{aligned}$$

where $h(t)=\sum_{j=1}^J 1_{\mathcal{T}_{0j}}(t)\,\frac{\partial}{\partial\overline{\mu}}\int\psi_0\left(\frac{y-\overline{\mu}}{\sigma_j}\right)\frac{1}{\sigma_j}f\left(\frac{y-\mu}{\sigma_j}\right)\lambda(dy)/_{\overline{\mu}=\mu}$, we have

$$\frac{\partial}{\partial\overline{\theta}}\int\tilde{\psi}(y,t,\overline{\theta})\,P_{\theta,\delta}(dy,dt)/_{\overline{\theta}=\theta} = \begin{pmatrix} H\,A_d'D_h\,A_d & 0_{r\times J}\\ 0_{J\times r} & D_b\end{pmatrix}$$

which is nonsingular.

Condition (3.9): Let be $\overline{\Theta}$ a compact subset of Θ with θ in its interior and $M:=\sup\{|\psi(y,t)|;\ y\in\mathbb{R};\ t\in\mathcal{T}_0\}$. From the location-scale case (see Clarke (1986), p. 203) we have that for all $\epsilon_1>0$ there exists $0<\epsilon_2<\frac{\epsilon_1}{2I}\min\{\frac{1}{M},\frac{1}{r+J}\}$ such that for all $t\in\mathcal{T}_0$ and Q_t with

$$\max_{y\in\mathbb{R}}|G_t(y)-F_{\theta,t}(y)|\leq\frac{2\epsilon_2}{\min\{\delta(\{t\});\ t\in\mathcal{T}_0\}} \tag{3.19}$$

we have

$$\sum_{j=1}^J 1_{\mathcal{T}_{0j}}(t)\max_{\overline{\theta}\in\overline{\Theta}}\left|\int\psi\left(\frac{y-a(t)'\overline{\beta}}{\overline{\sigma}_j}\right)(Q_t-P_{\theta,t})(dy)\right|\leq\frac{\epsilon_1}{2}$$

and

$$\sum_{j=1}^J 1_{\mathcal{T}_{0j}}(t)\frac{\partial}{\partial\overline{\theta}}\int\psi\left(\frac{y-a(t)'\overline{\beta}}{\overline{\sigma}_j}\right)Q_t(dy)/_{\overline{\theta}=\tilde{\theta}}$$
$$\subset \sum_{j=1}^J 1_{\mathcal{T}_{0j}}(t)\frac{\partial}{\partial\overline{\theta}}\int\psi\left(\frac{y-a(t)'\overline{\beta}}{\overline{\sigma}_j}\right)P_{\theta,t}(dy)/_{\overline{\theta}=\tilde{\theta}}+\frac{\epsilon_1}{2I}B_{(r+J)}$$

for all $\tilde{\theta} \in \overline{\Theta}$. Lemma 3.1 provides that $d_K(Q_\xi, P_{\theta,\delta}) \leq \epsilon_2$ implies (3.19) and $\max_{t\in\mathcal{T}_0} |\xi(\{t\}) - \delta(\{t\})| \leq \epsilon_2$. Hence we get for all Q_ξ with $d_K(Q_\xi, P_{\theta,\delta}) \leq \epsilon_2$

$$\begin{aligned}
&\max_{\overline{\theta}\in\overline{\Theta}} \Big| \int \tilde{\psi}(y,t,\overline{\theta})\, Q_t(dy)\, \xi(dt) - \int \tilde{\psi}(y,t,\overline{\theta})\, P_{\theta,t}(dy)\, \delta(dt) \Big| \\
&\leq \max_{\overline{\theta}\in\overline{\Theta}} \sum\nolimits_{t\in\mathcal{T}_0} \int |\tilde{\psi}(y,t,\overline{\theta})|\, Q_t(dy)\, |\xi(\{t\}) - \delta(\{t\})| \\
&\quad + \max_{\overline{\theta}\in\overline{\Theta}} \sum\nolimits_{t\in\mathcal{T}_0} \delta(\{t\}) \Big| \int \tilde{\psi}(y,t,\overline{\theta})\, (Q_t - P_{\theta,t})(dy) \Big| \\
&\leq M\, I\, \epsilon_2 + \frac{\epsilon_1}{2} < \epsilon_1
\end{aligned}$$

and

$$\begin{aligned}
&\frac{\partial}{\partial\overline{\theta}} \int \tilde{\psi}(y,t,\overline{\theta})\, Q_\xi(dy,dt)/_{\overline{\theta}=\tilde{\theta}} \\
&= \sum\nolimits_{t\in\mathcal{T}_0} \frac{\partial}{\partial\overline{\theta}} \int \tilde{\psi}(y,t,\overline{\theta})\, Q_t(dy)/_{\overline{\theta}=\tilde{\theta}}\, \xi(\{t\}) \\
&\subset \sum\nolimits_{t\in\mathcal{T}_0} \left(\frac{\partial}{\partial\overline{\theta}} \int \tilde{\psi}(y,t,\overline{\theta})\, P_{\theta,t}(dy)/_{\overline{\theta}=\tilde{\theta}} + \frac{\epsilon_1}{2\,I}\, B_{(r+J)} \right) \xi(\{t\}) \\
&\subset \sum\nolimits_{t\in\mathcal{T}_0} \left(\frac{\partial}{\partial\overline{\theta}} \int \tilde{\psi}(y,t,\overline{\theta})\, P_{\theta,t}(dy)/_{\overline{\theta}=\tilde{\theta}}\, \delta(\{t\}) \right. \\
&\quad \left. + \epsilon_2 (r+J)\, B_{(r+J)} + \frac{\epsilon_1}{2\,I}\, B_{(r+J)} \right) \\
&\subset \sum\nolimits_{t\in\mathcal{T}_0} \frac{\partial}{\partial\overline{\theta}} \int \tilde{\psi}(y,t,\overline{\theta})\, P_{\theta,t}(dy)/_{\overline{\theta}=\tilde{\theta}}\, \delta(\{t\}) \\
&\quad + \left(I\,(r+J)\,\epsilon_2 + \frac{\epsilon_1}{2} \right) B_{(r+J)} \\
&\subset \frac{\partial}{\partial\overline{\theta}} \int \tilde{\psi}(y,t,\overline{\theta})\, P_{\theta,\delta}(dy,dt)/_{\overline{\theta}=\tilde{\theta}} + \epsilon_1\, B_{(r+J)}.
\end{aligned}$$

Condition (3.10): The location-scale case provides that we have for all $t \in \mathcal{T}_0$

$$\Big| \int \tilde{\psi}(y,t,\theta)(Q_t - P_{\theta,t})(dy) \Big| \leq K \max_{y\in\mathbb{R}} |G_t(y) - F_{\theta,t}(y)|$$

for some $K \in \mathbb{R}$ (see Clarke (1986), p. 204). Hence Lemma 3.1 provides with $M := \max_{(y,t)\in\mathbb{R}\times\mathcal{T}_0} |\psi(y,t)|$

$$\begin{aligned}
&\left|\int \tilde{\psi}(y,t,\theta)(Q_\xi - P_{\theta,\delta})(dy,dt)\right| \\
&\leq \sum\nolimits_{t\in\mathcal{T}_0} \left|\int \tilde{\psi}(y,t,\theta)(Q_t - P_{\theta,t})(dy)\right| \xi(\{t\}) \\
&\quad + \sum\nolimits_{t\in\mathcal{T}_0} \int |\tilde{\psi}(y,t,\theta)|\, P_{\theta,t}(dy)\; |\xi(\{t\}) - \delta(\{t\})| \\
&\leq \frac{K\,2}{\min\{\delta(\{t\});\ t\in\mathcal{T}_0\}} d_K(Q_\xi, P_{\theta,\delta}) + M\, I\, d_K(Q_\xi, P_{\theta,\delta}).
\end{aligned}$$

Condition (3.11): Assume that there exists $\epsilon > 0$ and $(\theta_n)_{n\in\mathbb{N}} = ((\beta_n', \sigma_{1n}, ..., \sigma_{Jn})')_{n\in\mathbb{N}}$ with $|\theta_n - \theta| > \epsilon$ for $n \in \mathbb{N}$ such that

$$\lim_{n\to\infty} \nu(P_{\theta,\delta}, \theta_n) - \nu(P_{\theta,\delta}, \theta) = 0. \tag{3.20}$$

Then we have

$$\lim_{n\to\infty} \sum\nolimits_{t\in\mathcal{T}_0} \left(|a(t)'\beta_n - a(t)'\beta| + \sum\nolimits_{j=1}^{J} 1_{\mathcal{T}_{0j}}(t)|\sigma_{jn} - \sigma_j| \right) = 0$$

and in particular because of $\mathcal{T}_{0j} \neq \emptyset$ for all $j = 1, ..., J$

$$\lim_{n\to\infty} \sum\nolimits_{j=1}^{J} |\sigma_{jn} - \sigma_j| = 0$$

and

$$\lim_{n\to\infty} \sum\nolimits_{t\in\mathcal{T}_0} \left| a(t)' \frac{\beta_n - \beta}{\theta_n - \theta} \right| = 0$$

because of $\frac{\epsilon}{|\theta_n - \theta|} < 1$. Then there exists n_0 such that for all $n \geq n_0$

$$\epsilon < |\theta_n - \theta| \leq |\beta_n - \beta| + \sum\nolimits_{j=1}^{J} |\sigma_{jn} - \sigma_j| \leq |\beta_n - \beta| + \frac{\epsilon}{2}.$$

Hence $\frac{\beta_n - \beta}{|\theta_n - \theta|}$ is an element of the compact set $C := \{\tilde{\theta} \in \mathbb{R}^r;\ \frac{1}{2} \leq |\tilde{\theta}| \leq 1\}$ for $n \geq n_0$ so that there exists $\tilde{\theta}_0 \in C$ with $\sum_{t\in\mathcal{T}_0} |a(t)'\tilde{\theta}_0| = 0$ which is a contradiction to the identifiability of β at δ.

Condition (3.12): From the location-scale case (see Clarke (1983), Lemma 7.1, p. 1204) we have that for all $\epsilon_1 > 0$ there exists $\epsilon_2 > 0$ such that for all $t \in \mathcal{T}_0$

$$\max\nolimits_{y\in\mathbb{R}} |G_t(y) - F_{\theta,t}(y)| \leq \frac{2\,\epsilon_2}{\min\{\delta(\{t\});\ t\in\mathcal{T}_0\}}$$

implies

$$\max\nolimits_{\mu\in\mathbb{R}} \nu_1(Q_\xi, t, \mu) \leq \frac{\epsilon_1}{2\,I}$$

and

$$\max_{\sigma \in I\!R^+} \nu_2(Q_\xi, t, \sigma) \leq \frac{\epsilon_1}{2\,I},$$

where

$$\begin{aligned}
\nu_1(Q_\xi, t, \mu) &:= \left| |G_t^{-1}\left(\frac{1}{2}\right) - \mu| - |F_{\theta,t}^{-1}\left(\frac{1}{2}\right) - \mu| \right| \\
&\leq \left| G_t^{-1}\left(\frac{1}{2}\right) - F_{\theta,t}^{-1}\left(\frac{1}{2}\right) \right|, \\
\nu_2(Q_\xi, t, \sigma) & \\
&:= \left| | \left[G_t^{-1}\left(\frac{3}{4}\right) - G_t^{-1}\left(\frac{1}{4}\right) \right] \left[F^{-1}\left(\frac{3}{4}\right) - F^{-1}\left(\frac{1}{4}\right) \right]^{-1} - \sigma | \right. \\
&\quad \left. - | \left[F_{\theta,t}^{-1}\left(\frac{3}{4}\right) - F_{\theta,t}^{-1}\left(\frac{1}{4}\right) \right] \left[F^{-1}\left(\frac{3}{4}\right) - F^{-1}\left(\frac{1}{4}\right) \right]^{-1} - \sigma | \right| \\
&\leq \left[\left| G_t^{-1}\left(\frac{3}{4}\right) - F_{\theta,t}^{-1}\left(\frac{3}{4}\right) \right| + \left| G_t^{-1}\left(\frac{1}{4}\right) - F_{\theta,t}^{-1}\left(\frac{1}{4}\right) \right| \right] \cdot \\
&\qquad \left[F^{-1}\left(\frac{3}{4}\right) - F^{-1}\left(\frac{1}{4}\right) \right]^{-1}.
\end{aligned}$$

Then Lemma 3.1 provides for all Q_ξ with $d_K(Q_\xi, P_{\theta,\delta}) \leq \epsilon_2$ and $\mathrm{supp}(\xi) \subset \mathrm{supp}(\delta) = \mathcal{T}_0$

$$\begin{aligned}
&\max_{\overline{\theta} \in \Theta} |\nu(Q_\xi, \overline{\theta}) - \nu(P_{\theta,\delta}, \overline{\theta})| \\
&\quad \leq \sum_{t \in \mathcal{T}_0} \max_{\beta \in I\!R^r} \nu_1(Q_\xi, t, a(t)'\beta) \\
&\qquad + \sum_{t \in \mathcal{T}_0} \max_{\sigma \in I\!R^+} \nu_2(Q_\xi, t, \sigma) \leq 2\,I\,\frac{\epsilon_1}{2I} = \epsilon_1. \square
\end{aligned}$$

4

Robustness Measures: Bias and Breakdown Points

In Chapter 3 robustness properties are derived by regarding the behaviour of estimators and corresponding functionals in infinitesimal neighbourhoods of the ideal distribution $P_{\theta,\delta}$. Now, in Section 4.1 we will derive robustness properties of estimators by regarding the behaviour of their corresponding functionals in neighbourhoods which are not infinitesimal small. This behaviour is important for situations in which the amount of outliers is not decreasing to zero when the sample size is increasing to ∞. Moreover this robustness concept provides robustness measures, namely the bias and the breakdown point, which can be transferred to finite samples as is done in Section 4.2. In Section 4.3 the breakdown point for finite samples is derived for trimmed weighted L_p estimators in linear models, and in Section 4.4 it is investigated for nonlinear problems.

4.1 Asymptotic Bias and Breakdown Points

Here we will derive robustness properties of estimators by regarding the behaviour of their corresponding functionals in given neighbourhoods of $P_{\theta,\delta}$ which are not infinitesimal small. The maximal deviation from $\widehat{\zeta}(P_{\theta,\delta})$ in a given neighbourhood $\mathcal{U}(P_{\theta,\delta},\epsilon)$ of $P_{\theta,\delta}$ was introduced as a robustness measure by Huber (1964) (see also Huber (1981), p. 11). Here it will be called the *asymptotic bias for a given neighbourhood* to distinguish it from the asymptotic bias for shrinking neighbourhoods which is introduced in Section 5.1 and regarded in Chapter 7 and Chapter 8. The largest radius ϵ so that in the neighbourhood $\mathcal{U}(P_{\theta,\delta},\epsilon)$ the functional $\widehat{\zeta}$ is bounded, i.e. has no singularity, is also a wellknown robustness measure (see for example Huber (1981), Martin et al. (1989), Davies (1993)) and is called the *asymptotic breakdown point*. Both, the asymptotic bias and the asymptotic breakdown point, can be defined for arbitrary neighbourhoods (see in particular for different appropriate neighbourhoods Davies (1993)). But for designed experiments it is sufficient to define them for the contamination neighbourhood $\mathcal{U}_c(P_\delta, \epsilon)$ and the Kolmogorov neighbourhood $\mathcal{U}_K(P_\delta, \epsilon)$ around some probability measure $P_\delta \in \mathcal{P}^*$. For outlier robustness of estimators for a parametric family $\{P_{\theta,\delta};\ \theta \in \Theta\}$ the contamination neighbourhoods $\mathcal{U}_c(P_{\theta,\delta}, \epsilon)$ around $P_{\theta,\delta}$ should be regarded. But for robustness considerations at finite samples it also will be useful to regard Kolmogorov neighbourhoods around empirical distributions P_{y_N,d_N} (see Section 4.2).

Definition 4.1 (Asymptotic bias for a given neighbourhood)
The maximum asymptotic bias of a functional $\widehat{\zeta}$ at the contamination neighbourhood $\mathcal{U}_c(P_\delta, \epsilon)$ is defined as

$$b_{\epsilon,c}(\widehat{\zeta}, P_\delta) := \max\{|\widehat{\zeta}(P_\delta) - \widehat{\zeta}(Q_\delta)|;\ Q_\delta \in \mathcal{U}_c(P_\delta, \epsilon)\},$$

and at the Kolmogorov neighbourhood $\mathcal{U}_K(P_\delta, \epsilon)$ it is defined as

$$b_{\epsilon,K}(\widehat{\zeta}, P_\delta) := \max\{|\widehat{\zeta}(P_\delta) - \widehat{\zeta}(Q_\xi)|;\ Q_\xi \in \mathcal{U}_K(P_\delta, \epsilon)\}.$$

Definition 4.2 (Asymptotic breakdown point)
The asymptotic breakdown point of a functional $\widehat{\zeta}$ at P_δ is defined for contamination neighbourhoods as

$$\epsilon_c^*(\widehat{\zeta}, P_\delta) := \min\{\epsilon;\ b_{\epsilon,c}(\widehat{\zeta}, P_\delta) = b_{1,c}(\widehat{\zeta}, P_\delta)\}$$

and for Kolmogorov neighbourhoods as

$$\epsilon_K^*(\widehat{\zeta}, P_\delta) := \min\{\epsilon;\ b_{\epsilon,K}(\widehat{\zeta}, P_\delta) = b_{1,K}(\widehat{\zeta}, P_\delta)\}.$$

The following obvious lemma shows that alternative definitions of the breakdown point are possible and that usually the asymptotic breakdown point can be interpreted as the largest radius ϵ of a neighbourhood without singularity.

Lemma 4.1 *If $b_{1,c}(\widehat{\zeta}, P_\delta) = \infty$, then*

$$\begin{aligned} \epsilon_c^*(\widehat{\zeta}, P_\delta) &:= \min\{\epsilon;\ b_{\epsilon,c}(\widehat{\zeta}, P_\delta) = \infty\} \\ &= \min\{\epsilon;\ \max\{|\widehat{\zeta}(Q_\delta)|;\ Q_\delta \in \mathcal{U}_c(P_\delta, \epsilon)\} = \infty\}. \end{aligned}$$

The same holds for the asymptotic breakdown point for Kolmogorov neighbourhoods.

The next lemma, which follows at once from Lemma 3.2, shows how the bias and breakdown points for contamination and Kolmogorov neighbourhoods are related.

Lemma 4.2 *We always have $b_{\epsilon,c}(\widehat{\zeta}, P_\delta) \le b_{\epsilon,K}(\widehat{\zeta}, P_\delta)$ and in particular $\epsilon_c^*(\widehat{\zeta}, P_\delta) \ge \epsilon_K^*(\widehat{\zeta}, P_\delta)$.*

In Section 3.2 general asymptotic robustness was defined as continuity of the functional ζ with respect to the generalized Kolmogorov metric d_K.

The next obvious lemma shows that positive breakdown points for the Kolmogorov neighbourhoods are necessary for general asymptotic robustness.

Lemma 4.3 *If $\widehat{\zeta}$ is generally asymptotically robust, then $\epsilon_K^*(\widehat{\zeta}, P_{\theta,\delta}) > 0$ for all $\theta \in \Theta$.*

In Huber (1981), p.12, alternative definitions of the maximum asymptotic bias and the asymptotic breakdown point are given which are based on the distributions of sequences of estimators and therefore have more connection with the large sample robustness as defined in Section 3.2. Also basing on the distributions of sequences of estimators in Hampel et al. (1986), p. 97, another alternative definition of the asymptotic breakdown point is given. But the bias and the breakdown points given by these alternative definitions are difficult to calculate and often they only can be calculated if they coincide with those defined as in Definition 4.1 and 4.2.

Because the concepts of bias and breakdown point can be used also for finite samples here the concepts of bias and breakdown point will be mainly investigated for finite samples (see Section 4.2). In this section we only will give an upper bound for the asymptotic breakdown point (as defined in Definition 4.2) for regression equivariant functionals in linear models. In linear models the response function is given by $\mu(t,\beta) = a(t)'\beta$ and we regard functionals $\widehat{\zeta}$ of the form $\widehat{\zeta} = (\widehat{\varphi}', \widehat{\rho})'$ for estimating $\zeta(\theta) = \zeta((\beta', \sigma)') = (\varphi(\beta)', \rho(\sigma))'$. The regression equivariance is defined for linear transformations of the form

$$H_\gamma : I\!R \times \mathcal{T} \ni (y,t) \to (y + a(t)'\gamma, t) \in I\!R \times \mathcal{T}.$$

Definition 4.3 (Regression equivariance)
In a linear model the functional $\widehat{\zeta} = (\widehat{\varphi}', \widehat{\rho})'$ for ζ with $\zeta(\theta) = \zeta((\beta', \sigma)') = (\varphi(\beta)', \rho(\sigma))'$ and $\varphi(\beta) = L\beta$ is regression equivariant if for all $\gamma \in I\!R^r$ and all $Q_\delta \in \mathcal{P}^*$

$$\widehat{\zeta}(Q_\delta^{H_\gamma}) = (\widehat{\varphi}(Q_\delta^{H_\gamma})', \widehat{\rho}(Q_\delta^{H_\gamma}))' = ((\widehat{\varphi}(Q_\delta) + L\beta)', \widehat{\rho}(Q_\delta))'.$$

For regression equivariant functionals $\widehat{\zeta} = (\widehat{\varphi}', \widehat{\rho})'$ for $\zeta(\theta) = (\varphi(\beta)', \rho(\sigma))'$ the asymptotic breakdown point depends only on the maximum mass that δ puts on a set $\mathcal{D} \subset \mathcal{T}$ at which φ is not identifiable, i.e. on the maximum mass on a nonidentifying set (see Section 1.3).

Definition 4.4 (Maximum mass on a nonidentifying set)
The maximum mass of δ on a nonidentifying set for φ is defined as

$$\mathcal{M}_\varphi(\delta) := \max\{\delta(\mathcal{D});\ \mathcal{D} \textit{ is nonidentifying for } \varphi\}.$$

If φ is the identity, i.e. $\varphi(\beta) = \beta$, then $\mathcal{M}_\varphi(\delta) =: \mathcal{M}_\beta(\delta)$ coincides with $\Delta(P_\delta)$ which is regarded in Davies (1993, 1994). In particular in this case we have

$$\mathcal{M}_\beta(\delta) = \max\{\delta(\mathcal{D});\ \{a(t);\ t \in \mathcal{D}\} \text{ is a subset of a subspace of } I\!R^r\}.$$

Lemma 4.4 *If $\varphi(\beta) = L\beta$, then*

$$\begin{aligned} \mathcal{M}_{\varphi}(\delta) &= \max\{\ \delta(\{t \in \mathcal{T};\ a(t)'\beta = 0\}); \\ &\quad \beta \in I\!R^r \ \text{with}\ |L\beta| = 1\ \}. \end{aligned} \tag{4.1}$$

Proof. Lemma 1.1 provides the assertion: In particular, if $\mathcal{D} = \{t \in \mathcal{T};\ a(t)'\beta = 0\}$ with $|L\beta| = 1$, then φ is not identifiable at $\mathcal{D}$. Conversely, let $\mathcal{D}$ any nonidentifying set. Then there exists β with $L\beta \neq 0$ and $a(t)'\beta = 0$ for all $t \in \mathcal{D}$. Using $\frac{1}{|L\beta|}\beta$ instead of β provides the assertion. □

Because $\{\beta \in I\!R^r;\ |L\beta| = 1\}$ is a compact set the maximum in (4.1) is attained. Hence we have the following corollary.

Corollary 4.1 *If $\varphi(\beta) = L\beta$, then there exists a nonidentifying set $\mathcal{D}$ with $\delta(\mathcal{D}) = \mathcal{M}_{\varphi}(\delta)$.*

The following theorem, which gives the upper bound for the asymptotic breakdown point of regression equivariant estimators, is in the case $\varphi(\beta) = \beta$ an analogy to the Theorem 3.1 in Davies (1993) which is given for neighbourhoods based on linearly invariant metrics. For deriving the upper bound for the breakdown point for the Kolmogorov neighbourhood we could extend the proof of Theorem 3.1 in Davies (1993) because the generalized Kolmogorov metric is invariant with repect to the transformations H_γ, i.e. we always have $d_K(P_\delta^{H_\gamma}, Q_\xi^{H_\gamma}) = d_K(P_\delta, Q_\xi)$ because of $F_t^{H_\gamma}(y) = F_t(y - a(t)'\gamma)$. But the same upper bound also holds for the breakdown point for contamination neighbourhoods which is larger than the breakdown point for Kolmogorov neighbourhoods (see Lemma 4.2). Hence, even in the case of $\varphi(\beta) = \beta$ the following theorem does not follow from Davies Theorem 3.1. But at once it implies the upper bound for the breakdown point for Kolmogorov neighbourhoods.

Theorem 4.1 *If in a linear model the functional $\widehat{\zeta} = (\widehat{\varphi}', \widehat{\rho})'$ for ζ with $\zeta(\theta) = \zeta((\beta', \sigma)') = (\varphi(\beta)', \rho(\sigma))'$ is regression equivariant, then for all $P_\delta \in \mathcal{P}^*$*

$$\epsilon_c^*(\widehat{\zeta}, P_\delta) \leq \frac{1 - \mathcal{M}_{\varphi}(\delta)}{2}.$$

Proof. Let $\mathcal{D}$ be a nonidentifying set for φ with $\delta(\mathcal{D}) = \mathcal{M}_{\varphi}(\delta)$. Then Lemma 1.1 provides the existence of γ with $a(t)'\gamma = 0$ for all $t \in \mathcal{D}$ and $L\gamma \neq 0$. Define $Q_{v,\delta} \in \mathcal{P}^*$ for $v \in I\!R$ by

$$Q_{v,t} := \begin{cases} P_t & \text{for } t \in \mathcal{D}, \\ \frac{1}{2}P_t + \frac{1}{2}P_t^{H_{v\gamma}} & \text{for } t \notin \mathcal{D}. \end{cases}$$

Then with $\epsilon := \frac{1 - \mathcal{M}_{\varphi}(\delta)}{2}$ and

$$c(t) := \begin{cases} 0 & \text{for } t \in \mathcal{D} \\ \frac{1}{1 - \mathcal{M}_{\varphi}(\delta)} & \text{for } t \notin \mathcal{D} \end{cases}$$

we have $Q_{v,t} = (1-\epsilon\, c(t))\, P_t + \epsilon\, c(t)\, P_t^{H_{v\gamma}}$ for all $t \in \mathcal{D}$ with $\int c(t)\, \delta(dt) = \frac{\delta(\mathcal{T}\setminus\mathcal{D})}{1-\mathcal{M}_\varphi(\delta)} = 1$. Hence $Q_{v,\delta} \in \mathcal{U}_c(P_\delta, \epsilon)$ for all $v \in \mathbb{R}$. Moreover we have

$$Q_{v,t}^{H_{-v\gamma}} := \begin{cases} P_t^{H_{-v\gamma}} = P_t = Q_{-v,t} & \text{for } t \in \mathcal{D}, \\ \frac{1}{2} P_t^{H_{-v\gamma}} + \frac{1}{2} P_t = Q_{-v,t} & \text{for } t \notin \mathcal{D}. \end{cases}$$

Then the regression equivariance provides

$$\begin{aligned} &|\widehat{\zeta}(P_\delta) - \widehat{\zeta}(Q_{v,\delta})| + |\widehat{\zeta}(P_\delta) - \widehat{\zeta}(Q_{-v,\delta})| \geq |\widehat{\zeta}(Q_{v,\delta}) - \widehat{\zeta}(Q_{v,\delta}^{H_{-v\gamma}})| \\ &\quad = |(\widehat{\varphi}(Q_{v,\delta})', \widehat{\rho}(Q_{v,\delta}))' - ((\widehat{\varphi}(Q_{v,\delta}) - vL\gamma)', \widehat{\rho}(Q_{v,\delta}))'| \geq |v\, L\, \gamma| \to \infty \end{aligned}$$

for $|v| \to \infty$ so that $\max_{v \in \mathbb{R}} |\widehat{\zeta}(P_\delta) - \widehat{\zeta}(Q_{v,\delta})| = \infty$. □

4.2 Bias and Breakdown Points for Finite Samples

In Section 4.1 estimators $\widehat{\zeta}_N$ were regarded which are derived from functionals $\widehat{\zeta}$ by $\widehat{\zeta}_N(y_N, d_N) = \widehat{\zeta}(P_{y_N, d_N})$, where P_{y_N, d_N} is the empirical distribution at the sample (y_N, d_N). For these estimators the asymptotic bias and the asymptotic breakdown point were given by the behaviour of the functional $\widehat{\zeta}$ in a neighbourhood of the ideal distribution $P_{\theta,\delta}$. But instead of the ideal distribution $P_{\theta,\delta}$ we can also regard the empirical distribution P_{y_N, d_N} to get the bias and the breakdown point for finite samples (y_N, d_N). Thereby we should regard only those members of the contamination neighbourhood of P_{y_N, d_N} which are also empirical distributions so that they are also given by a finite sample. For example these empirical distributions are given by samples (y_{N+M}, d_{N+M}), where M arbitrary observations $\overline{y}_1, ..., \overline{y}_M$ at experimental conditions $\overline{t}_1, ..., \overline{t}_M$ are added to the original sample (y_N, d_N). Then these empirical distributions are lying in the simple contamination neigbourhood $\mathcal{U}_c^s(P_{y_N, d_N}, \epsilon)$ with radius $\epsilon = \frac{M}{N+M}$ (see also Lemma 4.5).

Because here we regard designed experiments the sequence $(d_N)_{N \in \mathbb{N}}$ of designs is given by the experimenter so that the M additional experimental conditions $\overline{t}_1, ..., \overline{t}_M$ are given by $(d_N)_{N \in \mathbb{N}}$ and in particular by d_{N+M}. Only the additional observations $\overline{y}_1, ..., \overline{y}_M$ at these experimental conditions are arbitrary and may contain outliers. Hence the bias and the breakdown point for finite samples can be based on the following set of alternative samples

$$\mathcal{Y}_M^+(y_N) := \{y_{N+M} \in \mathbb{R}^{N+M};\; y_{nN} = y_{n(N+M)} \text{ for } n = 1, ..., N\}.$$

Now the *bias by adding M observations* can also be defined for estimators which can not be given by functionals.

Because we also will regard estimators which are not always unique (see Section 4.3) we here allow for finite sample considerations estimators of the form $\widehat{\zeta}_N : \mathbb{R}^N \times \mathcal{T}^N \to \{S;\ S \subset \mathbb{R}^q\}$ which maps $\mathbb{R}^N \times \mathcal{T}^N$ on subsets of $\mathbb{R}^q$. Then by $|S_1 - S_2|$ we will denote the maximal distance between elements of S_1 and S_2, i.e.

$$|S_1 - S_2| := \max\{|s_1 - s_2|;\ s_1 \in S_1,\ s_2 \in S_2\}.$$

In particular $|S_1 - S_1|$ is the maximal distance inside S_1 and $|\{s_1\} - \{s_2\}| = |s_1 - s_2|$.

Definition 4.5 (Bias by adding M observations)
If $(\widehat{\zeta}_N)_{N\in\mathbb{N}}$ is a sequence of estimators at $(d_N)_{N\in\mathbb{N}}$, then the bias of $\widehat{\zeta}_N$ at d_N by adding M observations to y_N is defined as

$$\begin{aligned} b_M^+(\widehat{\zeta}_N, d_N, y_N) &:= \max\{|\widehat{\zeta}_N(y_N, d_N) - \widehat{\zeta}_{N+M}(y_{N+M}, d_{N+M})|; \\ &\qquad y_{N+M} \in \mathcal{Y}_M^+(y_N)\} \end{aligned}$$

The *breakdown point by adding observations* is the minimal fraction of added observations which can cause an arbitrarily large bias. Thereby at first we define the breakdown point for a given observation vector y_N. Because the observation vector y_N is not known before realizing the experiment we also define an over all breakdown point as the minimum breakdown point within all $y_N \in \mathbb{R}^N$.

Definition 4.6 (Breakdown point by adding observations)
If $(\widehat{\zeta}_N)_{N\in\mathbb{N}}$ is a se- quence of estimators at $(d_N)_{N\in\mathbb{N}}$, then the breakdown point of $\widehat{\zeta}_N$ at d_N by adding observations to y_N is defined as

$$\epsilon^+(\widehat{\zeta}_N, d_N, y_N) := \min\{\tfrac{M}{N+M};\ M \in \mathbb{N}_0 \text{ with } b_M^+(\widehat{\zeta}_N, d_N, y_N) = \infty\},$$

and the breakdown point of $\widehat{\zeta}_N$ at d_N by adding observations is defined as

$$\epsilon^+(\widehat{\zeta}_N, d_N) := \min\{\epsilon^+(\widehat{\zeta}_N, d_N, y_N);\ y_N \in \mathbb{R}^N\}.$$

The following lemma gives the relation between the breakdown point by adding observations and the asymptotic breakdown point at the empirical distribution P_{y_N,d_N} as defined in Section 4.1.

Lemma 4.5 *If $(\widehat{\zeta}_N)_{N\in\mathbb{N}}$ is given by the functional $\widehat{\zeta}$ and $b_{1,K}(\widehat{\zeta}, P_{y_N,d_N}) = \infty$, then*

$$\epsilon^+(\widehat{\zeta}_N, d_N, y_N) \geq \epsilon_K^*(\widehat{\zeta}, P_{y_N,d_N}).$$

Proof. The assertion follows with Lemma 3.2 at once from the fact that

$$\begin{aligned} P_{y_{N+M},d_{N+M}} &= \frac{1}{N+M}\sum_{n=1}^{N+M} e_{y_{n(N+M)},t_{n(N+M)}} \\ &= \frac{N}{N+M}\frac{1}{N}\sum_{n=1}^{N} e_{y_{nN},t_{nN}} + \frac{M}{N+M}\frac{1}{M}\sum_{n=N+1}^{N+M} e_{y_{n(N+M)},t_{n(N+M)}} \end{aligned}$$

is an element of $\mathcal{U}_c^s(P_{y_N,d_N},\epsilon) \subset \mathcal{U}_K(P_{y_N,d_N},\epsilon)$ with $\epsilon = \frac{M}{N+M}$. □

Instead of adding M observations to the given observations y_N we can substitute M observations of y_N. In particular we can substitute M observations by outliers. This leads to a breakdown point which was regarded by He, Jurečková et al. (1990), Ellis and Morgenthaler (1993) and Müller (1995b). For that define the set of alternative observations $\overline{y}_N$ as

$$\begin{aligned}\mathcal{Y}_M(y_N) \quad := \quad & \{\overline{y}_N \in I\!R^N;\ \text{there exist } m(1) < ... < m(N-M) \\ & \text{with } \overline{y}_{m(i)N} = y_{m(i)N} \text{ for all } i \in \{1,...,N-M\}\ \}.\end{aligned}$$

Definition 4.7 (Bias by replacing M observations)
The bias of an estimator $\widehat{\zeta}_N$ at d_N by replacing M observations of y_N is defined as

$$b_M(\widehat{\zeta}_N, d_N, y_N) := \max\{|\widehat{\zeta}_N(y_N, d_N) - \widehat{\zeta}_N(\overline{y}_N, d_N)|;\ \overline{y}_N \in \mathcal{Y}_M(y_N)\}.$$

Definition 4.8 (Breakdown point by replacing observations)
The breakdown point of an estimator $\widehat{\zeta}_N$ at d_N by replacing observations of y_N is defined as

$$\epsilon^*(\widehat{\zeta}_N, d_N, y_N) := \min\{\frac{M}{N};\ M \in I\!N_0 \ \text{with}\ b_M(\widehat{\zeta}, d_N, y_N) = \infty\},$$

and the breakdown point of $\widehat{\zeta}_N$ at d_N by replacing observations is defined as

$$\epsilon^*(\widehat{\zeta}_N, d_N) := \min\{\epsilon^*(\widehat{\zeta}_N, d_N, y_N);\ y_N \in I\!R^N\}.$$

As the breakdown point for adding observations the breakdown point for replacing observations is related to the asymptotic breakdown point at the empirical distribution P_{y_N,d_N}.

Lemma 4.6 *If $(\widehat{\zeta}_N)_{N\in I\!N}$ is given by the functional $\widehat{\zeta}$ and $b_{1,K}(\widehat{\zeta}, P_{y_N,d_N}) = \infty$, then*

$$\epsilon^*(\widehat{\zeta}_N, d_N, y_N) \geq \epsilon_K^*(\widehat{\zeta}, P_{y_N,d_N}).$$

Proof. Regard any $M \in \{1,...,N\}$ and any $\overline{y}_N \in \mathcal{Y}_M(y_N)$. Without loss of generality we can assume that $y_{nN} = \overline{y}_{nN}$ for $n = M+1,...,N$. Set $\mathcal{T}_0 = \{t_{1N},...,t_{NN}\}$, $N_t = \sum_{n=1}^N 1_{\{t\}}(t_{nN})$, $M_t = \sum_{n=1}^M 1_{\{t\}}(t_{nN})$,

$$\begin{aligned}P_{M,t} \quad &= \quad \frac{1}{M_t}\sum_{n=1}^{M} 1_{\{t\}}(t_{nN})\, e_{(y_{nN},t)} \\ \overline{P}_{M,t} \quad &= \quad \frac{1}{M_t}\sum_{n=1}^{M} 1_{\{t\}}(t_{nN})\, e_{(\overline{y}_{nN},t)} \\ P_{N-M,t} \quad &= \quad \frac{1}{N_t - M_t}\sum_{n=M+1}^{N} 1_{\{t\}}(t_{nN})\, e_{(y_{nN},t)}\end{aligned}$$

and let $F_{M,t}$, $\overline{F}_{M,t}$, $F_{N-M,t}$ the corresponding distribution functions. Then we have

$$\begin{aligned}
P_{\overline{y}_N,d_N} &= \frac{1}{N}\sum\nolimits_{n=1}^{N} e_{(\overline{y}_{nN},t_{nN})} \\
&= \sum\nolimits_{t\in\mathcal{T}_0} \frac{N_t}{N}\left(\frac{M_t}{N_t}\frac{1}{M_t}\sum\nolimits_{n=1}^{M} 1_{\{t\}}(t_{nN})\, e_{(\overline{y}_{nN},t)} \right.\\
&\qquad \left. + \frac{N_t-M_t}{N_t}\frac{1}{N_t-M_t}\sum\nolimits_{n=M+1}^{N} 1_{\{t\}}(t_{nN})\, e_{(y_{nN},t)}\right) \\
&= \sum\nolimits_{t\in\mathcal{T}_0} \frac{N_t}{N}\left(\frac{M_t}{N_t}\overline{P}_{M,t} + \frac{N_t-M_t}{N_t}P_{N-M,t}\right)
\end{aligned}$$

and similarly

$$P_{y_N,d_N} = \sum\nolimits_{t\in\mathcal{T}_0} \frac{N_t}{N}\left(\frac{M_t}{N_t}P_{M,t} + \frac{N_t-M_t}{N_t}P_{N-M,t}\right).$$

This implies

$$\begin{aligned}
&d_K(P_{y_N,d_N}, P_{\overline{y}_N,d_N}) \\
&= \max\nolimits_{(y,t)\in\mathbb{R}\times\mathcal{T}_0}\left|\frac{N_t}{N}\left(\frac{M_t}{N_t}F_{M,t}(y) + \frac{N_t-M_t}{N_t}F_{N-M,t}(y)\right)\right. \\
&\qquad \left. - \frac{N_t}{N}\left(\frac{M_t}{N_t}\overline{F}_{M,t}(y) + \frac{N_t-M_t}{N_t}F_{N-M,t}(y)\right)\right| \\
&= \max\nolimits_{(y,t)\in\mathbb{R}\times\mathcal{T}_0}\frac{M_t}{N}|F_{M,t}(y) - \overline{F}_{M,t}(y)| \\
&\le \max\nolimits_{t\in\mathcal{T}_0}\frac{M_t}{N} \le \frac{M}{N}.
\end{aligned}$$

Hence $P_{\overline{y}_N,d_N} \in \mathcal{U}_K^0(P_{y_N,d_N}, \frac{M}{N})$ which provides the assertion. □

The following lemma describes the relation between the breakdown point by adding observations and the breakdown point by replacing observations.

Lemma 4.7 *If* $\epsilon^+(\widehat{\zeta}_N, d_N) = \frac{M}{N+M}$, *then* $\epsilon^+(\widehat{\zeta}_N, d_N) \ge \epsilon^*(\widehat{\zeta}_{N+M}, d_{N+M})$, *and if* $\epsilon^*(\widehat{\zeta}_N, d_N) = \frac{M}{N}$, *then* $\epsilon^*(\widehat{\zeta}_N, d_N) \ge \epsilon^+(\widehat{\zeta}_{N-M}, d_{N-M})$.

Proof. Let be $\epsilon^+(\widehat{\zeta}_N, d_N, y_N) = \frac{M}{N+M}$ and $y_{N+M} \in \mathcal{Y}_M^+(y_N)$ arbitrary. Then for every $K \in \mathbb{N}$ there exists $y_{N+M}^K \in \mathcal{Y}_M^+(y_N)$ with $|\widehat{\zeta}_{N+M}(y_{N+M}^K, d_{N+M})| > K$. Because $y_{N+M}^K \in \mathcal{Y}_M(y_{N+M})$ the first assertion follows. Now let be $\epsilon^*(\widehat{\zeta}_N, d_N, y_N) = \frac{M}{N}$. Then for every $K \in \mathbb{N}$ there exists $y_N^K \in \mathcal{Y}_M(y_N)$ with $|\widehat{\zeta}_N(y_N^K, d_N)| > K$. Because $y_N^K \in \mathcal{Y}_M(y_N)$ there exists y_{N-M} and a subsequence $(y_N^{K_n})_{n\in\mathbb{N}}$ of $(y_N^K)_{K\in\mathbb{N}}$ such that $y_N^{K_n} \in \mathcal{Y}_M^+(y_{N-M})$ for all $n \in \mathbb{N}$ which provides the second assertion. □

The bias and the breakdown point by replacing observations has the advantage that only the estimator $\widehat{\zeta}_N$ and the design d_N at the sample size N has to be given while for the bias and the breakdown point by adding observations the estimators and designs for all sample sizes N, $N+1$, $N+2, \ldots$ must be given. Moreover, usually in the literatur the breakdown point is defined by replacing the observations. Hence, in the following we only regard the bias and the breakdown point by replacing observations.

In the definition of the breakdown point mainly used in the literatur not only the observations are replaced arbitrarily but also the corresponding experimental conditions. See in particular Rousseeuw and Leroy (1987), Coakley (1991), Coakley and Mili (1993), Mili and Coakley (1993) and the original definition of Donoho and Huber (1983). This is due to the assumption that the experimental conditions are random and therefore also can contain outliers. For comparisons we also give this original definition of the breakdown point which we call the *breakdown point for contaminated experimental conditions*. This breakdown point is based on

$$\begin{aligned}\mathcal{Y}_M(y_N, d_N) \;:=\; & \{(\overline{y}_N, \overline{d}_N) \in I\!R^N \times \mathcal{T}^N;\ \text{there exist} \\ & m(1) < \ldots < m(N-M) \text{ with} \\ & (\overline{y}_{m(i)N}, \overline{t}_{m(i)N}) = (y_{m(i)N}, t_{m(i)N}) \text{ for all} \\ & i \in \{1, \ldots, N-M\}\ \}.\end{aligned}$$

Definition 4.9 (Bias for contaminated experimental conditions)
For contaminated experimental conditions the bias of an estimator $\widehat{\zeta}_N$ at (y_N, d_N) by replacing M observations and experimental conditions is defined as

$$\begin{aligned}\tilde{b}_M(\widehat{\zeta}_N, d_N, y_N) \;:=\; & \max\{|\widehat{\zeta}_N(y_N, d_N) - \widehat{\zeta}_N(\overline{y}_N, \overline{d}_N)|; \\ & (\overline{y}_N, \overline{d}_N) \in \mathcal{Y}_M(y_N, d_N)\}.\end{aligned}$$

Definition 4.10 (Breakdown point for contaminated experimental conditions)
For contaminated experimental conditions the breakdown point of an estimator $\widehat{\zeta}_N$ at (y_N, d_N) is defined as

$$\tilde{\epsilon}^*(\widehat{\zeta}_N, d_N, y_N) := \min\{\frac{M}{N};\ \tilde{b}_M(\widehat{\zeta}, d_N, y_N) = \infty\}.$$

The following lemma which already was given in Müller (1995b) is obvious.

Lemma 4.8 *For any estimator $\widehat{\zeta}_N$ we have*

$$\tilde{\epsilon}^*(\widehat{\zeta}_N, d_N, y_N) \leq \epsilon^*(\widehat{\zeta}_N, d_N, y_N)$$

for all $(y_N, d_N) \in I\!R^N \times \mathcal{T}^N$.

4.3 Breakdown Points in Linear Models

In this section we present calculations of the finite sample breakdown point by replacing observations as defined in Section 4.2. For that we assume a linear model, i.e. a model where the response function $\mu(t,\beta)$ has the form $\mu(t,\beta) = a(t)'\beta$. At first we derive an upper bound for the breakdown point of regression equivariant estimators.

For regression equivariant estimators for β Rousseeuw and Leroy (1987) derived an upper bound for the breakdown point for contaminated experimental conditions at designs d_N which provides regressors $a(t_{1N}),...,a(t_{NN})$ in general position which means that any $r-1$ regressors are linearly independent. In Mili and Coakley (1993) and Müller (1995b) this result was generalized for every design d_N, where Mili and Coakley (1993) regarded contaminated experimental conditions while Müller (1995b) assumed that the experimental conditions are given by the experimenter so that they are without contamination. Here we will generalize the result of Müller (1995b) for regression equivariant estimators $\widehat{\varphi}_N$ for general linear aspects $\varphi(\beta) = L\beta$ of β. The result is analogous to the result given in Section 4.1 for the asymptotic breakdown point of regression equivariant functionals. For simplicity here we drop the estimation of $\rho(\sigma)$, a function of the variance of the errors. I.e. we have $\widehat{\zeta}_N = \widehat{\varphi}_N$ and $\zeta(\theta) = \zeta((\beta',\sigma)') = \varphi(\beta)$.

Definition 4.11 (Regression equivariant estimators)
An estimator $\widehat{\varphi}_N$ for $\varphi(\beta) = L\beta$ is regression equivariant at d_N if for all $\gamma \in I\!R^r$ and all $y_N \in I\!R^N$

$$\widehat{\varphi}_N(y_N + A_{d_N}\gamma, d_N) \quad = \quad \widehat{\varphi}_N(y_N, d_N) + L\gamma.$$

For regression equivariant estimators for $\varphi(\beta)$ the breakdown point depends only on the maximum number of experimental conditions in a set $\mathcal{D} \subset \mathcal{T}$ at which φ is not identifiable, i.e. on the maximum number of experimental conditions in a nonidentifying set (see Section 1.3).

Definition 4.12 (Maximum number of experimental conditions in a nonidentifying set)
The maximum number of experimental conditions of the design d_N in a nonidentifying set for φ is defined as

$$\mathcal{N}_\varphi(d_N) := \max\{\sum_{n=1}^N 1_\mathcal{D}(t_{nN});\ \mathcal{D} \text{ is nonidentifying for } \varphi\}.$$

The maximum number of experimental conditions of the design d_N in a nonidentifying set corresponds to the maximal mass of the corresponding design measure $\delta_N = \frac{1}{N}\sum_{n=1}^N e_{t_{nN}}$ on a nonidentifying set as defined in Section 4.1. I.e. we have $\frac{1}{N}\mathcal{N}_\varphi(d_N) = \mathcal{M}_\varphi(\delta_N)$.

If φ is the identity, i.e. $\varphi(\beta) = \beta$, then $\mathcal{N}_\varphi(d_N) =: \mathcal{N}_\beta(d_N)$ coincides with $N(X)$ used in Müller (1995b) and satisfies

$$\mathcal{N}_\beta(d_N) = \max\{\textstyle\sum_{n=1}^N 1_\mathcal{V}(a(t_{nN}));\ \mathcal{V} \text{ is a subset of a subspace of } I\!R^r\}.$$

In particular for $\varphi(\beta) = \beta$ we always have $\mathcal{N}_\varphi(d_N) \geq r-1$ and if $\mathcal{N}_\varphi(d_N) = r-1$ is satisfies, then the regressors $a(t_{1N}), ..., a(t_{NN})$ are in general position (see Rousseeuw and Leroy (1987)).

For general φ we always have $\mathcal{N}_\varphi(d_N) \geq s-1$ but often it is impossible to achieve $\mathcal{N}_\varphi(d_N) = s-1$.

The following theorem, which gives the upper bound for the breakdown point of regression equivariant estimators, generalizes the result in Müller (1995b) which is given for $\varphi(\beta) = \beta$ and is the analogy to Theorem 4.1. For that set $\lfloor z \rfloor := \max\{n \in \mathbb{N};\ n \leq z\}$ for any $z \in \mathbb{R}^+$.

Theorem 4.2 *If at d_N the estimator $\widehat{\varphi}_N$ for $\varphi(\beta) = L\beta$ is regression equivariant, then for all $y_N \in \mathbb{R}^N$*

$$\epsilon^*(\widehat{\varphi}_N, d_N, y_N) \leq \frac{1}{N}\left\lfloor \frac{N - \mathcal{N}_\varphi(d_N) + 1}{2} \right\rfloor.$$

Proof. Without loss of generality we can assume that $t_{(N-\mathcal{N}_\varphi(d_N)+1)N}, ..., t_{NN}$ lie in a nonidentifying set. Then Lemma 1.1 provides the existence of $\gamma \in \mathbb{R}^r$ with $a(t_{nN})'\gamma = 0$ for $n = N - \mathcal{N}_\varphi(d_N) + 1, ..., N$ and $L\gamma \neq 0$. Now assume $\epsilon^*(\widehat{\varphi}_N, d_N, y_N) > \frac{1}{N} M$ with $M = \lfloor \frac{N-\mathcal{N}_\varphi(d_N)+1}{2} \rfloor$. Then there exists a $k \in \mathbb{R}$ with $|\widehat{\varphi}_N(y_N, d_N) - \widehat{\varphi}_N(\overline{y}_N, d_N)| \leq k$ for all $\overline{y}_N \in \mathcal{Y}_M(y_N)$. In particular this holds for

$$\overline{y}_N^1 := (y_{1N} + a(t_{1N})'v\gamma, ..., y_{MN} + a(t_{MN})'v\gamma, y_{(M+1)N}, ..., y_{NN})'$$

and

$$\overline{y}_N^2 := (y_{1N}, ..., y_{MN}, y_{(M+1)N} - a(t_{(M+1)N})'v\gamma, ..., y_{NN} - a(t_{NN})'v\gamma)',$$

where $v \in \mathbb{R}$. Thereby $\overline{y}_N^2$ lies in $\mathcal{Y}_M(y_N)$ because $N - \mathcal{N}_\varphi(d_N) - M = \lfloor \frac{N-\mathcal{N}_\varphi(d_N)}{2} \rfloor \leq M$. Because of $\overline{y}_N^1 = \overline{y}_N^2 + v\, A_{d_N}\gamma$ and the regression equivariance of $\widehat{\varphi}_N$ we have $\widehat{\varphi}_N(\overline{y}_N^1, d_N) = \widehat{\varphi}_N(\overline{y}_N^2, d_N) + v\, L\gamma$ for all $v \in \mathbb{R}$. But this provides the contradiction

$$\begin{aligned} |v|\,|L\gamma| &= |\widehat{\varphi}_N(\overline{y}_N^2, d_N) - \widehat{\varphi}_N(\overline{y}_N^1, d_N)| \\ &\leq |\widehat{\varphi}_N(\overline{y}_N^2, d_N) - \widehat{\varphi}_N(y_N, d_N)| + |\widehat{\varphi}_N(y_N, d_N) - \widehat{\varphi}_N(\overline{y}_N^1, d_N)| \\ &\leq 2\,k \end{aligned}$$

for all $v \in \mathbb{R}$. □

The following corollary of Theorem 4.2 shows drastically that if φ is not identifiable at d_N, i.e. $\mathcal{N}_\varphi(d_N) = N$, then a regression equivariant estimator can not be unique and the set of estimated values is so large that arbitrary large distances inside the set appear.

Corollary 4.2 *If φ is not identifiable at d_N and the estimator $\widehat{\varphi}_N$ for $\varphi(\beta)$ is regression equivariant, then for all $y_N \in I\!R^N$*

$$\epsilon^*(\widehat{\varphi}_N, d_N, y_N) = 0.$$

Now we define a class of regression equivariant estimators for which also lower bounds for the breakdown point and bounds for the bias by replacing observations can be calculated. In particular this class of estimators includes the estimators which maximize the breakdown point within all regression equivariant estimators (see Section 9.1). This class of estimators is the class of *h-trimmed weighted L_p estimators.* For estimating β the breakdown points of these estimators were already investigated in Müller (1995b). Here these estimators will be generalized to estimators of general linear aspect and the breakdown points of the generalized estimators will be given.

To define these estimators at first we define the h-trimmed L_p estimators for the whole parameter β by using the following notation: If

$$r_1(y_N, d_N, \beta) := |y_{1N} - a(t_{1N})'\beta|, ..., r_N(y_N, d_N, \beta) := |y_{NN} - a(t_{NN})'\beta|$$

are the absolute values of the residuals, then $r_{(1)}(y_N, d_N, \beta), ...,$ $r_{(N)}(y_N, d_N, \beta)$ denote the ordered residuals, i.e. we have $r_{(1)}(y_N, d_N, \beta) \leq ... \leq r_{(N)}(y_N, d_N, \beta)$.

Definition 4.13 (Trimmed weighted L_p estimator for β)
An estimator $\widehat{\beta}_{h,p}$ is called a h-trimmed weighted L_p estimator for β at d_N with weights $w_l, ..., w_h > 0$ if it satisfies

$$\widehat{\beta}_{h,p}(y_N, d_N) = \arg\min\{\sum\nolimits_{n=l}^{h} w_n\, r_{(n)}(y_N, d_N, \beta)^p;\ \beta \in I\!R^r\} \tag{4.2}$$

for all $y_N \in I\!R^N$, where $l, h \in \{1, ..., N\}$ and $p > 0$.

Note that because (4.2) may have several solutions a h-trimmed L_p estimator is not unique so that it is defined as a mapping into subsets of $I\!R^r$.

Often the bias bounds and the breakdown points of h-trimmed L_p estimtors does not depend on p, l and the weights $w_l, ..., w_h$ so that usually we do not mention explicitly the weights and l.

A special h-trimmed L_p estimator is the L_1 *estimator*, the least absolute value estimator, with $h = N$, $p = 1$, $l = 1$ and $w_1 = ... = w_N = 1$. Because we will regard this estimator separately we also define it separately.

Definition 4.14 (L_1 estimator for β)
An estimator $\widehat{\beta}_1$ is called a L_1 estimator for β if it is a h-trimmed L_p estimator for β with $h = N$, $p = 1$, $l = 1$ and $w_1 = ... = w_N = 1$.

Besides the L_1 estimator the class of trimmed weighted L_p estimators includes many relevant estimators as the least squares (LS) estimator, the least median of squares (LMS) estimator and least trimmed squares (LTS) estimators. For the definitions of the LMS estimator and the LTS estimator see Rousseeuw (1984) and Rousseeuw and Leroy (1987). Thereby the LS estimator is a N-trimmed weighted L_2 estimator with $l = 1$ and $w_n = 1$. The LMS estimator is a $\lfloor \frac{N+2}{2} \rfloor$-trimmed weighted L_2 estimator with $l = \frac{N+1}{2}$ and $w_n = 1$ for odd N and $l = \frac{N}{2}$ and $w_n = \frac{1}{2}$ for even N, and a LTS estimator is a h-trimmed weighted L_2 estimator with $l = 1$ and $w_n = 1$. In Mili and Coakley (1993) more general estimators, the so-called D-estimators, are regarded. But the examples of the D-estimators which they give are the h-trimmed weighted L_1 estimators and the h-trimmed weighted L_2 estimators so that it is an open question whether the class of D-estimators contains a relevant estimator which is not a h-trimmed weighted L_p estimator. A special class of trimmed weighted L_1 estimators are the rank-based estimators which Hössjer (1994) regarded and which in particular are asymptotically normal with convergence rate of $\sqrt{N}$.

By generalizing the Theorem 1 on p. 113 in Rousseeuw and Leroy (1987) which concerns only the LMS estimator we can show that in all relevant cases a solution of (4.2) exists so that always the h-trimmed weighted L_p estimator exists. Moreover this result shows how h-trimmed L_p estimators can be used to provide high breakdown point estimators for linear aspects. Thereby we also get a bound for the bias by replacing observations. For that set

$$\begin{aligned} R(y_N) &:= \max\{|y_{nN}|;\ n = 1, ..., N\}, \\ S(d_N) &:= \max\{|a(t_{nN})|;\ n = 1, ..., N\}, \\ W &:= \left(\frac{2 \max\{w_n;\ n = l, ..., h\}\,(h - l + 1)}{\min\{w_n;\ n = l, ..., h\}} \right)^{\frac{1}{p}} \end{aligned}$$

and

$$\begin{aligned} \rho_\varphi(d_N) := \frac{1}{2} \inf\{\tau > 0;\ & \text{there exists a nonidentifying set } \mathcal{D} \text{ so} \\ & \text{that } V(\mathcal{D})^\tau \text{ covers at least } \mathcal{N}_\varphi(d_N) + 1 \text{ of the } a(t_{nN})\}. \end{aligned}$$

Thereby for any set $\mathcal{D} \subset \mathcal{T}$ we define $V(\mathcal{D})$ as the linear hull of $\{a(t) \in I\!R^r;\ t \in \mathcal{D}\}$ and

$$V^\tau := \{x \in I\!R^r;\ |x - x_0| \le \tau \text{ for some } x_0 \in V\}.$$

Because of the definition of $\mathcal{N}_\varphi(d_N)$ (see Section 4.2) we have $\rho_\varphi(d_N) > 0$. Now we have the following lemma which already was included in the Lemmata 8.1 and 8.2 of Müller (1995b).

Lemma 4.9 *Let be* $M = \min\{N-h+1, h-\mathcal{N}_\varphi(d_N)\} \geq 1$, $w_l, ..., w_h, p > 0$ *and* $y \in I\!R^N$. *Then for all* $\overline{y} \in \mathcal{Y}_{M-1}(y_N)$ *and all* $\beta \in I\!R^r$ *with* $|\beta| > \frac{R(y_N)\, S(d_N)\,(W+1)}{\rho_\varphi(d_N)^2}$ *and* $\varphi(\beta) = L\,\beta \neq 0$ *we have*

$$\sum_{n=l}^{h} w_n\, r_{(n)}(\overline{y}_N, d_N, \beta)^p > 2 \sum_{n=l}^{h} w_n\, r_{(n)}(\overline{y}_N, d_N, 0_r)^p.$$

Proof. At first we prove the following assertion for

$$V_\beta := \{x \in I\!R^r;\ x'\beta = 0\}$$

and $\rho > 0$: If $\beta \neq 0$, then any $x \notin V_\beta^\rho$ with $|x| \leq k$ satisfies $|x'\beta| > \frac{\rho^2}{k}\, |\beta|$. To prove this let x_V denote the projection of x on V_β. Then we have $|x|^2 = |x - x_V|^2 + |x_V|^2$, i.e. $|x_V|^2 = |x|^2 - |x - x_V|^2 < k^2 - \rho^2$. This implies

$$\begin{aligned} \frac{|x'\beta|}{|x|\,|\beta|} &= \cos(\beta, x) = \frac{1}{\sqrt{1 + (\tan(\beta, x))^2}} \\ &= \frac{1}{\sqrt{1 + \frac{|x_V|^2}{|x - x_V|^2}}} \geq \frac{1}{\sqrt{1 + \frac{k^2 - \rho^2}{\rho^2}}} = \frac{\rho}{k}. \end{aligned}$$

Hence we get $|x'\beta| \geq \frac{\rho}{k}\, |x - 0|\, |\beta| > \frac{\rho^2}{k}\, |\beta|$.

Now we are going to prove the assertion of Lemma 4.9: Let be $\overline{y}_N = (\overline{y}_{1N}, ..., \overline{y}_{NN})'$ any corrupted sample by retaining at least $N - M + 1 = N - \min\{N - h + 1, h - \mathcal{N}_\varphi(d_N)\} + 1 \geq h$ observations. Because at least h observations are retained we have

$$\begin{aligned} &\sum_{n=l}^{h} w_n\, r_{(n)}(\overline{y}_N, d_N, 0_r)^p \qquad (4.3) \\ &\leq \max\{w_n;\ n = l, ..., h\}\, (h - l + 1)\, R(y_N)^p. \end{aligned}$$

Now regard any $\beta \in I\!R^r$ with $|\beta| > \frac{R(y_N)\, S(d_N)\,(W+1)}{\rho_\varphi(d_N)^2}$ and $L\,\beta \neq 0$. Then $\{t \in \mathcal{T};\ a(t)'\beta = 0\} = \{t \in \mathcal{T};\ a(t) \in V_\beta\}$ is a nonidentifying set for φ. Then the subspace $V_\beta^{\rho_\varphi(d_N)}$ contains at most $\mathcal{N}_\varphi(d_N)$ experimental conditions $a(t_{nN})$. Thus at least $N - \min\{N-h+1, h-\mathcal{N}_\varphi(d_N)\} + 1 - \mathcal{N}_\varphi(d_N) \geq N - h + 1$ experimental conditions at which the observations are retained do not lie in $V_\beta^{\rho_\varphi(d_N)}$. According to the first part of the proof they satisfy $|a(t_n)'\beta| > \frac{\rho_\varphi(d_N)^2}{S(d_N)}\, |\beta| \geq (W + 1)\, R(y_N)$, and therefore for any w_n they satisfy

$$\begin{aligned} & w_n\, |\overline{y}_{nN} - a(t_{nN})'\beta|^p = w_n\, |y_{nN} - a(t_{nN})'\beta|^p \\ \geq\ & w_n\, (|a(t_{nN})'\beta| - |y_{nN}|)^p \\ >\ & w_n\, ((W + 1) R(y_N) - R(y_N))^p = w_n\, W^p\, R(y_N)^p \\ \geq\ & \min\{w_n;\ n = l, ..., h\} \frac{2 \max\{w_n;\ n = l, ..., h\}\, (h - l + 1)}{\min\{w_n;\ n = l, ..., h\}}\, R(y_N)^p \\ =\ & 2 \max\{w_n;\ n = l, ..., h\}\, (h - l + 1)\, R(y_N)^p. \end{aligned}$$

In particular at least one of $r_{(l)}(\overline{y}_N, d_N, \beta), ..., r_{(h)}(\overline{y}_N, d_N, \beta)$ satisfies

$$w_n\, r_{(n)}(\overline{y}_N, d_N, \beta)^p > 2 \max\{w_n;\ n = l, ..., h\}\,(h - l + 1)\, R(y_N)^p$$

so that with (4.3) we have

$$\sum\nolimits_{n=l}^{h} w_n\, r_{(n)}(\overline{y}_N, d_N, \beta)^p > 2 \sum\nolimits_{n=l}^{h} w_n\, r_{(n)}(\overline{y}_N, d_N, 0_r)^p$$

for all $\beta \in I\!R^r$ with $|\beta| > \frac{R(y_N)\, S(d_N)\,(W+1)}{\rho_\varphi(d_N)^2}$. □

A consequence of Lemma 4.9 is the following central lemma which provides the existence and a bias bound for h-trimmed L_p estimators and which coincides with Lemma 8.2 in Müller (1995b).

Lemma 4.10 *Let be $M = \min\{N - h + 1, h - \mathcal{N}_\beta(d_N)\} \geq 1$ and $y_N \in I\!R^N$. Then for all $\overline{y}_N \in \mathcal{Y}_{M-1}(y_N)$ every h-trimmed weighted L_p estimator $\widehat{\beta}_{h,p}(\overline{y}_N, d_N)$ for β exists and satisfies*

$$|\beta| \leq \frac{R(y_N)\, S(d_N)\,(W+1)}{\rho_\beta(d_N)^2}$$

for all $\beta \in \widehat{\beta}_{h,p}(\overline{y}_N, d_N)$.

Proof. According to Lemma 4.9 only a sequence in the compact set given by all

$$|\beta| \leq \frac{R(y_N)\, S(d_N)\,(W+1)}{\rho_\beta(d_N)^2}$$

can approach the minimum of $\sum_{n=l}^{h} w_n\, r_{(n)}(\overline{y}_N, d_N, \beta)^p$. Because $\sum_{n=l}^{h} w_n\, r_{(n)}(\overline{y}_N, d_N, \beta)^p$ is continuous in β the infimum is also attained. □

Now the existence of h-trimmed L_p estimators follows at once from Lemma 4.10 (see also Theorem 3.1 in Müller (1995b)).

Theorem 4.3 *If $\mathcal{N}_\beta(d_N) < h$, then for all $y_N \in I\!R^N$ every h-trimmed weighted L_p estimator $\widehat{\beta}_{h,p}(y_N, d_N)$ for β exists and satisfies*

$$|\beta| \leq \frac{R(y_N)\, S(d_N)\,(W+1)}{\rho_\beta(d_N)^2}$$

for all $\beta \in \widehat{\beta}_h(y_N, d_N)$.

If $h \leq \mathcal{N}_\beta(d_N)$, then there are observations so that the set of solutions of (4.2) is unbounded. For example, if t_{nN}, $n = 1, ..., \mathcal{N}_\beta(d_N)$, of d_N are lying in a nonidentifying set and $y_N = A_{d_N}\beta$, then

$$\widehat{\beta}_h(y_N, d_N) \supset \{\beta + v;\ a(t_{nN})'v = 0 \text{ for } n = 1, ..., \mathcal{N}_\beta(d_N)\}$$

so that $\widehat{\beta}_h(y_N, d_N)$ is an unbounded set. (compare also with Theorem 6.4 in Mili and Coakley (1993)). Therefore for estimating β it makes no sense to regard h-trimmed L_p estimators in the case $h \leq \mathcal{N}_\beta(d_N)$.

But for estimating a linear aspect $\varphi(\beta) = L\,\beta$ of β it can be useful to regard h-trimmed L_p estimators for β also in situations where $h \leq \mathcal{N}_\beta(d_N)$ because the set $\{L\,\beta;\ \beta \in \widehat{\beta}_h(y_N, d_N)\}$ may be still bounded. To include also the case that no solution of (4.2) exist we define generally h-trimmed L_p estimators for a linear aspect $\varphi(\beta)$ as follows.

Definition 4.15 (Trimmed weighted L_p estimator for $\varphi(\beta) = L\,\beta$) *An estimator $\widehat{\varphi}_{h,p}$ is called a h-trimmed weighted L_p estimator for $\varphi(\beta)$ at d_N with weights $w_l, ..., w_h > 0$ if it satisfies*

$$\begin{aligned} \widehat{\varphi}_{h,p}(y_N, d_N) \quad = \{\widehat{\varphi} \in I\!R^s;\ & \text{there exists } (\beta_m)_{m \in I\!N} \text{ with} \\ & \widehat{\varphi} = \lim_{m\to\infty} L\,\beta_m \text{ and} \\ & \lim_{m\to\infty} \sum\nolimits_{n=l}^{h} w_n\, r_{(n)}(y_N, d_N, \beta_m)^p \\ & = \min_{\beta \in I\!R^r} \big(\sum\nolimits_{n=l}^{h} w_n\, r_{(n)}(y_N, d_N, \beta)^p\big)\} \end{aligned} \tag{4.4}$$

for all $y_N \in I\!R^N$, where $l, h \in \{1, ..., N\}$ and $p > 0$.

The following lemma generalizes Lemma 4.10 which concerns only estimators of β.

Lemma 4.11 *Let be $\varphi(\beta) = L\,\beta = (L_1, ...L_s)'\beta$, $M = \min\{N - h + 1, h - \mathcal{N}_\varphi(d_N)\} \geq 1$ and $y_N \in I\!R^N$. Then for all $\overline{y}_N \in \mathcal{Y}_{M-1}(y_N)$ every h-trimmed weighted L_p estimator $\widehat{\varphi}_{h,p}(\overline{y}_N, d_N)$ for φ exists and satisfies*

$$|\widehat{\varphi}| \leq \sqrt{\sum\nolimits_{i=1}^{s} |L_i|^2}\ \frac{R(y_N)\,S(d_N)\,(W+1)}{\rho_\varphi(d_N)^2}$$

for all $\widehat{\varphi} \in \widehat{\varphi}_{h,p}(\overline{y}_N, d_N)$.

Proof. Let be β_m a sequence approaching

$$\min_{\beta \in I\!R^r} \big(\sum\nolimits_{n=l}^{h} w_n\, r_{(n)}(\overline{y}_N, d_N, \beta)^p\big).$$

Then there exists a subsequence β_{m_k} such that $L\,\beta_{m_k} = 0$ for all $k \in I\!N$ or $L\,\beta_{m_k} \neq 0$ for all $k \in I\!N$. If $L\,\beta_{m_k} = 0$ for all $k \in I\!N$, then

$$\widehat{\varphi} = 0 = \lim_{k\to\infty} L\,\beta_{m_k} \in \widehat{\varphi}_{h,p}(\overline{y}_N, d_N).$$

Otherwise, according to Lemma 4.9, there exists k_0 such that $(\beta_{m_k})_{k>k_0}$ is a sequence in the compact set given by all β with $|\beta| < \frac{R(y_N)\,S(d_N)\,(W+1)}{\rho_\varphi(d_N)^2}$.

Hence there exists a subsequence β_{m_l} of β_{m_k} with $\lim_{l\to\infty} \beta_{m_l} = \beta_0$ and $|\beta_0| < \frac{R(y_N)\,S(d_N)\,(W+1)}{\rho_\varphi(d_N)^2}$ which implies

$$\widehat{\varphi} = L\,\beta_0 = \lim_{k\to\infty} L\,\beta_{m_k} \in \widehat{\varphi}_{h,p}(\overline{y}_N, d_N)$$

and

$$|\widehat{\varphi}| \le |L\,\beta_0| \le \sqrt{\sum\nolimits_{i=1}^{s} |L_i|^2}\;\frac{R(y_N)\,S(d_N)\,(W+1)}{\rho_\varphi(d_N)^2}.\square$$

Now Lemma 4.11 provides the generalization of Theorem 4.3.

Theorem 4.4 *If $\mathcal{N}_\varphi(d_N) < h$, then for all $y_N \in I\!R^N$ every h-trimmed weighted L_p estimator $\widehat{\varphi}_{h,p}(y_N, d_N)$ for φ exists and satisfies*

$$|\widehat{\varphi}| \le \sqrt{\sum\nolimits_{i=1}^{s} |L_i|^2}\;\frac{R(y_N)\,S(d_N)\,(W+1)}{\rho_\varphi(d_N)^2}$$

for all $\widehat{\varphi} \in \widehat{\varphi}_{h,p}(y_N, d_N)$.

As consequences of Lemma 4.10 and Lemma 4.11 we also get for h-trimmed L_p estimators for β and $\varphi(\beta)$ the following bounds for the bias by replacing observations.

Theorem 4.5
a) If $M = \min\{N-h+1, h-\mathcal{N}_\beta(d_N)\} \ge 1$, then for every h-trimmed L_p estimator $\widehat{\beta}_{h,p}$ for β at d_N we have for all $y_N \in I\!R^N$

$$b_M(\widehat{\beta}_{h,p}, d_N, y_N) \le 2\,\frac{R(y_N)\,S(d_N)\,(W+1)}{\rho_\beta(d_N)^2}.$$

b) If $\varphi(\beta) = L\,\beta = (L_1, ...L_s)'\beta$ and $M = \min\{N-h+1, h-\mathcal{N}_\varphi(d_N)\} \ge 1$, then for every h-trimmed L_p estimator $\widehat{\varphi}_{h,p}$ for φ at d_N we have for all $y_N \in I\!R^N$

$$b_M(\widehat{\varphi}_{h,p}, d_N, y_N) \le 2\sqrt{\sum\nolimits_{i=1}^{s} |L_i|^2}\;\frac{R(y_N)\,S(d_N)\,(W+1)}{\rho_\varphi(d_N)^2}.$$

Now we derive the breakdown point by replacing observations (see Section 4.2) for h-trimmed L_p estimators of general linear aspects $\varphi(\beta) = L\,\beta$. These results generalize the results given in Müller (1995b) for estimators concerning only the whole parameter vector β.

In Section 4.2 Theorem 4.2 provides an upper bound for the breakdown point for every regression equivariant estimator. Because h-trimmed L_p estimators are regression equivariant (see Theorem 4.1 in Mili and Coakley (1993)) we get as a consequence of Theorem 4.2 an upper bound for the breakdown point of h-trimmed L_p estimators.

Corollary 4.3 *Any h-trimmed weighted L_p estimator $\widehat{\varphi}_{h,p}$ for φ at d_N satisfies*

$$\epsilon^*(\widehat{\varphi}_{h,p}, d_N, y_N) \leq \frac{1}{N} \left\lfloor \frac{N - \mathcal{N}_\varphi(d_N) + 1}{2} \right\rfloor$$

for all $y_N \in I\!R^N$.

Besides the existence of h-trimmed L_p estimators and a bias bound, Lemma 4.11 also provides the following theorem which gives a lower bound for h-trimmed L_p estimators and which generalizes Theorem 4.1 in Müller (1995b) concerning only estimators for β.

Theorem 4.6 *Any h-trimmed weighted L_p estimator $\widehat{\varphi}_{h,p}$ for φ at d_N with $h > \mathcal{N}_\varphi(d_N)$ satisfies*

$$\epsilon^*(\widehat{\varphi}_{h,p}, d_N, y_N) \geq \frac{1}{N} \min\{N - h + 1, h - \mathcal{N}_\varphi(d_N)\}$$

for all $y_N \in I\!R^N$, and in particular

$$\epsilon^*(\widehat{\varphi}_{h,p}, d_N) \geq \frac{1}{N} \min\{N - h + 1, h - \mathcal{N}_\varphi(d_N)\}.$$

For $h - \mathcal{N}_\varphi(d_N) \leq N - h + 1$, i.e. $h \leq \frac{N+\mathcal{N}_\varphi(d_N)+1}{2}$, the lower bound is attained by observations satisfying an exact fit, i.e. for observations y_N satisfying $y_N = A_{d_N}\beta$.

Theorem 4.7 *If $\mathcal{N}_\varphi(d_N) < h \leq \frac{N+\mathcal{N}_\varphi(d_N)+1}{2}$, then any h-trimmed weighted L_p estimator $\widehat{\varphi}_{h,p}$ for φ at d_N satisfies*

$$\epsilon^*(\widehat{\varphi}_{h,p}, d_N, A_{d_N}\beta) = \frac{1}{N} (h - \mathcal{N}_\varphi(d_N))$$

for all $\beta \in I\!R^r$, and in particular

$$\epsilon^*(\widehat{\varphi}_{h,p}, d_N) = \frac{1}{N} (h - \mathcal{N}_\varphi(d_N)).$$

Proof. Without loss of generality we can assume that t_{nN} with $n = 1, ..., \mathcal{N}_\varphi(d_N)$ lie in a nonidentifying set. Then there exists a vector γ with $L\gamma \neq 0$ which is orthogonal to $\{a(t_{nN});\ n = 1, ..., \mathcal{N}_\varphi(d_N)\}$, i.e. $a(t_{nN})'\gamma = 0$ for $n \leq \mathcal{N}_\varphi(d_N)$. Define for any $v \in I\!R$ alternative observation vectors $\overline{y}_N$ by $\overline{y}_{nN} = a(t_{nN})'(\beta + v\gamma)$ for $n \leq h$ and $\overline{y}_{nN} = a(t_{nN})'\beta$ for $n > h$ which ensures $\overline{y}_N \in \mathcal{Y}_{h-\mathcal{N}_\varphi(d_N)}(A_{d_N}\beta)$. Because

$$\sum\nolimits_{n=1}^{h} w_n\, r_{(n)}(\overline{y}_N, d_N, \beta + v\gamma)^p = 0$$

we have $\beta + v\gamma \in \widehat{\beta}_{h,p}(\overline{y}_N, d_N)$ as well as $\beta \in \widehat{\beta}_{h,p}(A_{d_N}\beta, d_N)$. Hence we get

$$\begin{aligned}
&\sup\{|\widehat{\varphi}_{h,p}(A_{d_N}\beta, d_N) - \widehat{\varphi}_{h,p}(\overline{y}_N, d_N)|;\ \overline{y}_N \in \mathcal{Y}_{h-\mathcal{N}_\varphi(d_N)}(A_{d_N}\beta)\} \\
&\quad \geq \sup\{|L\,\widehat{\beta}_{h,p}(A_{d_N}\beta, d_N) - L\,\widehat{\beta}_{h,p}(\overline{y}_N, d_N)|;\ \overline{y}_N \in \mathcal{Y}_{h-\mathcal{N}_\varphi(d_N)}(A_{d_N}\beta)\} \\
&\quad \geq \sup\{|L\gamma|\,|v|;\ v \in I\!R\} = \infty
\end{aligned}$$

so that with Theorem 4.6 the assertion is proved. Compare also the proofs of Theorem 4.2 in Müller (1995b) and Theorem 6.3 in Mili and Coakley (1993). □

In particular for the LMS estimator for β, i.e. for $h = \lfloor \frac{N+2}{2} \rfloor$, we get $\epsilon^*(\widehat{\beta}_{h,p}, d_N) = \frac{1}{N}\left(\lfloor \frac{N}{2} \rfloor - \mathcal{N}_\beta(d_N) + 1\right)$. If the experimental conditions are in general position, i.e. $\mathcal{N}_\beta(d_N) = r-1$, then we have $\epsilon^*(\widehat{\varphi}_{h,p}, d_N) = \frac{1}{N}\left(\lfloor \frac{N}{2} \rfloor - r + 2\right)$ which coincides with the result given by Theorem 2 in Rousseeuw and Leroy (1987) on p. 118 for contaminated experimental conditions.

Another case, where the breakdown point $\epsilon^*(\widehat{\varphi}_{h,p}, d_N)$ can be given explicitely, appears when the lower and the upper bound for the breakdown point coincide. This is the case if and only if h satisfies $\lfloor \frac{N+\mathcal{N}_\varphi(d_N)+1}{2} \rfloor \leq h \leq \lfloor \frac{N+\mathcal{N}_\varphi(d_N)+2}{2} \rfloor$. Hence for those values h we get the h-trimmed weighted L_p estimators with the highest breakdown point. This provides the following theorem which coincides with Theorem 4.3 in Müller (1995b) if β shall be estimated. Moreover for estimating β it also coincides with Theorem 6.1 given in Mili and Coakley (1993) for contaminated experimental conditions.

Theorem 4.8 *Any h-trimmed weighted L_p estimator $\widehat{\varphi}_{h,p}$ for φ at d_N with $\lfloor \frac{N+\mathcal{N}_\varphi(d_N)+1}{2} \rfloor \leq h \leq \lfloor \frac{N+\mathcal{N}_\varphi(d_N)+2}{2} \rfloor$ satisfies*

$$\epsilon^*(\widehat{\varphi}_{h,p}, d_N, y_N) = \frac{1}{N}\left\lfloor \frac{N - \mathcal{N}_\varphi(d_N) + 1}{2} \right\rfloor$$

for all $y_N \in I\!R^N$.

Theorem 4.7 and Theorem 4.8 show that the breakdown point $\epsilon^*(\widehat{\varphi}_{h,p}, d_N)$ increases when $\mathcal{N}_\varphi(d_N)$ decreases. For estimating β this means that the highest breakdown appears at designs d_N with $\mathcal{N}_\beta(d_N) = r - 1$, i.e. at designs with experimental conditions in general position. Then any h-trimmed weighted L_p estimator $\widehat{\beta}_{h,p}$ for β with $\lfloor \frac{N+r}{2} \rfloor \leq h \leq \lfloor \frac{N+r+1}{2} \rfloor$ satisfies $\epsilon^*(\widehat{\beta}_{h,p}, d_N) = \frac{1}{N}\lfloor \frac{N-r+2}{2} \rfloor$. Because $h = \lfloor \frac{N}{2} \rfloor + \lfloor \frac{r+1}{2} \rfloor \in [\lfloor \frac{N+r}{2} \rfloor, \lfloor \frac{N+r+1}{2} \rfloor]$ this result coincides with Theorem 6 on p. 132 in Rousseeuw and Leroy (1987) which is given for the LTS estimator and contaminated experimental conditions.

The breakdown point for $h > \lfloor \frac{N+\mathcal{N}_\varphi(d_N)+2}{2} \rfloor$ is much more difficult to calculate. In particular sometimes it does not only depend on $\mathcal{N}_\varphi(d_N)$ but also on the position of the experimental conditions $t_1, ..., t_N$ (see He,

Jurečková et al. (1990) and Ellis and Morgenthaler (1992)). For example the L_1 estimator $\widehat{\beta}_1$ in a linear regression model (where $a(t) = (1,t)'$ and $\beta = (\beta_0, \beta_1)' \in I\!R^2$) can have a breakdown point greater than $\frac{1}{4}$ but also equal to $\frac{1}{N}$ for designs with N different experimental conditions, i.e. for $\mathcal{N}_\varphi(d_N) = 1$. The following two theorems provide upper bounds for the breakdown point of the L_1 estimator in a linear regression model. Thereby the first theorem provides an upper bound which shows how the breakdown point depends on the position of the experimental conditions while the second theorem provides that $\frac{1}{4}$ is the upper bound for all designs with $N = 4K$, $K \in I\!N$. The upper bound of $\frac{1}{4}$ is attained by several designs (see Section 9.1 and Ellis and Morgenthaler (1992)).

Theorem 4.9 *Let be $t_{1N} \leq ... \leq t_{NN}$. Then the breakdown point of the L_1 estimator $\widehat{\beta}_1$ for β at $d_N = (t_{1N}, ..., t_{NN})'$ in a linear regression model satisfies*

$$\begin{aligned}\epsilon^*(\widehat{\beta}_1, d_N) \leq \tfrac{1}{N} \min\{M;\ & \textstyle\sum_{n=1}^{N-M}(t_{nN} - t_{1N}) \leq \sum_{n=N-M+1}^{N}(t_{nN} - t_{1N}) \\ & \text{or } \textstyle\sum_{n=M+1}^{N}(t_{NN} - t_{nN}) \leq \sum_{n=1}^{M}(t_{NN} - t_{nN})\}\end{aligned}$$

Proof. For y_N given by $y_{nN} = 0$ for $n = 1, ..., N$ we have $\widehat{\beta}_1(y_N, d_N) = \{0\}$. Now assume $\sum_{n=1}^{N-M}(t_{nN} - t_{1N}) \leq \sum_{n=N-M+1}^{N}(t_{nN} - t_{1N})$. Then for any $\overline{y}_N^0$ given by $\overline{y}_{nN}^0 = 0$ for $n = 1, ..., N-M$ and $\overline{y}_{nN}^0 = \beta_0 + \beta_1 t_{nN}$ for $n = N-M+1, ..., N$ with $\beta_0 + \beta_1 t_{1N} = 0$ we get

$$\begin{aligned}\sum_{n=1}^{N} |\overline{y}_{nN}^0| &= \sum_{n=N-M+1}^{N} |\beta_0 + \beta_1 t_{nN} - \beta_0 - \beta_1 t_{1N}| \\ &= |\beta_1| \sum_{n=N-M+1}^{N} (t_{nN} - t_{1N}) \geq |\beta_1| \sum_{n=1}^{N-M} (t_{nN} - t_{1N}) \\ &= \sum_{n=1}^{N-M} |\beta_0 + \beta_1 t_{nN} - \beta_0 - \beta_1 t_{1N}| = \sum_{n=1}^{N} |\overline{y}_{nN}^0 - \beta_0 - \beta_1 t_{nN}|.\end{aligned}$$

Hence there exists $\overline{\beta}$ with $0 \neq \overline{\beta} \in \widehat{\beta}_1(\overline{y}_N^0, d_N)$. Because $\widehat{\beta}_1(v\overline{y}_N^0, d_N) = v\widehat{\beta}_1(\overline{y}_N^0, d_N)$ for all $v \in I\!R$ it follows

$$\begin{aligned}&\sup\{|\widehat{\beta}_1(y_N, d_N) - \widehat{\beta}_1(\overline{y}_N, d_N)|;\ \overline{y}_N \in \mathcal{Y}_M(y_N)\} \\ \geq\ &\sup\{|v\,\widehat{\beta}_1(\overline{y}_N^0, d_N)|;\ v \in I\!R\} = \infty.\end{aligned}$$

The same holds in the case $\sum_{n=M+1}^{N}(t_{nN} - t_{NN}) \leq \sum_{n=1}^{M}(t_{nN} - t_{NN})$ so that the assertion follows. □

Theorem 4.10 *If $N = 4K$ with $K \in I\!N$, then the breakdown point of the L_1 estimator $\widehat{\beta}_1$ for β at d_N in a linear regression model satisfies*

$$\epsilon^*(\widehat{\beta}_1, d_N) \leq \frac{1}{4}.$$

Proof. See Ellis and Morgenthaler (1992). □

4.4 Breakdown Points for Nonlinear Problems

At first we regard the problem of estimating a nonlinear aspect $\varphi(\beta) \in I\!R^s$ in a linear model. For example the nonlinear aspect may be $\varphi((\beta_1, \beta_2)') = \frac{\beta_1}{\beta_2}$ or $\varphi(\beta) = \Phi(\beta)$, where Φ is some distribution function. In particular for examples like $\varphi(\beta) = \Phi(\beta)$ the set $\varphi(\mathcal{B}) = \{\varphi;\ \varphi = \varphi(\beta)$ for some $\beta \in \mathcal{B}\}$ is bounded although $\mathcal{B}$ is unbounded. For the example $\varphi(\beta) = \Phi(\beta)$ we have $\varphi(\mathcal{B}) = [0, 1]$ although $\mathcal{B} = I\!R$. Hence every estimator for φ should be also bounded by 0 and 1 so that there will be no observations with estimated values tending to infinity. In such situations the estimator would never break down if a definition of Section 4.2 is used for the breakdown point. But the definition of the breakdown point by replacing observations can be generalized in the sense as was already used in Section 4.1 for defining the asymptotic breakdown point. Namely the breakdown point can be defined as the smallest fraction of replaced observations so that the bias caused by this replaced observations is equal to the bias by replacing all observations.

Definition 4.16 (Breakdown point of estimators for a general aspect)
The breakdown point of an estimator $\widehat{\varphi}_N$ for a general aspect φ at d_N by replacing observations of y_N is defined as

$$\begin{aligned}\epsilon^*(\widehat{\varphi}_N, d_N, y_N) \quad &:= \quad \min\{\frac{M}{N};\ M \in I\!N_0 \text{ with} \\ &b_M(\widehat{\varphi}_N, d_N, y_N) = b_N(\widehat{\varphi}_N, d_N, y_N)\},\end{aligned}$$

and the breakdown point of $\widehat{\varphi}_N$ at d_N by replacing observations is defined as

$$\epsilon^*(\widehat{\varphi}_N, d_N) := \min\{\epsilon^*(\widehat{\varphi}_N, d_N, y_N);\ y_N \in I\!R^N\}.$$

Because for linear aspects we always have $b_N(\widehat{\varphi}_N, d_N, y_N) = \infty$ the above definition is really a generalization of the Definition 4.8 in Section 4.2.

In some situations the breakdown point of an estimator of a nonlinear aspect can be estimated by the breakdown points of estimators for linear aspects. This is shown by the following theorem.

Theorem 4.11 *If $\widehat{\varphi}_N = \rho(\widehat{\varphi}_{N,1}, ..., \widehat{\varphi}_{N,J})$, $\rho : I\!R^J \to I\!R$ is strictly increasing in all components, $\max\{\widehat{\varphi}_N(y_N, d_N);\ y_N \in I\!R^N\} = \max\{\rho(v);\ v \in I\!R^J\}$ and $\min\{\widehat{\varphi}_N(y_N, d_N);\ y_N \in I\!R^N\} = \min\{\rho(v);\ v \in I\!R^J\}$, then*

$$\epsilon^*(\widehat{\varphi}_N, d_N, y_N) \geq \min\{\epsilon^*(\widehat{\varphi}_{N,j}, d_N, y_N);\ j = 1, ..., J\}.$$

Proof. Let be $M = \min\{\epsilon^*(\widehat{\varphi}_{N,j}, d_N, y_N);\ j = 1, ..., J\}$. Then there exists $K \in I\!N$ such that for all $\overline{y}_N \in \mathcal{Y}_{M-1}(y_N)$ we have $|\widehat{\varphi}_{N,j}(\overline{y}_N, d_N)| \leq$

K for $j = 1, ..., J$. The assumption that ρ is increasing in all components provides

$$\rho(-K, ..., -K) \leq \rho(\widehat{\varphi}_{N,1}(\overline{y}_N, d_N), ..., \widehat{\varphi}_{N,J}(\overline{y}_N, d_N)) \leq \rho(K, ..., K)$$

for all $\overline{y}_N \in \mathcal{Y}_{M-1}(y_N)$. Because ρ is strictly increasing in all components and $M_\infty := \max\{\widehat{\varphi}_N(y_N, d_N);\ y_N \in I\!R^N\} = \max\{\rho(v);\ v \in I\!R^J\}$ and $M_{-\infty} := \min\{\widehat{\varphi}_N(y_N, d_N);\ y_N \in I\!R^N\} = \min\{\rho(v);\ v \in I\!R^J\}$ we have

$$\begin{aligned} & b_N(\widehat{\varphi}_N, d_N, y_N) \\ = \;& \max\{|\widehat{\varphi}_N - M_\infty|, |\widehat{\varphi}_N - M_{-\infty}|\} \\ > \;& \max\{|\widehat{\varphi}_N - \rho(K, \ldots, K)|, |\widehat{\varphi}_N - \rho(-K, \ldots, -K)|\} \\ \geq \;& b_{M-1}(\widehat{\varphi}_N, d_N, y_N) \end{aligned}$$

so that $\epsilon^*(\widehat{\varphi}_N, d_N, y_N) \geq M - 1$. □

If additionally to the assumptions of Theorem 4.11 $\rho(v_1, ..., v_J) = v_1 \cdot ... \cdot v_J$ for all $(v_1, ..., v_J)' \in I\!R^J$ and $\widehat{\varphi}_{N,j}$, $j = 1, ..., J$, are regression and scale equivariant, then also an upper bound can be derived. As for estimating linear aspects this upper bound depends on $\mathcal{N}_\varphi(d_N)$ which also for nonlinear aspects is defined as in Section 4.3. Because regression equivariance was also already defined in Section 4.3 we here only have to define scale equivariance.

Definition 4.17 (Scale equivariant estimator)
An estimator $\widehat{\varphi}_N$ for φ is scale equivariant at d_N if for all $\eta \in I\!R$ and all $y_N \in I\!R^N$

$$\widehat{\varphi}_N(\eta y_N, d_N) = \eta \widehat{\varphi}_N(y_N, d_N).$$

To show the upper bound we need the following lemma.

Lemma 4.12 *Let be $\varphi_1(\beta) = (l_1, ..., l_J)'\beta$ and $\varphi_2(\beta) = l_1'\beta \cdot ... \cdot l_J'\beta$ with $l_1, ..., l_J \in I\!R^r \setminus \{0_r\}$. Then φ_1 is identifiable at $\mathcal{D}$ if and only if φ_2 is identifiable at $\mathcal{D}$.*

Proof. Assume that φ_1 is identifiable at $\mathcal{D}$. Then for all $\beta, \overline{\beta} \in I\!R^r$ we have that $a(t)'\beta = a(t)'\overline{\beta}$ for $t \in \mathcal{D}$ implies $\varphi_1(\beta) = \varphi_1(\overline{\beta})$, i.e. $l_j'\beta = l_j'\overline{\beta}$ for $j = 1, ..., J$. But then we have $l_1'\beta \cdot ... \cdot l_J'\beta = l_1'\overline{\beta} \cdot ... \cdot l_J'\overline{\beta}$ so that also φ_2 is identifiable at $\mathcal{D}$.

Now assume that φ_2 is identifiable at $\mathcal{D}$. Then for all $\beta \in I\!R$ with $a(t)'\beta = 0$ for $t \in \mathcal{D}$ we have $\varphi_2(\beta) = l_1'\beta \cdot ... \cdot l_J'\beta = 0$. Now assume that there exists $J_0 \geq 1$ so that without loss of generality

$$\begin{aligned} & l_j'\beta \neq 0 \quad \text{for } j = 1, ..., J_0 \text{ and} \\ & l_j'\beta = 0 \quad \text{for } j = J_0 + 1, ..., J. \end{aligned} \tag{4.5}$$

Because of $l_1'\beta \cdot ... \cdot l_J'\beta = 0$ we have $J_0 < J$ so that a $l \in \mathbb{R}^r$ exists such that $l_j'l \neq 0$ for $j = J_0 + 1, ..., J$. Then for every $k \in \mathbb{R}$ we have $a(t)'(\beta + k\,l) = a(t)'k\,l$ for $t \in \mathcal{D}$ which, because of the identifiability of φ_2, implies

$$\prod_{j=1}^{J} k\,l_j'l = \prod_{j=1}^{J}(l_j'\beta + k\,l_j'l) = \prod_{j=1}^{J_0}(l_j'\beta + k\,l_j'l) \cdot \prod_{j=J_0+1}^{J} k\,l_j'l.$$

Hence we have

$$\begin{aligned}\prod_{j=1}^{J_0} k\,l_j'l &= \prod_{j=1}^{J_0}(l_j'\beta + k\,l_j'l) \\ &= \sum\nolimits_{j=0}^{J_0} \sum\nolimits_{\{i_1,...,i_j\}\subset\{1,...,J_0\}} k^j \prod_{i\in\{i_1,...,i_j\}} l_i'l \cdot \prod_{i\notin\{i_1,...,i_j\}} l_i'\beta \\ &= \prod_{j=1}^{J_0} l_j'\beta + \sum\nolimits_{j=1}^{J_0-1} k^j\, v_j + \prod_{j=1}^{J_0} k\,l_j'l\end{aligned}$$

with

$$v_j := \sum\nolimits_{\{i_1,...,i_j\}\subset\{1,...,J_0\}} \prod_{i\in\{i_1,...,i_j\}} l_i'l \cdot \prod_{i\notin\{i_1,...,i_j\}} l_i'\beta.$$

Then $\prod_{j=1}^{J_0} l_j'\beta + \sum_{j=1}^{J_0-1} k^j\, v_j$ is a polynomial in k with degree of $J_0 - 1$ which is for all $k \in \mathbb{R}$ equal to zero. Hence all coefficients of the polynomial are equal to zero. In particular we have $\prod_{j=1}^{J_0} l_j'\beta = 0$ which is a contradiction to (4.5). Therefore we have $l_j'\beta = 0$ for all $j = 1, ..., J$ so that φ_1 is identifiable at $\mathcal{D}$. □

Theorem 4.12 *Let be $\varphi(\beta) = l_1'\beta \cdot ... \cdot l_J'\beta$ with $l_1, ..., l_J \in \mathbb{R}^r \setminus \{0_r\}$, and $\widehat{\varphi}_N = \widehat{\varphi}_{N,1} \cdot ... \cdot \widehat{\varphi}_{N,J}$, where $\widehat{\varphi}_{N,j}$ is a regression and scale equivariant estimator for $l_j'\beta$ for $j = 1, \ldots, J$. Then*

$$\epsilon^*(\widehat{\varphi}_N, d_N) \leq \frac{1}{N}\left\lfloor \frac{N - \mathcal{N}_\varphi(d_N) + 1}{2} \right\rfloor.$$

Proof. We have to show that by replacing $\left\lfloor \frac{N-\mathcal{N}_\varphi(d_N)+1}{2} \right\rfloor$ observations the estimator $\widehat{\varphi}_N$ breaks down. For that we have to ensure that at least one component $\widehat{\varphi}_j$ is going to infinity while no another component is going to zero. To show this take any nonidentifying set $\mathcal{D}$ for φ with $\sum_{n=1}^N 1_{\mathcal{D}}(t_{nN}) = \mathcal{N}_\varphi(d_N)$ and set $M = \left\lfloor \frac{N-\mathcal{N}_\varphi(d_N)+1}{2} \right\rfloor$. Then there exists $\beta, \overline{\beta} \in \mathbb{R}^r$ with $a(t)'\beta = a(t)'\overline{\beta}$ for all $t \in \mathcal{D}$ and $\varphi(\beta) \neq \varphi(\overline{\beta})$ (see Section 1.3). Then without loss of generality we have $t_{nN} \in \mathcal{D}$ for $n = 1, ..., \mathcal{N}_\varphi(d_N)$

and $l_1'(\beta - \overline{\beta}) \neq 0$ because of Lemma 4.12. Define $\overline{y}_N^1$ by $\overline{y}_{nN}^1 = 0$ for $n = 1, ..., N - M$ and $\overline{y}_{nN}^1 = a(t_{nN})'(\beta - \overline{\beta})$ for $n = N - M + 1, ..., N$. Define also $\overline{y}_N^2$ by $\overline{y}_{nN}^2 = 0$ for $n = 1, ..., \mathcal{N}_\varphi(d_N)$ and $n = N - M + 1, ..., N$ and $\overline{y}_{nN}^2 = -a(t_{nN})'(\beta - \overline{\beta})$ for $n = \mathcal{N}_\varphi(d_N) + 1, ..., N - M$. Then $K\,\overline{y}_N^1, K\,\overline{y}_N^2 \in \mathcal{Y}_M(0_N)$ for all $K \in \mathbb{N}$ and $\overline{y}_N^1 = \overline{y}_N^2 + A_{d_N}(\beta - \overline{\beta})$ so that with the regression equivariance of $\widehat{\varphi}_{N1}$ we have

$$
\begin{aligned}
&|\widehat{\varphi}_{N,1}(\overline{y}_N^1, d_N)| + |\widehat{\varphi}_{N,1}(\overline{y}_N^2, d_N)| \\
&\geq \; |\widehat{\varphi}_{N,1}(\overline{y}_N^1, d_N) - \widehat{\varphi}_{N,1}(\overline{y}_N^2, d_N)| \\
&= \; |\widehat{\varphi}_{N,1}(\overline{y}_N^2 + A_{d_N}(\beta - \overline{\beta}), d_N) - \widehat{\varphi}_{N,1}(\overline{y}_N^2, d_N)| \\
&= \; |\widehat{\varphi}_{N,1}(\overline{y}_N^2, d_N) + l_1'(\beta - \overline{\beta}) - \widehat{\varphi}_{N,1}(\overline{y}_N^2, d_N)| \\
&= \; |l_1'(\beta - \overline{\beta})| \neq 0.
\end{aligned}
$$

Then $|\widehat{\varphi}_{N,1}(\overline{y}_N^1, d_N)| \neq 0$ or $|\widehat{\varphi}_{N,1}(\overline{y}_N^2, d_N)| \neq 0$. Assume without loss of generality $|\widehat{\varphi}_{N,1}(\overline{y}_N^1, d_N)| \neq 0$. Take any γ with $l_j'\gamma \neq 0$ for $j = 2, ..., J$. Then with the regression and scale equivariance we have

$$
|\widehat{\varphi}_{N,j}(A_{d_N}\gamma + K\overline{y}_N^1, d_N)| = |l_j'\gamma + K\widehat{\varphi}_{N,j}(\overline{y}_N^1, d_N)| \to \infty
$$

for $K \to \infty$ if $\widehat{\varphi}_j(\overline{y}_N^1, d_N) \neq 0$. At least this holds for $\widehat{\varphi}_{N,1}$. If $\widehat{\varphi}_j(\overline{y}_N^1, d_N) = 0$, then

$$
|\widehat{\varphi}_{N,j}(A_{d_N}\gamma + K\overline{y}_N^1, d_N)| = |l_j'\gamma + K\widehat{\varphi}_{N,j}(\overline{y}_N^1, d_N)| = |l_j'\gamma| \neq 0
$$

for all $K \in \mathbb{N}$. Hence $|\widehat{\varphi}_N(A_{d_N}\gamma + K\overline{y}_N^1, d_N)|$ is converging to infinity for $K \to \infty$. Because $A_{d_N}\gamma + K\overline{y}_N^1 \in \mathcal{Y}_M(A_{d_N}\gamma)$ for all $K \in \mathbb{N}$ the assertion is proved. □

As for linear aspects the upper bound of Theorem 4.12 is attained by estimators based on h-trimmed L_p estimators which obviously are regression and scale equivariant.

Theorem 4.13 *Let be $\varphi(\beta) = l_1'\beta \cdot ... \cdot l_J'\beta$, where $l_1, ..., l_J \in \mathbb{R}^r$ are linearly independent, and $\widehat{\varphi}_N = \widehat{\varphi}_{N,1} \cdot ... \cdot \widehat{\varphi}_{N,J}$, where $\widehat{\varphi}_{N,j}$ is a h-trimmed L_p estimator for $l_j'\beta$ with $\lfloor \frac{N + \mathcal{N}_\varphi(d_N) + 1}{2} \rfloor \leq h \leq \lfloor \frac{N + \mathcal{N}_\varphi(d_N) + 2}{2} \rfloor$ for $j = 1, ..., J$. Then*

$$
\epsilon^*(\widehat{\varphi}_N, d_N) = \frac{1}{N} \left\lfloor \frac{N - \mathcal{N}_\varphi(d_N) + 1}{2} \right\rfloor .
$$

Proof. Lemma 4.12 provides $\mathcal{N}_\varphi(d_N) = \max\{\mathcal{N}_{\varphi_j}(d_N);\ j = 1, .., J\}$ so that $h > \mathcal{N}_{\varphi_j}(d_N)$ for all $j = 1, ..., J$. Hence Theorem 4.6 yields

$$
\begin{aligned}
&\min\{\epsilon^*(\widehat{\varphi}_{N,j}, d_N);\ j = 1, ..., J\} \\
&\geq \; \frac{1}{N} \min\{N - h + 1, h - \mathcal{N}_{\varphi_1}(d_N), ..., h - \mathcal{N}_{\varphi_J}(d_N)\}
\end{aligned}
$$

$$
\begin{aligned}
&= \frac{1}{N}\min\{N-h+1, h-\mathcal{N}_\varphi(d_N)\} \\
&\geq \tfrac{1}{N}\min\left\{N - \left\lfloor \tfrac{N+\mathcal{N}_\varphi(d_N)+2}{2}\right\rfloor + 1, \left\lfloor \tfrac{N+\mathcal{N}_\varphi(d_N)+1}{2}\right\rfloor - \mathcal{N}_\varphi(d_N)\right\} \\
&= \tfrac{1}{N}\min\left\{\left\lfloor \tfrac{2N-N-\mathcal{N}_\varphi(d_N)-1}{2}\right\rfloor + 1, \left\lfloor \tfrac{N+\mathcal{N}_\varphi(d_N)+1-2\mathcal{N}_\varphi(d_N)}{2}\right\rfloor\right\} \\
&= \tfrac{1}{N}\left\lfloor \tfrac{N-\mathcal{N}_\varphi(d_N)+1}{2}\right\rfloor .
\end{aligned}
$$

Moreover, because $l_1, \ldots, l_J$ are linearly independent

$$
\begin{aligned}
\max\{\widehat{\varphi}_N(y_N, d_N);\ y_N \in I\!R^N\} &\geq \max\{\widehat{\varphi}_N(A_{d_N}\beta, d_N);\ \beta \in I\!R^r\} \\
&\geq \max\{\prod_{j=1}^J l_j'\beta;\ \beta \in I\!R^r\} = \infty \\
&= \max\{\prod_{j=1}^J v_j;\ (v_1, \ldots, v_J)' \in I\!R^J\},
\end{aligned}
$$

and similarly

$$
\min\{\widehat{\varphi}_N(y_N, d_N);\ y_N \in I\!R^N\} = \min\{\prod_{j=1}^J v_j;\ (v_1, \ldots, v_J)' \in I\!R^J\}.
$$

Hence, together with Theorem 4.11 and Theorem 4.12 the assertion is proved. □

Example 4.1 (Volume of a cube)
By measuring the length of the edges of a cube its volume shall be estimated. Thereby we assume that the length of the edges can be measured only with errors, for example because the cube is very small. Then the length of the edges $\beta_1, \beta_2, \beta_3$ are unknown and the observations Y_{nN} are the measured length of the edges, i.e. $Y_{nN} = \beta_i + Z_{nN}$, if in the nth observation the edge i is measured. Then we have $Y_{nN} = a(t_{nN})'\beta + Z_{nN}$ with $\beta = (\beta_1, \beta_2, \beta_3)'$ and $a(t) = (1_1(t), 1_2(t), 1_3(t))'$ with $\mathcal{T} = \{1, 2, 3\}$, i.e. a one-way lay-out model. The interesting aspect is $\varphi(\beta) = \beta_1 \cdot \beta_2 \cdot \beta_3 = l_1'\beta \cdot l_2'\beta \cdot l_3'\beta$, where $l_1 = (1,0,0)'$, $l_2 = (0,1,0)'$ and $l_3 = (0,0,1)'$ are linearly independent. Hence, according to Theorem 4.13 an estimator $\widehat{\varphi}_N$ of the form $\widehat{\varphi}_N = \widehat{\varphi}_{N,1} \cdot \widehat{\varphi}_{N,2} \cdot \widehat{\varphi}_{N,3}$, where $\widehat{\varphi}_{N,j}$ are h-trimmed L_p estimators for $\varphi_j(\beta) = l_j'\beta$ with $\lfloor \frac{N+\mathcal{N}_\varphi(d_N)+1}{2} \rfloor \leq h \leq \lfloor \frac{N+\mathcal{N}_\varphi(d_N)+2}{2} \rfloor$ for $j = 1, 2, 3$, is breakdown point maximizing with maximum breakdown point of

$$
\epsilon^*(\widehat{\varphi}_N, d_N) = \frac{1}{N}\left\lfloor \frac{N - \mathcal{N}_\varphi(d_N) + 1}{2} \right\rfloor .
$$

If we use a design, where the edges 1,2,3 are measured with equal numbers of repetitions, then according to Theorem 9.2 and Lemma 4.12 the design is breakdown point maximizing. □

In opposite to nonlinear aspects of product type for nonlinear aspects given by a quotient only a very negative result can be shown which provides only the smallest breakdown point of $\frac{1}{N}$ for estimators of these aspects. This is caused by the fact that the estimated denominator easily can become equal to zero.

Theorem 4.14 *If the aspect is given by*

$$\varphi(\beta) = \frac{c_1 + \varphi_1(\beta)}{c_2 + \varphi_2(\beta)}$$

and the estimator is given by

$$\widehat{\varphi}_N(y_N, d_N) = \frac{c_1 + \widehat{\varphi}_{N,1}(y_N, d_N)}{c_2 + \widehat{\varphi}_{N,2}(y_N, d_N)},$$

where $\varphi_j(\beta) = l_j'\beta$ *for* $j = 1, 2$, l_1 *and* l_2 *are linearly independent and* $\widehat{\varphi}_{N,1}$ *and* $\widehat{\varphi}_{N,2}$ *are regression equivariant estimators for* $\varphi_1(\beta)$ *and* $\varphi_2(\beta)$, *respectively, then*

$$\epsilon^*(\widehat{\varphi}_N, d_N) \leq \frac{1}{N}.$$

Proof. Take any $y_N \in I\!R^N$ and $\gamma_1 \in I\!R^r$ with $l_2'\gamma_1 \neq 0$. Then there exists $\gamma_2 \in I\!R^r$ with $l_2'\gamma_2 = c_2 + \widehat{\varphi}_{N,2}(y_N, d_N)$. Set $y_N^0 = y_N - A_{d_N}\gamma_2$ and $y_N^1 = y_N^0 + A_{d_N}\gamma_1$. Then the regression equivariance of $\widehat{\varphi}_{N,2}$ provides $c_2 + \widehat{\varphi}_{N,2}(y_N^0, d_N) = 0$ and $c_2 + \widehat{\varphi}_{N,2}(y_N^1, d_N) \neq 0$. Define

$$y_N^M = (y_{1N}^1, ..., y_{MN}^1, y_{(M+1)N}^0, ..., y_{NN}^0)'$$

and

$$M_0 = \max\{M \in \{1, ..., N\};\ c_2 + \widehat{\varphi}_{N,2}(y_N^M, d_N) = 0\}.$$

Then we have $M_0 < N$, $c_2 + \widehat{\varphi}_{N,2}(y_N^{M_0}, d_N) = 0$, $c_2 + \widehat{\varphi}_{N,2}(y_N^{M_0+1}, d_N) \neq 0$ and $y_N^{M_0} \in \mathcal{Y}_1(y_N^{M_0+1})$. If $c_1 + \widehat{\varphi}_{N,1}(y_N^{M_0}, d_N) \neq 0$, then $|\widehat{\varphi}_N(y_N^{M_0}, d_N) - \widehat{\varphi}_N(y_N^{M_0+1}, d_N)| = \infty$. If $c_1 + \widehat{\varphi}_{N,1}(y_N^{M_0}, d_N) = 0$, then because l_1 and l_2 are linearly independent there exists $\gamma_3 \in I\!R^r$ with $l_2'\gamma_3 = 0$ and $l_1'\gamma_3 \neq 0$. Regard now instead of $y_N^{M_0}$ and $y_N^{M_0+1}$ the observation vectors $\tilde{y}_N^{M_0} = y_N^{M_0} + A_{d_N}\gamma_3$ and $\tilde{y}_N^{M_0+1} = y_N^{M_0+1} + A_{d_N}\gamma_3$. Then still it holds $\tilde{y}_N^{M_0} \in \mathcal{Y}_1(\tilde{y}_N^{M_0+1})$ and the regression equivariance provides $c_2 + \widehat{\varphi}_{N,2}(\tilde{y}_N^{M_0}, d_N) = 0$, $c_2 + \widehat{\varphi}_{N,2}(\tilde{y}_N^{M_0+1}, d_N) \neq 0$ and $c_1 + \widehat{\varphi}_{N,1}(\tilde{y}_N^{M_0}, d_N) \neq 0$ so that $|\widehat{\varphi}_N(\tilde{y}_N^{M_0}, d_N) - \widehat{\varphi}_N(\tilde{y}_N^{M_0+1}, d_N)| = \infty$. Hence the assertion is proved. □

Example 4.2 (Linear calibration)
In the linear calibration problem we have a simple linear regression model $Y_{nN} = \beta_0 + \beta_1 t_{nN} + Z_{nN}$ with $\beta = (\beta_0, \beta_1)' \in I\!R^2$ and the interesting aspect

is $\varphi(\beta) = \frac{y-\beta_0}{\beta_1}$ (see Example 2.3). Then we have $\varphi(\beta) = \frac{y+l_1'\beta}{l_2'\beta}$, where $l_1 = (-1,0)'$ and $l_2 = (0,1)'$ are linearly independent. Hence according to Theorem 4.14 any estimator of the form $\widehat{\varphi}_N(y_N, d_N) = \frac{y+\widehat{\varphi}_{N,1}(y_N,d_N)}{\widehat{\varphi}_{N,2}(y_N,d_N)}$, where $\widehat{\varphi}_{N,1}$ and $\widehat{\varphi}_{N,2}$ are regression equivariant estimators for $l_1'\beta$ and $l_2'\beta$, respectively, has at most a breakdown point of $\frac{1}{N}$. □

Theorem 4.14 provides a very low breakdown point for estimators of nonlinear aspects given by a quotient. Because the breakdown point $\epsilon^*(\widehat{\varphi}_N, d_N)$ at a design d_N is the minimum of all breakdown points $\epsilon^*(\widehat{\varphi}_N, d_N, y_N)$ at a given observation vector y_N the low breakdown point $\epsilon^*(\widehat{\varphi}_N, d_N)$ may be caused by observation vectors for which the probability of occurrence is equal to zero or which occur only with a very small probability. Hence we should use a more probabilistic approach like the approach regarded in Section 4.1. For estimating a linear aspect a more probabilistic approach is not necessary because often the breakdown point $\epsilon^*(\widehat{\varphi}_N, d_N, y_N)$ is independent of y_N. In particular this holds if the lower and the upper bound given by Theorem 4.6 and Corollary 4.3 coincide which is the case for breakdown point maximizing estimators.

Sometimes nonlinear models can be transformed in a linear model with an interesting nonlinear aspect. See for example the rhythmometry model which was studied in Kitsos et al. (1988). But often such a transformation is not possible. Breakdown points for really nonlinear models were only studied by Christmann (1993) for logistic regression and by Stromberg and Ruppert (1992) and Sakata and White (1995) by using new definitions of breakdown points. But in these definitions also the experimental conditions are contaminated so that it cannot be used for designed experiments. Moreover, these definitions concern only the breakdown behaviour of the response function and not of the parameter itself. Therefore the investigation of breakdown points for designed nonlinear models is still an open problem (see also the outlook).

5

Asymptotic Robustness for Shrinking Contamination

In this chapter we show that the robustness concept based on influence functions (see Section 3.2) is closely connected with a robustness concept based on shrinking neighbourhoods. In particular, the concepts coincide for estimators given by Fréchet differentiable functionals as is shown in Section 5.1. But Section 5.1 also shows that for robustness concepts based on shrinking neighbourhoods also the larger class of asymptotically linear estimators can be used instead of the class of estimators which satisfy some differentiability condition. Therefore the robustness properties are given for asymptotically linear estimators. Especially, these properties are derived in Section 5.2 for estimating a linear aspect in a linear model, in Section 5.3 for estimating a nonlinear aspect in a linear model and in Section 5.4 for estimation in a nonlinear model.

5.1 Asymptotic Behaviour of Estimators in Shrinking Neighbourhoods

At first we show that estimators derived from Fréchet differentiable functionals $\widehat{\zeta}$ by $\widehat{\zeta}_N(y_N, d_N) = \widehat{\zeta}(P_{y_N,d_N})$ are asymptotically linear for distributions $(Q_{N,\theta,\xi})_{N\in\mathbb{N}}$ of a shrinking Kolmogorov neighbourhood $\mathcal{U}_{K,\epsilon}(P_{\theta,\delta})$ (for the shrinking Kolmogorov neighbourhood see Section 3.1). The asymptotic linearity provides that the estimators are asymptotically normally distributed so that robustness and optimality criteria can be derived from the asymptotic normal distribution.

To derive the asymptotic linarity we assume that the Fréchet derivative at $P_{\theta,\delta}$ is of form (3.2), where $\psi_\theta : \mathbb{R} \times \mathcal{T} \to \mathbb{R}^q$ and $P_{\theta,\delta}$ satisfy the following conditions:

- $\widehat{\zeta}(P_{\theta,\delta}) = \zeta(\theta)$ for all $\theta \in \Theta$. (5.1)
- $\int \psi_\theta(y,t)\, P_{\theta,\delta}(dy, dt) = 0.$ (5.2)
- $\text{supp}(\delta) = \{t_1, ..., t_I\}.$ (5.3)

- There exist $K \in I\!R$ and $\epsilon > 0$ such that:

$$d_K(P_{\overline{\theta},\delta}, P_{\theta,\delta}) \leq K|\overline{\theta} - \theta|, \tag{5.4}$$

$$\lim_{y\to\infty} \psi_{\overline{\theta}}(y,t) = \psi_{\overline{\theta}}(\infty,t) \text{ and } \lim_{y\to-\infty} \psi_{\overline{\theta}}(y,t) = \psi_{\overline{\theta}}(-\infty,t), \tag{5.5}$$

$$\int \left| \frac{\partial}{\partial \overline{y}} \psi_{\overline{\theta}}(\overline{y},t)/_{\overline{y}=y} - \frac{\partial}{\partial \overline{y}} \psi_{\theta}(\overline{y},t)/_{\overline{y}=y} \right| \lambda(dy) \leq K|\overline{\theta} - \theta| \tag{5.6}$$

$$\text{and } |\psi_{\overline{\theta}}(y,t) - \psi_{\theta}(y,t)| \leq K|\overline{\theta} - \theta| \tag{5.7}$$

for all $(y,t) \in I\!R \times \text{supp}(\delta)$ and $\overline{\theta}$ with $|\overline{\theta} - \theta| \leq \epsilon$.

- $\max_{t\in\text{supp}(\delta)} \sqrt{N}|\delta_N(\{t\}) - \delta(\{t\})| = O(1)$, i.e. (5.8)

$\max_{t\in\text{supp}(\delta)} \sqrt{N}|\delta_N(\{t\}) - \delta(\{t\})|$ is bounded.

Under these conditions we have the following result which extends a result of Bednarski et al. (1991) to planned experiments. This result is stronger than the results of Huber (1981) and Rieder (1994) concerning asymptotic linearity of estimators given by Fréchet or Hadamard differentiable functionals because the asymptotic linearity holds in the large shrinking Kolmogorov neighbourhood $\mathcal{U}_{K,\epsilon}(P_{\theta,\delta})$.

Theorem 5.1 *Let $\widehat{\zeta}$ be Fréchet differentiable at $P_{\theta,\delta}$ with derivative $\tilde{\zeta}(Q_\xi - P_{\theta,\delta}) = \int \psi_\theta(y,t)Q_\xi(dy,dt)$, where $P_{\theta,\delta}$ and ψ_θ satisfy conditions (5.1) - (5.7) and $Q_{N,\theta,\xi_N} \in \mathcal{U}_K(P_{\theta_N,\delta}, \frac{1}{\sqrt{N}}\epsilon)$ for some $(\theta_N)_{N\in I\!N} \in J_\theta$ for $N \in I\!N$, i.e. $(Q_{N,\theta,\xi_N})_{N\in I\!N} \in \mathcal{U}_{K,\epsilon}(P_{\theta,\delta})$. For deterministic designs additionally assume condition 5.8. Then the estimators $\widehat{\zeta}_N$ given by $\widehat{\zeta}$ satisfy*

$$\lim_{N\to\infty} Q_\theta^N\left(\left\{(y_N,d_N);\ \sqrt{N}\left|\widehat{\zeta}_N(y_N,d_N) - \zeta(\theta_N) - \frac{1}{N}\sum_{n=1}^N \psi_{\theta_N}(y_{nN},t_{nN})\right| > \epsilon_1\right\}\right) = 0 \tag{5.9}$$

for all $\epsilon_1 > 0$.

Proof. See the Appendix A.1. □

An estimator satisfying (5.9) for all $(Q_{N,\theta,\xi_N})_{N\in I\!N} \in \mathcal{U}_{K,\epsilon}(P_{\theta,\delta})$ is called *strongly asymptotically linear*. Under the conditions of Theorem 5.1 the function ψ_θ is the influence function of the estimator $\widehat{\zeta}_N$ as defined in Section 3.2. If additionally $\widehat{\zeta}$ is continuous in a neighbourhood of $P_{\theta,\delta}$, then according to Lemma 3.3 the influence function ψ_θ is continuous and bounded. Moreover, if $\widehat{\zeta}(P_{\theta,\xi}) = \zeta(\theta)$ for all ξ with $\text{supp}(\xi) = \text{supp}(\delta)$, then according to Lemma 3.3 the function ψ_θ satisfies

- ψ_θ is continuous and bounded. (5.10)
- $\int \psi_\theta(y,t)P_{\theta,t}(dy) = 0$ for all $t \in \mathcal{T}$, $\theta \in \Theta$. (5.11)

Thereby the property $\widehat{\zeta}(P_{\theta,\xi}) = \zeta(\theta)$ for all designs ξ with $\text{supp}(\xi) = \text{supp}(\delta)$ is important for designed experiments because then the estimators given by $\widehat{\zeta}$ can be used for different designs so that after all the best design for these estimators can be chosen.

Theorem 3.2 provides that the influence function of a functional which is based on a M-functional additionally satisfies

$$- \frac{\partial}{\partial \overline{\theta}} \int \psi_{\overline{\theta}}(y,t) P_{\theta,\delta}(dy,dt) /_{\overline{\theta}=\theta} = \dot{\zeta}(\theta) := \frac{\partial}{\partial \overline{\theta}} \zeta(\overline{\theta}) /_{\overline{\theta}=\theta}. \tag{5.12}$$

Bednarski et al. (1991) have shown that under some regularity conditions for every estimator which is strongly asymptotically linear in ψ_θ, i.e. an estimator satisfying (5.9), there exists a Fréchet differentiable M-functional with score function ψ_θ. Hence usually for designed experiments the function ψ_θ of a strongly asymptotically linear estimator can be interpreted as an influence function and will satisfy the conditions (5.10), (5.11) and (5.12). Therefore in general a function ψ_θ which gives the linear representation and satisfies conditions (5.10) - (5.12) is called the *influence function* of the strongly asymptotically linear estimator.

Definition 5.1 (Strong asymptotic linearity)
An estimator $\widehat{\zeta}_N$ for $\zeta(\theta)$ is strongly asymptotically linear at $P_{\theta,\delta}$ with influence function ψ_θ if ψ_θ satisfies conditions (5.10) - (5.12) and for all sequences $(Q_{N,\theta,\xi_N})_{N\in\mathbb{N}} \in \mathcal{U}_{K,\epsilon}(P_{\theta,\delta})$, i.e. $Q_{N,\theta,\xi_N} \in \mathcal{U}_K(P_{\theta_N,\delta}, \frac{1}{\sqrt{N}}\epsilon_1)$ for some $(\theta_N)_{N\in\mathbb{N}} \in J_\theta$ for $N \in \mathbb{N}$, we have

$$\begin{aligned} \lim_{N\to\infty} Q_\theta^N \Bigg(\Bigg\{ (y_N, d_N);\ \sqrt{N} \bigg| & \widehat{\zeta}_N(y_N, d_N) - \zeta(\theta_N) \\ & - \frac{1}{N} \sum_{n=1}^N \psi_{\theta_N}(y_{nN}, t_{nN}) \bigg| > \epsilon \Bigg\} \Bigg) \\ = 0 \end{aligned}$$

for all $\epsilon > 0$.

Example 5.1 (M-functionals)
Let be $P_{\theta,\delta}$ and $\tilde{\psi}$ as in Example 3.1. In particular $\tilde{\psi}$ satisfies

$$\begin{aligned} \tilde{\psi}(y,t,\theta) &= (\tilde{\psi}_0(y,t,\theta), ..., \tilde{\psi}_J(y,t,\theta)) \\ &= \psi\left(\frac{y - a(t)'\beta}{\sigma(t)}, t \right) \in \mathbb{R}^{r+J}, \end{aligned}$$

where ψ is defined by

$$\psi(y,t) = (a(t)' H' \psi_0(y,t), \psi_1(y,t), ..., \psi_J(y,t))'$$

with

$$\psi_0(y,t) := w(t) \operatorname{sgn}(y) \min\{|y|, b(t)\}$$

and

$$\psi_j(y,t) := \left[\psi_0(y,t)^2 - \int \psi_0(y,t)^2 P(dy)\right] 1_{\mathcal{T}_{0j}}(t).$$

In Example 3.1 it was shown that the M-functional with this score function is Fréchet differentiable. Additionally we have

$$\frac{\partial}{\partial \overline{y}}\tilde{\psi}_0(\overline{y},t,\theta)/_{\overline{y}=y} = H\, a(t)\, w(t)\, 1_{[-b(t)\sigma(t)+a(t)'\beta,b(t)\sigma(t)+a(t)'\beta]}$$

which implies condition 5.6 because of

$$\begin{aligned}
&\int \left|\frac{\partial}{\partial \overline{y}}\tilde{\psi}(\overline{y},t,\theta)/_{\overline{y}=y} - \frac{\partial}{\partial \overline{y}}\tilde{\psi}(\overline{y},t,\overline{\theta})/_{\overline{y}=y}\right| \lambda(dy)\\
&\leq \; |H\, a(t)\, w(t)|\; [|-b(t)\sigma(t)+a(t)'\beta+b(t)\overline{\sigma}(t)a(t)'\overline{\beta}|\\
&\qquad + |b(t)\sigma(t)+a(t)'\beta - b(t)\overline{\sigma}(t)a(t)'\overline{\beta}|]\\
&\leq \; 2\,|H\, a(t)\, w(t)|\; [b(t)\,|\sigma(t)-\overline{\sigma}(t)| + |a(t)|\,|\beta-\overline{\beta}|]\\
&\leq \; K\,|\theta-\overline{\theta}|.
\end{aligned}$$

Because $\tilde{\psi}$ is Lipschitz continuous also condition (5.7) holds. Hence a M-estimator with this score function is strongly asymptotically linear. □

The following theorem shows that strongly asymptotically linear estimators are asymptotically normally distributed in the shrinking Kolmogorov neighbourhood and therefore also in the full contamination neighbourhood. Compare also with Huber (1981), Bickel (1981, 1984), Rieder (1985, 1987, 1994) who derived similar results, but mainly in contiguous neighbourhoods.

Theorem 5.2 *Let $(\widehat{\zeta}_N)_{N\in\mathbb{N}}$ be strongly asymptotically linear at $P_{\theta,\delta}$ with influence function ψ_θ and for deterministic designs assume $\delta_N(\{t\}) \to \delta(\{t\})$ for $N \to \infty$ for all $t \in supp(\delta)$. Then $\widehat{\zeta}_N$ is asymptotically normally distributed in the shrinking Kolmogorov neighbourhood, i.e.*

$$\begin{aligned}
&\mathcal{L}\left(\sqrt{N}\left(\widehat{\zeta}_N - \zeta(\theta) - b_N\right)\middle| Q_\theta^N\right)\\
&\quad \xrightarrow{N\to\infty} \mathcal{N}\left(0, \int \psi_\theta(y,t)\,\psi_\theta(y,t)'\, P_{\theta,\delta}(dy,dt)\right)
\end{aligned}$$

for all $(Q_{N,\theta,\xi_N})_{N\in\mathbb{N}} \in \mathcal{U}_{K,\epsilon}(P_{\theta,\delta})$, where

$$b_N := \begin{cases} \int \psi_\theta(y,t) Q_{N,\theta,\xi_N}(dy,dt) & \textit{for random designs,}\\ \frac{1}{N}\sum_{n=1}^N \int \psi_\theta(y,t_{nN}) Q_{N,\theta,t_{nN}}(dy) & \textit{for deterministic designs.}\end{cases}$$

Proof. At first regard deterministic designs and assume that ψ_θ is univariate, i.e. $\psi_\theta : \mathbb{R} \times \mathcal{T} \to \mathbb{R}$. Set

$$\mathcal{T}_0 := \{t_1, ..., t_I\} := \text{supp}(\delta),$$

$$M := \max_{(y,t)\in\mathbb{R}\times\mathcal{T}_0} |\psi_\theta(y,t)|,$$
$$\mu_{N,t} := \int \psi_\theta(y,t)\, Q_{N,\theta,t}(dy),$$
$$\sigma_{N,t}^2 := \int (\psi_\theta(y,t) - \mu_{N,t})^2\, Q_{N,\theta,t}(dy),$$
$$\sigma_t^2 := \int \psi_\theta(y,t)^2\, P_{\theta,t}(dy).$$

Because $Q_{N,\theta,t}$ converges weakly to $P_{\theta,t}$ according to Lemma 3.1, ψ_θ is continuous and bounded and $\int \psi_\theta(y,t)\, P_{\theta,t}(dy) = 0$ according to condition (5.11), for every $\epsilon_1 > 0$ and $\epsilon_2 > 0$ there exists $N_0 \in \mathbb{N}$ such that

$$\int 1_{\{4|\psi_\theta(y,t)|\geq \epsilon_1 \sqrt{N_0 \delta(\{t\})}\sigma_t\}}(y)\ \psi(y,t,\theta)^2\, \frac{1}{\sigma_t^2}\, P_{\theta,t}(dy) < \frac{\epsilon_2}{16\, I},$$
$$|\sigma_t - \sigma_{N,t}| < \frac{\sigma_t}{2},$$
$$\int |(\mu_{N,t} - 2\psi_\theta(y,t))\, \mu_{N,t}|\, Q_{N,\theta,t}(dy) \leq 3\, M\, |\mu_{N,t}| < \frac{\epsilon_2\, \sigma_t^2}{8\, I},$$
$$|\delta(\{t\}) - \delta_N(\{t\})| < \frac{\delta(\{t\})}{2}$$

for all $t \in \mathcal{T}_0$ and $N \geq N_0$. Because $1_{\{4|\psi_\theta(y,t)|\geq \epsilon_1 \sqrt{N_0 \delta(\{t\})}\sigma_t\}}(y)\ \psi(y,t,\theta)^2$ is upper semicontinuous there exists $N_1 \geq N_0$ with

$$\begin{aligned}
&\int 1_{\{4|\psi_\theta(y,t)|\geq \epsilon_1 \sqrt{N_0 \delta(\{t\})}\sigma_t\}}(y)\ \psi_\theta(y,t)^2 \frac{1}{\sigma_t^2}\, Q_{N,\theta,t}(dy) \\
\leq\ & \int 1_{\{4|\psi_\theta(y,t)|\geq \epsilon_1 \sqrt{N_0 \delta(\{t\})}\sigma_t\}}(y)\ \psi_\theta(y,t)^2\, \frac{1}{\sigma_t^2}\, P_{\theta,t}(dy) + \frac{\epsilon_2}{16\, I} \\
<\ & \frac{\epsilon_2}{8\, I}
\end{aligned}$$

for all $t \in \mathcal{T}_0$ and $N \geq N_1$ (see Billingsley (1968), p. 17, Problem 7). Hence with

$$\begin{aligned}
&\frac{1}{\sum_{t\in\mathcal{T}} N\delta_N(\{t\})\sigma_{N,t}^2} \sum_{t\in\mathcal{T}} N\, \delta_N(\{t\}) \cdot \\
&\quad \int 1_{\{|\psi_\theta(y,t)|\geq \epsilon_1 \sqrt{\sum_{t\in\mathcal{T}} N\delta_N(\{t\})\sigma_{N,t}^2}\}}(y)\ (\psi_\theta(y,t) - \mu_{N,t})^2\ Q_{N,\theta,t}(dy) \\
\leq\ & \sum_{t\in\mathcal{T}_0} \frac{4}{\sigma_t^2} \\
&\quad \cdot \int 1_{\{|\psi_\theta(y,t)|\geq \epsilon_1 \sqrt{\sum_{t\in\mathcal{T}} N\delta_N(\{t\})\sigma_{N,t}^2}\}}(y)\ \psi_\theta(y,t)^2\ Q_{N,\theta,t}(dy) \\
&+ \sum_{t\in\mathcal{T}_0} \frac{4}{\sigma_t^2} \int (\mu_{N,t} - 2\psi_\theta(y,t))\, \mu_{N,t}\, Q_{N,\theta,t}(dy)
\end{aligned}$$

$$\leq \sum_{t\in\mathcal{T}_0} \int 1_{\{4|\psi_\theta(y,t)|\geq \epsilon_1\sqrt{N_0\delta(\{t\})}\sigma_t\}}(y)\ \psi_\theta(y,t)^2\ \frac{4}{\sigma_t^2}\ Q_{N,\theta,t}(dy)$$
$$+ I\,\frac{4}{\sigma_t^2}\,\frac{\epsilon_2\,\sigma_t^2}{8\,I}$$
$$\leq\ 4\,I\,\frac{\epsilon_2}{8\,I} + \frac{\epsilon_2}{2} = \epsilon_2$$

the Lindeberg condition is satisfied so that with Lindeberg's theorem (see for example Billingsley (1968), p. 42)

$$\mathcal{L}\left(\frac{\sqrt{N}\frac{1}{N}\sum_{n=1}^N(\psi_\theta(y_{nN},t_{nN})-\mu_{N,t_{nN}})}{\frac{1}{N}\sum_{n=1}^N\sigma_{N,t_{nN}}}\,\middle|\,Q_\theta^N\right)\ \overset{N\to\infty}{\longrightarrow}\ \mathcal{N}(0,1).$$

Because

$$\frac{1}{N}\sum_{n=1}^N \sigma_{N,t_{nN}} = \sum_{t\in\mathcal{T}} \delta_N(\{t\})\sigma_{N,t}$$
$$\overset{N\to\infty}{\longrightarrow}\ \sum_{t\in\mathcal{T}} \delta(\{t\})\sigma_t = \int \psi_\theta(y,t)^2\,P_{\theta,\delta}(dy,dt)$$

the assertion follows for the univariate case for deterministic designs. For random designs in the univariate case the proof is more simple. We only have to drop in the above proof the dependencies on t and to substitute $Q_{N,\theta,t}$ and $P_{\theta,t}$ by Q_{N,θ,ξ_N} and $P_{\theta,\delta}$. The multivariate case follows similarly by using additionally the central limit theorem of Cramér and Wold (see Billingsley (1968), p. 49). □

Because the full shrinking contamination neighbourhood is a subset of the shrinking Kolmogorov neighbourhood (see Lemma 3.2) we get from Theorem 5.2 the following corollary.

Corollary 5.1 *Let $(\widehat{\zeta}_N)_{N\in\mathbb{N}}$ be strongly asymptotically linear at $P_{\theta,\delta}$ with influence function ψ_θ and for deterministic designs assume $\delta_N(\{t\})\to\delta(\{t\})$ for $N\to\infty$ for all $t\in supp(\delta)$. Then*

$$\mathcal{L}\left(\sqrt{N}\left(\widehat{\zeta}_N-\zeta(\theta)\right)\middle|\,Q_\theta^N\right)$$
$$\overset{N\to\infty}{\longrightarrow}\ \mathcal{N}\left(b(\psi_\theta,(Q_{N,\theta,\delta})_{N\in\mathbb{N}}),\int \psi_\theta(y,t)\,\psi_\theta(y,t)'\,P_{\theta,\delta}(dy,dt)\right)$$

for all $(Q_{N,\theta,\delta})_{N\in\mathbb{N}}\in\mathcal{U}_{c,\epsilon}(P_{\theta,\delta})$, where

$$b(\psi_\theta,(Q_{N,\theta,\delta})_{N\in\mathbb{N}}) := \lim_{N\to\infty}\sqrt{N}\int \psi_\theta(y,t)Q_{N,\theta,\delta}(dy,dt).$$

Proof. Let be $(Q_{N,\theta,\delta})_{N\in\mathbb{N}}\in\mathcal{U}_{c,\epsilon}(P_{\theta,\delta})$ arbitrary. Then $Q_{N,\theta,t} = (1-N^{-1/2}\epsilon\,c(t))P_{\theta,t}+N^{-1/2}\epsilon\,c(t)\tilde{P}_t$ for all $N\in\mathbb{N}$ such that because of Condition 5.11

$$\sqrt{N}\int \psi_\theta(y,t)Q_{N,\theta,\delta}(dy,dt) = \epsilon\int \psi_\theta(y,t)\,c(t)\,\tilde{P}_\delta(dy,dt)$$

for all $N \in \mathbb{N}$ is satisfied. Hence for random designs the assertion follows from Theorem 5.2. For deterministic designs additionally note that

$$\begin{aligned}
&\sqrt{N}\,\frac{1}{N}\sum\nolimits_{n=1}^{N} \int \psi_\theta(y,t_{nN})Q_{N,\theta,t_{nN}}(dy) \\
= \quad & \sum\nolimits_{t\in\mathcal{T}} \delta_N(\{t\}) \int \psi_\theta(y,t)Q_{N,\theta,t}(dy) \\
= \quad & \sum\nolimits_{t\in\mathcal{T}} \delta_N(\{t\})\,\epsilon \int \psi_\theta(y,t)\,c(t)\,\tilde{P}_t(dy) \\
\overset{N\to\infty}{\longrightarrow} \quad & \sum\nolimits_{t\in\mathcal{T}} \delta(\{t\})\,\epsilon \int \psi_\theta(y,t)\,c(t)\,\tilde{P}_t(dy) \\
= \quad & \epsilon \int \psi_\theta(y,t)\,c(t)\,\tilde{P}_\delta(dy,dt).\square
\end{aligned}$$

As already mentioned every strongly asymptotically linear estimator provides a Fréchet differentiable M-functional. Hence strong asymptotic linearity and Fréchet differentiability are closely connected. Moreover, Bednarski et al. (1991) showed in their Lemma 3.1 that the median is not strongly asymptotically linear although the median is very outlier robust. But the median is very sensitive with respect to small changes of the data which are close to the value given by the median, i.e. the median is not inlier robust. This means that Fréchet differentiability describes besides outlier robustness also other robustness properties. Because here mainly outlier robustness should be regarded we refer for other robustness properties in particular to Lindsay (1993) and Basu et al. (1993) and regard estimators which satisfy a weaker asymptotic linearity condition. This asymptotic linearity condition is called *weak asymptotic linearity* and is satisfied by the median and corresponding estimators in higher dimensions as well as by Gauss-Markov estimators. Thereby the function ψ_θ which gives the linear representation is still called the influence function although the relation to the real influence functions as defined in Section 3.2 is not always obvious. To include estimators as the median and the Gauss-Markov etimator we can use instead of condition (5.10) the following very weak condition on the influence function:

$$\int |\psi_\theta(y,t)|^2 P_{\theta,\delta}(dy,dt) < \infty. \tag{5.13}$$

Definition 5.2 (Weak asymptotic linearity)
An estimator $\widehat{\zeta}_N$ for $\zeta(\theta)$ is weakly asymptotically linear at $P_{\theta,\delta}$ with influence function ψ_θ if ψ_θ satisfies conditions (5.11) - (5.13) and

$$\lim_{N\to\infty} P_\theta^N \left(\left\{ (y_N, d_N);\ \sqrt{N} \left| \widehat{\zeta}_N(y_N, d_N) - \zeta(\theta) - \frac{1}{N} \sum_{n=1}^N \psi_\theta(y_{nN}, t_{nN}) \right| > \epsilon \right\} \right) = 0$$

for all $\epsilon > 0$.

Weak asymptotic linearity can be derived by Hadamard differentiable functionals (see Rieder (1994)) and by straightforward methods (see Section 5.2, Section 5.3 or Section 5.4). Moreover an estimator which is weakly asymptotically linear is also asymptotically linear for all distributions which are contiguous to P_θ^N, in particular for all distributions of the restricted shrinking neighbourhood $\mathcal{U}^0_{c,\epsilon}(P_{\theta,\delta})$. This follows in particular from the proof of the following theorem which shows the asymptotic normality of weakly asymptotically linear estimators in the restricted shrinking contamination neighbourhood.

Theorem 5.3 *Let $(\widehat{\zeta}_N)_{N\in\mathbb{N}}$ be weakly asymptotically linear at $P_{\theta,\delta}$ with influence function ψ_θ and for deterministic designs assume $\delta_N(\{t\}) \to \delta(\{t\})$ for $N \to \infty$ for all $t \in supp(\delta)$. Then $\widehat{\zeta}_N$ is asymptotically normally distributed for shrinking contamination, i.e.*

$$\mathcal{L}\left(\sqrt{N}\left(\widehat{\zeta}_N - \zeta(\theta)\right) \Big| Q_\theta^N \right) \stackrel{N\to\infty}{\longrightarrow} \mathcal{N}\left(b(\psi_\theta, (Q_{N,\theta,\delta})_{N\in\mathbb{N}}),\ \int \psi_\theta(y,t)\, \psi_\theta(y,t)'\, P_{\theta,\delta}(dy,dt) \right)$$

for all $(Q_{N,\theta,\delta})_{N\in\mathbb{N}} \in \mathcal{U}^0_{c,\epsilon}(P_{\theta,\delta})$, where

$$b(\psi_\theta, (Q_{N,\theta,\delta})_{N\in\mathbb{N}}) := \lim_{N\to\infty} \sqrt{N} \int \psi_\theta(y,t) Q_{N,\theta,\delta}(dy,dt).$$

Proof. The assertion follows by the third lemma of LeCam (see Hájek and Šidák (1967), p. 208) and for deterministic designs additionally with Lindeberg's theorem. See also Bickel (1981, 1984), Rieder (1985, 1987, 1994), Kurotschka and Müller (1992), Müller (1994e). □

Any sequence $(Q_{N,\theta,\delta})_{N\in\mathbb{N}}$ of the restricted or the full shrinking contamination neighbourhood has elements $Q_{N,\theta,\delta}$ which satisfy $Q_{N,\theta,t} = (1 - N^{-1/2}\,\epsilon\, c(t)) P_{\theta,\delta} + N^{-1/2}\,\epsilon\, c(t)\, \tilde{P}_t$. Hence because of Condition 5.11 the asymptotic bias $b(\psi_\theta, (Q_{N,\theta,\delta})_{N\in\mathbb{N}})$ of weakly asymptotically linear estimators in the restricted shrinking contamination neighbourhood as well as of strongly asymptotically linear estimators in the full shrinking contamination neighbourhood can be estimated by

$$\begin{aligned}
|b(\psi_\theta, (Q_{N,\theta,\delta})_{N\in\mathbb{N}})| &= \lim_{N\to\infty} |\sqrt{N} \int \psi_\theta(y,t) Q_{N,\theta,\delta}(dy,dt)| \\
&= |\epsilon \int \psi_\theta(y,t)\, c(t)\, \tilde{P}_\delta(dy,dt)| \\
&\leq \max_{(y,t)\in\mathbb{R}\times \mathrm{supp}(\delta)} \epsilon\, |\psi_\theta(y,t)|\, |\int c(t)\, \delta(dt)| \\
&\leq \max_{(y,t)\in\mathbb{R}\times \mathrm{supp}(\delta)} \epsilon\, |\psi_\theta(y,t)|.
\end{aligned}$$

This upper bound for the asymptotic bias is already attained in the restricted shrinking contamination neighbourhood. This shows the following lemma.

Lemma 5.1 *The asymptotic bias* $b(\psi_\theta, (Q_{N,\theta,\delta})_{N\in\mathbb{N}})$ *satisfies*

$$\begin{aligned}
&\max_{(y,t)\in\mathbb{R}\times \mathrm{supp}(\delta)} \epsilon\, |\psi_\theta(y,t)| \\
&\quad = \max\{|b(\psi_\theta, (Q_{N,\theta,\delta})_{N\in\mathbb{N}})|;\ (Q_{N,\theta,\delta})_{N\in\mathbb{N}} \in \mathcal{U}^0_{c,\epsilon}(P_{\theta,\delta})\} \\
&\quad = \max\{|b(\psi_\theta, (Q_{N,\theta,\delta})_{N\in\mathbb{N}})|;\ (Q_{N,\theta,\delta})_{N\in\mathbb{N}} \in \mathcal{U}_{c,\epsilon}(P_{\theta,\delta})\}.
\end{aligned}$$

Proof. See Rieder (1985, 1987, 1994, Section 7.3), Müller (1987), Kurotschka and Müller (1992). □

For $\epsilon = 1$ here the maximum asymptotic bias is also called *asymptotic bias for shrinking contamination*.

Definition 5.3 (Asymptotic bias of shrinking contamination) *The maximum asymptotic bias* $b(\psi_\theta, \delta)$ *of shrinking contamination of an asymptotically linear estimator with influence function* ψ_θ *at* $P_{\theta,\delta}$ *is defined as*

$$b(\psi_\theta, \delta) := \|\psi_\theta\|_\delta,$$

where

$$\|\psi_\theta\|_\delta := \max_{(y,t)\in\mathbb{R}\times \mathrm{supp}(\delta)} |\psi_\theta(y,t)|.$$

If the estimator $\widehat{\zeta}_N$ can be derived from a functional $\widehat{\zeta}$ which has a Fréchet derivative of the form $\dot{\zeta}(Q_\xi - P_{\theta,\delta}) = \int \psi_\theta(y,t) Q_\xi(dy,dt)$, where ψ_θ satisfies Conditions (5.10) - (5.12), then the asymptotic bias for shrinking contamination coincides with the gross-error-sensitivity as defined in Section 3.2.

This is, for example, the case for M-estimators (see Section 3.3). But if the estimator can not be derived from a functional which is Fréchet differentiable or at least Hadamard differentiable (see Rieder (1994)), then in general it is not obvious how the asymptotic bias for shrinking contamination and the gross-error-sensitivity are related. Hence it make sense to define an additional robustness criterion via the asymptotic bias for shrinking contamination.

Definition 5.4 (Asymptotic robustness for shrinking contamination)
A weakly asymptotically linear estimator with influence function ψ_θ *is asymptotically robust for shrinking contamination if* $b(\psi_\theta, \delta) < \infty$ *for all* $\theta \in \Theta$.

5.2 Robust Estimation in Contaminated Linear Models

As in Section 1.2 the linear model is given by $y_{nN} = a(t_{nN})'\beta + Z_{nN}$, where the ideal (central) distribution of Z_{nN} at t_{nN} is a symmetric distribution with mean 0, variance $\sigma(t_{nN})^2$ and bounded Lebesgue density. Then the distribution of $\frac{1}{\sigma(t_{nN})} Z_{nN}$ is a symmetric distribution P with mean 0, variance 1 and bounded Lebesgue density f. If we assume that all experimental conditions of a sequence of designs is lying in a finite set $\mathcal{T}_0$, i.e.

$$\mathrm{supp}(\delta_N) \subset \mathcal{T}_0 := \{\tau_1, ..., \tau_I\} \subset \mathcal{T}$$

for all $N \in I\!N$, then the ideal distribution of y_{nN} at t_{nN} depends only on the parameter $\theta = (\beta', \sigma(\tau_1), ..., \sigma(\tau_I))'$ so that the ideal distribution of y_{nN} at t_{nN} is denoted by $P_{\theta, t_{nN}}$ and the ideal distribution of y_N by P_θ^N (see Section 3.1 and Section 3.2). In such linear models a large class of weakly asymptotically linear estimators is the class of *one-step M-estimators* which generalize the M-estimators and which are easy to calculate. There are also other weakly asymptotically linear estimators as R-, L- and S-estimators which are based on ranks, the order statistic or on scale estimators (see for example Davies (1990), Jurečková (1992), Gutenbrunner and Jurečková (1992),

Hössjer (1994), Rieder (1994)). But the class of one-step M-estimators contains all relevant estimators for asymptotically optimal robust estimation. In particular, it contains the L_1 estimator and the least squares estimator which appear as limit cases of optimum estimation, namely as optimal solutions for minimum bias bound and without bias bound, respectively (see Section 7.2). Moreover, the one-step M-estimators which are robust for shrinking contamination have a simple characterization (see below), and there exist one-step M-estimators with high breakdown point (see Section 9.2). Therefore, here we regard only one-step M-estimators.

In Bickel (1975) simple one-step M-estimators in linear models were regarded where the score function is independent of the experimental conditions. By regarding also score functions which are dependent on the experimental conditions Simpson et al. (1992) generalized the simple one-step M-estimators to general one-step M-estimators based on a Newton-Raphson approach or a scoring approach. Here we regard the generalized one-step M-estimators based on the Newton-Raphson approach. These estimators can easily be extended for estimation of arbitrary linear aspects $\varphi(\beta) = L\beta$ of β (see Müller (1987) and Kurotschka and Müller (1992)). To include the case that the variances $\sigma^2(t)$ may be unknown and different and the case that also at designs, where β itself is not identifiable, a linear aspect shall be estimated, here one-step-M-estimators are defined as follows.

Definition 5.5 (One-step M-estimator)
An estimator $\widehat{\varphi}_N : I\!R^N \times \mathcal{T}^N \to I\!R^s$ *is called a one-step M-estimator for* $\varphi(\beta) = L\beta$ *with score function* $\psi : I\!R \times \mathcal{T} \to I\!R^s$, *initial estimator* $\widehat{\beta}_N : I\!R^N \times \mathcal{T}^N \to I\!R^r$ *for* β *and variance estimators* $\widehat{\sigma}^2_N(t) : I\!R^N \times \mathcal{T}^N \to I\!R^+$ *for* $\sigma^2(t)$, $t \in \mathcal{T}$, *if*

$$\widehat{\varphi}(y_N, d_N) = L\widehat{\beta}_N(y_N, d_N) + \frac{1}{N}\sum_{n=1}^N \psi\left(\frac{y_{nN} - a(t_{nN})'\widehat{\beta}_N(y_N,d_N)}{\widehat{\sigma}_N(t_{nN})(y_N,d_N)}, t_{nN} \right) \widehat{\sigma}_N(t_{nN})(y_N, d_N).$$

For example the least squares estimator for β is a one-step M-estimator with the least squares estimator as initial estimator and

$$\psi_\infty(z,t) = \mathcal{I}(\delta)^{-1}\, a(t)\, z \tag{5.14}$$

as score function which is independent of the choice of the variance estimator, i.e. which needs no variance estimator. The Gauss-Markov estimator for a linear aspect $\varphi(\beta) = L\beta$ is a one-step M-estimator with the least squares estimator as initial estimator and

$$\psi_\infty(z,t) = L\,\mathcal{I}(\delta)^{-}\, a(t)\, z \tag{5.15}$$

as score function. The L_1 estimator for β is a one-step M-estimator with the L_1 estimator as initial estimator and

$$\psi_{01}(z,t) = \sqrt{\frac{\pi}{2}}\,\mathcal{I}(\delta)^{-1}\,a(t)\,\mathrm{sgn}(z) \tag{5.16}$$

as score function, where sgn(·) is the signum function. i.e.

$$\mathrm{sgn}(z) := \begin{cases} 1 & \text{for } z > 0, \\ 0 & \text{for } z = 0, \\ -1 & \text{for } z < 0. \end{cases}$$

Note that because the score function is multiplied with the variance estimator and divided by the variance estimator within the score function a one-step M-estimator is scale equivariant, i.e.

$$\widehat{\varphi}_N(s \cdot y_N, d_N) = s \cdot \widehat{\varphi}_N(y_N, d_N) \text{ for all } s \in I\!R^+$$

if the variance estimators and the initial estimator are scale equivariant. In particular the least squares estimator and the corresponding empirical variance estimator

$$\widehat{\sigma}^2_N(t) := \frac{1}{N_t(d_N)} \sum_{n=1}^N 1_{\{t\}}(t_{nN})\,(y_{nN} - \overline{y}_t(y_N, d_N))^2 \tag{5.17}$$

are scale equivariant initial and variance estimators, where $N_t(d_N) := \sum_{n=1}^N 1_{\{t\}}(t_{nN})$, $\overline{y}_t(y_N, d_N) := \frac{1}{N_t(d_N)} \sum_{n=1}^N 1_{\{t\}}(t_{nN})\, y_{nN}$ and $\mathrm{supp}(\delta_N) \in \{\tau_1, ..., \tau_I\} \subset \mathcal{T}$ for all $N \in I\!N$. More robust estimators as a M-estimator for $\widehat{\beta}_N$ and Huber's Proposal 2 (Huber (1964), p. 96) for $\widehat{\sigma}^2_N(t)$ will improve the robustness of the one-step M-estimators for small sample sizes. In particular, if we use a high breakdown point estimator as initial estimator and a corresponding variance estimator, then also the one-step M-estimator has a high breakdown point for finite sample sizes (see Jurečková and Portnoy (1987), Rousseeuw (1984), Simpson et al. (1992) and Section 9.2). But the asymptotic behaviour in shrinking neighbourhoods does not depend on the robustness of the initial estimator and the variance estimators for finite samples. The initial estimator and the variance estimators should only satisfy the following regularity conditions:

- $\sqrt{N}(a(\tau_1), ..., a(\tau_I))'(\widehat{\beta}_N - \beta)$ is tight at the ideal model P_θ^N. (5.18)
- $\sqrt{N}(\widehat{\sigma}_N(t) - \sigma(t))$ is tight at the ideal model P_θ^N for all $t \in \mathcal{T}_0$. (5.19)

Condition (5.18) is due to the possibility that β itself may be not identifiable. For example, the least squares estimator and the empirical variance estimator given by (5.17) fulfill the conditions (5.18) and (5.19) for many ideal distributions P_θ^N and in particular if P is the standard normal distribution.

For the score functions the following simple conditions are assumed (compare the conditions given in Maronna and Yohai (1981)):

- $\psi(z,t) = \psi_0(z,t) + c(t)\text{sgn}(z)$, where for all $t \in \mathcal{T}_0$ (5.20)
  $\psi_0(\cdot,t)$ is antisymmetric, continuous and there exists
  $l_1(t), ..., l_L(t)$ so that $\psi_0(\cdot,t)$ has bounded and continuous
  derivatives on $\mathbb{R} \setminus \{l_1(t), ..., l_L(t)\}$,
- $\int \psi(z,t)a(t)'zP(dz)\delta(dt) = L.$ (5.21)

Condition (5.20) can be generalized, as Lemma 5.2 below shows, but in this form it is easy to check. In particular, it ensured that the score function satisfies condition (5.11) which is necessary for weak asymptotic linearity (see Section 5.1). The condition (5.21) ensures condition (5.12) which is also necessary for weak asymptotic linearity. It is due to the aspect which shall be estimated and can be often fulfilled for a given score function by multiplying with a suitable matrix. In particular the score functions of the least squares estimator, the Gauss-Markov estimator and the L_1 estimator, given by (5.14), (5.15) and (5.16), respectively, satisfy the conditions (5.20) and (5.21). Also the score functions of the Hampel-Krasker estimator (see Hampel (1978), Krasker (1980)), of the Krasker-Welsch estimator (see Krasker and Welsch (1982)) and of all optimal robust estimators for linear aspects fulfill condition (5.20) and (5.21) (see Section 7.2, Müller (1987), Kurotschka and Müller (1992) and Müller (1994e)).

The following lemma is a first step of proving the weak asymptotic linearity of one-step M-estimators and also shows how condition (5.20) can be generalized (compare also the conditions given in Bickel (1975), Yohai and Maronna (1979) and Rieder (1985, 1994, Section 6.4.2)). In particular Rieder (1985, 1994) shows weak asymptotic linearity of one-step M-estimators under relative general conditions. But he assumes that the initial estimator is a discretized version of a tight estimator, where the discretization decreases with $\sqrt{N}$, which is a very complicated assumption for the praxis and in particular for simulation studies.

Lemma 5.2 *If $\psi = (\psi_1, ..., \psi_s)' : I\!R \times T \to I\!R^s$ satisfies condition (5.20), then*

$$\bullet\ \lambda(b,q,t,\sigma) := \int \psi\left(\frac{z\sigma - a(t)'b}{\sigma + q}, t\right)(\sigma+q)P(dz) \text{ is for} \tag{5.22}$$

all $t \in \mathcal{T}_0, \sigma \in I\!R^+$ continuously differentiable in a neighbourhood of $(b,q) = (0,0)$, where

$$\frac{\partial}{\partial(b,q)}\lambda(b,q,t,\sigma)/_{(b,q)=(0,0)} = (-\int \psi(z,t)a(t)'zP(dz) \mid 0_{s\times 1}),$$

and there exists $K \in I\!R^+$ and antisymmetric and monotone increasing functions $\psi_j^+(\cdot,t) : I\!R \to I\!R$ and $\psi_j^-(\cdot,t) : I\!R \to I\!R$ with

$$\bullet\ \psi_j(\cdot,t) = \psi_j^+(\cdot,t) - \psi_j^-(\cdot,t), \tag{5.23}$$

$$\bullet\ \psi_j^{\pm}\left(\frac{z}{\sigma+q}, t\right)(\sigma+q) \text{ is as function of } q \text{ continuous on } \left[\frac{-\sigma}{4}, \frac{\sigma}{4}\right] \text{ for all } z \in I\!R, \tag{5.24}$$

$$\bullet\ \int \psi_j^{\pm}(z,t)^2 P(dz) < \infty, \tag{5.25}$$

$$\bullet\ \int \left|\psi_j^{\pm}\left(z + \frac{k+h}{\sigma}, t\right) - \psi_j^{\pm}\left(z + \frac{k}{\sigma}, t\right)\right|^2 P(dz) \leq K\frac{|h|}{\sigma}, \tag{5.26}$$

$$\bullet\ \left|\int\left(\psi_j^{\pm}\left(\frac{z\sigma+k+h}{\sigma+q}, t\right) - \psi_j^{\pm}\left(\frac{z\sigma+k}{\sigma+q}, t\right)\right)P(dz)\right| \leq K\frac{|h|}{\sigma}, \tag{5.27}$$

$$\bullet\ \int \left|\psi_j^{\pm}\left(\frac{z\sigma+h}{\sigma+q+\rho}, t\right)(\sigma+q+\rho) - \psi_j^{\pm}\left(\frac{z\sigma+h}{\sigma+q}, t\right)(\sigma+q)\right|^2 P(dz) \leq K|\rho|^2, \tag{5.28}$$

for all $j \in \{1, ..., s\}$, $t \in \mathcal{T}_0$, $k \in I\!R$, $\sigma \in I\!R^+$ and $h, q, \rho \in [\frac{-\sigma}{4}, \frac{\sigma}{4}]$.

Proof. See Müller (1994e), Lemma 4.1. □

The following theorem shows that one-step M-estimators are weakly asymptotically linear with influence function ψ_θ satisfying

$$\psi_\theta(y,t) = \psi\left(\frac{y - a(t)'\beta}{\sigma(t)}, t\right)\sigma(t) \tag{5.29}$$

for all $(y,t) \in I\!R \times \mathcal{T}_0$ and all $\theta = (\beta', \sigma(\tau_1), ..., \sigma(\tau_I))' \in I\!R^r \times (I\!R^+)^I$. Note that for estimating $\varphi(\beta) = L\beta$ in a linear model for influence functions ψ_θ of the form (5.29) the conditions (5.11) - (5.13), which concern $\int \psi_\theta(y,t)P_{\theta,t}(dy) = 0$ for all $t \in \mathcal{T}$, $\frac{\partial}{\partial\theta}\int \psi_{\overline{\theta}}(y,t)P_{\theta,\delta}(dy,dt)/_{\overline{\theta}=\theta} = -\dot{\zeta}(\theta)$

and $\int |\psi_\theta(y,t)|^2 P_{\theta,\delta}(dy,dt) < \infty$, are equivalent with

- $$\int \psi(z,t)P(dz) = 0 \text{ for all } t \in \mathcal{T}, \tag{5.30}$$
- $$\int \psi(z,t)\, a(t)'\, z\, P(dz)\delta(dt) = L, \tag{5.31}$$
- $$\int |\psi(z,t)|^2 P(dz)\delta(dt) < \infty. \tag{5.32}$$

The equivalence of the conditions holds because in linear models $\frac{Y_{nN}-a(t_{nN})'\beta}{\sigma(t_{nN})}$ is distributed according to P and because of property (5.22) and the fact that $\dot{\zeta}(\theta) = \dot{\varphi}(\beta) = L$. Hence Theorem 4.2 in Müller (1994e) provides the following theorem.

Lemma 5.3 *Let $\widehat{\varphi}_N$ be a one-step M-estimator for $\varphi(\beta) = L\beta$ with score function ψ satisfying conditions (5.20) and (5.21), or (5.21) - (5.28), initial estimator $\widehat{\beta}_N$ satisfying condition (5.18) and variance estimators $\widehat{\sigma}_N(t)$ satisfying condition (5.19) and for deterministic designs assume $\delta_N(\{t\}) \to \delta(\{t\})$ for $N \to \infty$ for all $t \in \mathcal{T}_0$. Then for all $\theta = (\beta', \sigma(\tau_1), ..., \sigma(\tau_I))' \in \mathbb{R}^r \times (\mathbb{R}^+)^I$ the estimator $\widehat{\varphi}_N$ is weakly asymptotically linear at $P_{\theta,\delta}$ with influence function ψ_θ satisfying*

$$\psi_\theta(y,t) = \psi\left(\frac{y - a(t)'\beta}{\sigma(t)}, t\right) \sigma(t)$$

for all $(y,t) \in \mathbb{R} \times \mathcal{T}$.

Proof. See Müller (1994e), Theorem 4.2. □

Because for estimating a linear aspect in a linear model the conditions (5.11) - (5.13) are equivalent with the conditions (5.30) - (5.32), if the influence function satisfies condition (5.29) with ψ satisfying (5.20) and (5.21), we can define in linear models weakly asymptotically linear estimators as follows.

Definition 5.6 (AL-estimator)
In a linear model an estimator $\widehat{\varphi}_N$ for $\varphi(\beta) = L\beta$ is called a weakly asymptotically linear estimator (briefly an AL-estimator) with influence function $\psi : \mathbb{R} \times \mathcal{T} \to \mathbb{R}^s$ if ψ satisfies conditions (5.30) - (5.32) and

$$\lim_{N\to\infty} P_\theta^N \Bigg(\Bigg\{(y_N, d_N);\ \sqrt{N}\, \Big|\widehat{\varphi}_N(y_N, d_N) - \varphi(\beta) - \sum\nolimits_{n=1}^N \psi\left(\frac{y_{nN} - a(t_{nN})'\beta}{\sigma(t_{nN})}, t_{nN}\right) \sigma(t_{nN})\Big| > \epsilon\Bigg\}\Bigg) = 0$$

for all $\epsilon > 0$ and all $\theta = (\beta', \sigma(\tau_1), ..., \sigma(\tau_I))' \in \mathbb{R} \times (\mathbb{R}^+)^I$.

For general weakly asymptotically linear estimators Theorem 5.3 provides their asymptotic normality in shrinking contamination neighbourhoods. For weakly asymptotically linear estimators for a linear aspect in a linear model, i.e. for AL-estimators, this theorem simplifies to the following theorem (see also Bickel (1981, 1984), Rieder (1985, 1987, 1994), Kurotschka and Müller (1992) and Müller (1994e)). This theorem in particular shows that for linear problems the asymptotic covariance matrix and the maximum asymptotic bias for shrinking contamination as defined in Definition 5.3 do not depend on the unknown parameter β. To see this property we only have to use the transformation theorem for integrals and the fact that the distribution Q_θ^N of Y_N and (Y_N, D_N), respectively, is completely determined by the distribution $Q^N := Q^N_{(0'_N,1)}$ of the normed errors Z_N and (Z_N, D_N), respectively. In particular, if $(Q_{N,\delta})_{N\in\mathbb{N}} \in \mathcal{U}^0_{c,\epsilon}(P_\delta)$ is the distribution of $(\frac{1}{\sigma(T_{nN})}(Z_{nN}, T_{nN}))_{N\in\mathbb{N}}$, then for every θ the distributions $(Q_{N,\theta,\delta})_{N\in\mathbb{N}}$ of $((Y_{nN}, T_{nN}))_{N\in\mathbb{N}}$ satisfy $(Q_{N,\theta,\delta})_{N\in\mathbb{N}} \in \mathcal{U}^0_{c,\epsilon}(P_{\theta,\delta})$, and vice versa.

Theorem 5.4 *Let $(\widehat{\varphi}_N)_{N\in\mathbb{N}}$ be an AL-estimator for $\varphi(\beta) = L\beta$ with influence function ψ and for deterministic designs assume $\delta_N(\{t\}) \to \delta(\{t\})$ for $N \to \infty$ for all $t \in \mathcal{T}_0$. Then for all $\theta = (\beta', \sigma(\tau_1), ..., \sigma(\tau_I))' \in \mathbb{R} \times (\mathbb{R}^+)^I$*

$$\mathcal{L}\left(\sqrt{N}\,(\widehat{\varphi}_N - \varphi(\beta))\middle|\, Q_\theta^N\right)$$
$$\overset{N\to\infty}{\longrightarrow} \mathcal{N}\left(b(\psi\,\sigma, (Q_{N,\delta})_{N\in\mathbb{N}}), \int \sigma(t)^2\, \psi(z,t)\, \psi(z,t)'\, P(dz)\, \delta(dt)\right)$$

with

$$b(\psi\,\sigma, (Q_{N,\delta})_{N\in\mathbb{N}}) = \lim_{N\to\infty} \sqrt{N} \int \psi(z,t)\,\sigma(t)\, Q_{N,\delta}(dz, dt)$$

for all $(Q_{N,\delta})_{N\in\mathbb{N}} \in \mathcal{U}^0_{c,\epsilon}(P_\delta)$, and the maximum asymptotic bias of shrinking contamination satisfies

$$b(\psi\,\sigma, \delta) = \max_{(z,t)\in\mathbb{R}\times\text{supp}(\delta)} \epsilon\, |\psi(z,t)\,\sigma(t)|.$$

Theorem 5.4 and Lemma 5.3 provide at once a simple characterization of robust one-step M-estimators.

Theorem 5.5 *Under the assumptions of Lemma 5.3 a one-step M-estimator for $\varphi(\beta) = L\beta$ with score function ψ is asymptotically robust for shrinking contamination if and only if ψ is bounded.*

Example 5.2 (One-way lay-out, continuation of Example 2.2)
Consider a one-way classification model with four treatments 1,2,3,4 where the observations are given by

$$\begin{aligned} Y_{nN} &= 1_{\{1\}}(t_{nN})\beta_1 + 1_{\{2\}}(t_{nN})\beta_2 + 1_{\{3\}}(t_{nN})\beta_3 + 1_{\{4\}}(t_{nN})\beta_4 + Z_{nN} \\ &= a(t_{nN})'\beta + Z_{nN} \end{aligned}$$

with $t_{nN} \in \{1,2,3,4\} = \mathcal{T}$, $a(t) = (1_{\{1\}}(t), 1_{\{2\}}(t), 1_{\{3\}}(t), 1_{\{4\}}(t))'$ and $\beta = (\beta_1, \beta_2, \beta_3, \beta_4)'$ (see also Example 2.2). Assume that at the ideal model the error variables Z_{nN} are normally distributed with mean 0 and unknown variance $\sigma(t_{nN})^2$. For estimating the linear aspect $\varphi(\beta) = (\beta_2 - \beta_1, \beta_3 - \beta_1, \beta_4 - \beta_1)'$, so that the level 1 is the control level, a simple design measure is $\delta = \frac{1}{4}(e_1 + e_2 + e_3 + e_4)$. A deterministic design sequence $(d_N)_{N \in \mathbb{N}}$ for which the corresponding design measures $(\delta_N)_{N \in \mathbb{N}}$ converges weakly to this design measure δ is for example given by $d_4 = (1,2,3,4)$, $d_5 = (1,2,3,4,1)$, $d_6 = (1,2,3,4,1,2)$, $d_7 = (1,2,3,4,1,2,3)$, $d_8 = (1,2,3,4,1,2,3,4)$, $d_9 = (1,2,3,4,1,2,3,4,1)$ and so forth.

A robust score function for estimating $\varphi(\beta)$ at the design measure δ is ψ_0 given by

$$\psi_0(z,t) = \begin{cases} (-1,-1,-1)' \operatorname{sgn}(z)\, 4\sqrt{\frac{\pi}{2}} & \text{for } t = 1, \\ (1_2(t), 1_3(t), 1_4(t))' \operatorname{sgn}(z) \frac{\min\{|z|, b\, y\}}{y} & \text{for } t \neq 1, \end{cases}$$

with $b = 4\sqrt{\frac{3\pi}{2}}$ and $y = (2\Phi(b\,y) - 1)\frac{1}{4} \approx 0.2409$, where Φ denotes the standard normal distribution function. Note that because of Lemma A.4 in the appendix the fixed point y exists. This score function obviously satisfies condition (5.20), and according to Lemma A.2 in the appendix it also satisfies (5.21). Hence, a one-step M-estimator with this score function, with initial estimator satisfying (5.18) and variance estimators satisfying (5.19) is according to Theorem 5.3 an AL-estimator with influence function ψ_0. Moreover, according to Theorem 5.4, in shrinking contamination neighbourhoods with radius ϵ this estimator is also asymptotically normally distributed with a maximum asymptotic bias equal to $b(\psi\,\sigma, \delta) = \epsilon \cdot 4 \cdot \sqrt{\frac{3\pi}{2}} \cdot \max\{\sigma(1), \sigma(2), \sigma(3)\}$. In Section 7.2 we will see that a weakly asymptotically linear estimator with this score function is optimal for minimum asymptotic bias if $\sigma(t) = \sigma$ for all $t \in \mathcal{T}_0$.

Another score function which also satisfies condition (5.20) and (5.21) and is optimal for robust estimation with bias bound $2b$ (see Section 7.2) is given by

$$\psi(z,t) = \begin{cases} (-1,-1,-1)' \operatorname{sgn}(z) \frac{\min\{|z|, 2\, b\, v\}}{\sqrt{3}\, v} & \text{for } t = 1, \\ (1_2(t), 1_3(t), 1_4(t))' \operatorname{sgn}(z) \frac{\min\{|z|, 2\, b\, w\}}{w} & \text{for } t \neq 1, \end{cases}$$

where again $b = 4\sqrt{\frac{3\pi}{2}}$ and $v = (2\Phi(2\,b\,v) - 1)\frac{1}{4\sqrt{3}} \approx 0.1424$ and $w = (2\Phi(2\,b\,w) - 1)\frac{1}{4} \approx 0.2500$. In a shrinking contamination neighbourhood with radius ϵ a one-step M-estimator with this score function is asymptotically normally distributed with maximum asymptotic bias equal to $\epsilon \cdot 6 \cdot \sqrt{\frac{3\pi}{2}} \cdot \max\{\sigma(1), \sigma(2), \sigma(3)\}$. □

5.3 Robust Estimation of Nonlinear Aspects

For efficient estimation of a nonlinear aspect $\varphi(\beta)$ in a linear model in Section 2.3 we regarded estimators of the form $\widehat{\varphi}_N^{LS}(y_N, d_N) = \varphi(\widehat{\beta}_N^{LS}(y_N, d_N))$, where $\widehat{\beta}_N^{LS}$ is the least squares estimator for β. Because of Lemma 1.4 we have $\varphi(\beta) = \varphi^*(A_{\mathcal{D}}\beta)$ for all $\beta \in \mathcal{B}$ if and only if φ is identifiable at the finite set $\mathcal{D} = \text{supp}(\delta)$. Then we have

$$\widehat{\varphi}_N^{LS}(y_N, d_N) = \varphi^*(A_{\mathcal{D}}\widehat{\beta}_N^{LS}(y_N, d_N)) \tag{5.33}$$

for $\widehat{\beta}_N^{LS}(y_N, d_N) \in \mathcal{B}$, where $\widehat{\eta}_N^{LS}(y_N, d_N) := A_{\mathcal{D}}\widehat{\beta}_N^{LS}(y_N, d_N)$ is the Gauss-Markov estimator for $\eta(\beta) := A_{\mathcal{D}}\beta$, and therefore is most efficient for estimating η. Because $A_{\mathcal{D}}\widehat{\beta}_N^{LS}$ is converging to $A_{\mathcal{D}}\beta$ almost everywhere for $N \to \infty$ at P_θ with $\theta = (\beta', \sigma(\tau_1), ..., \sigma(\tau_I))'$, we have $\widehat{\beta}_N^{LS}(y_N, d_N) \in \mathcal{B}$ and therefore (5.33) almost everywhere for $N \to \infty$ and $\beta \in \mathcal{B}$. In the case, where β itself is not identifiable at $\mathcal{D}$ and therefore $\widehat{\beta}_N^{LS}$ is not unique, this holds at least for one version of $\widehat{\beta}_N^{LS}$. Hence, we also can use (5.33) for defining $\widehat{\varphi}_N^{LS}$. This definition of $\widehat{\varphi}_N^{LS}$ has also the advantage that $\widehat{\varphi}_N^{LS}$ is uniquely defined also if β is not identifiable at $\mathcal{D}$. Moreover, for robust estimation we do not have that $A_{\mathcal{D}}\widehat{\beta}_N$ is optimal robust for estimating η if $\widehat{\beta}_N$ is optimal robust for estimating β. Therefore for robust estimation we define estimators $\widehat{\varphi}_N$ for a nonlinear aspect φ by replacing the Gauss-Markov estimator $\widehat{\eta}_N^{LS}$ for η in (5.33) by a robust estimator for η. In particular we can replace the Gauss-Markov estimator for η by an AL-estimator $\widehat{\eta}_N$ for η, i.e. by a weakly asymptotically linear estimator in the linear model with an influence function ψ_θ satisfying $\psi_\theta(y,t) = \psi(\frac{y-a(t)'\beta}{\sigma(t)}, t)\sigma(t)$ for $\theta = (\beta', \sigma(\tau_1), ..., \sigma(\tau_I))'$ (see Section 5.2). Then the resulting estimator $\widehat{\varphi}_N$ given by $\widehat{\varphi}_N(y_N, d_N) = \varphi^*(\widehat{\eta}_N(y_N, d_N))$ is also weakly asymptotically linear in the sense of Definition 5.2. Moreover, locally at θ the estimator $\widehat{\varphi}_N$ behaves asymptotically as the estimator $\dot{\varphi}_\beta^*\widehat{\eta}_N$ for the linear aspect φ_β given by $\varphi_\beta(\tilde{\beta}) = \dot{\varphi}_\beta^* A_{\mathcal{D}}\tilde{\beta} = \dot{\varphi}_\beta\tilde{\beta}$. Thereby we define

$$\dot{\varphi}_\beta^* = \frac{\partial}{\partial\eta}\varphi^*(\eta)/_{\eta=A_{\mathcal{D}}\beta},$$

where $\dot{\varphi}_\beta$ defined as in Section 1.3 satisfies $\dot{\varphi}_\beta = \dot{\varphi}_\beta^* A_{\mathcal{D}}$ for all $\beta \in \mathcal{B}$.

Lemma 5.4 *Let $\mathcal{B}$ be a convex and open subset of $I\!R^r$, $\varphi : I\!R^r \to I\!R^s$ be continuously differentiable on $\mathcal{B}$, supp(δ) $= \mathcal{D} \subset \{\tau_1, ..., \tau_I\} = \mathcal{T}_0$ and $\varphi(\beta) = \varphi^*(A_\mathcal{D}\beta)$ for all $\beta \in \mathcal{B}$. If $\widehat{\eta}_N$ is an AL-estimator for $\eta(\beta) = A_\mathcal{D}\beta$ with influence function ψ and $\widehat{\varphi}_N$ is an estimator for φ given by $\widehat{\varphi}_N(y_N, d_N) = \varphi^*(\widehat{\eta}_N(y_N, d_N))$, then*

$$\sqrt{N}[\widehat{\varphi}_N(y_N, d_N) - \varphi(\beta)] - \sqrt{N}[\dot{\varphi}_\beta^* \widehat{\eta}_N(y_N, d_N) - \dot{\varphi}_\beta \beta] \overset{N\to\infty}{\longrightarrow} 0$$

in probability for $(P_\theta^N)_{N \in I\!N}$ with $\theta = (\beta', \sigma(\tau_1), ..., \sigma(\tau_I))' \in \mathcal{B} \times (I\!R^+)^I$.

Proof. With φ also φ^* is continuously differentiable. Then, because of the asymptotic normality of $\widehat{\eta}_N$ (see Theorem 5.4) we have

$$\begin{aligned}
&\sqrt{N}[\widehat{\varphi}_N(y_N, d_N) - \varphi(\beta)] - \sqrt{N}[\dot{\varphi}_\beta^* \widehat{\eta}_N(y_N, d_N) - \dot{\varphi}_\beta \beta] \\
&= \sqrt{N}|\widehat{\eta}_N(y_N, d_N) - \eta(\beta)| \\
&\quad \cdot \frac{\varphi^*(\widehat{\eta}_N(y_N, d_N)) - \varphi^*(\eta(\beta)) - \dot{\varphi}_\beta^*(\widehat{\eta}_N(y_N, d_N) - \eta(\beta))}{|\widehat{\eta}_N(y_N, d_N) - \eta(\beta)|} \\
&\overset{N\to\infty}{\longrightarrow} 0
\end{aligned}$$

in probability for $(P_\theta^N)_{N \in I\!N}$. □

Lemma 5.5 *If $\widehat{\eta}_N$ is an AL-estimator for $\eta(\beta) = A_\mathcal{D}\beta$ at δ with influence function ψ, then under the assumptions of Lemma 5.4 for all $\theta = (\beta', \sigma(\tau_1), ..., \sigma(\tau_I))' \in \mathcal{B} \times (I\!R^+)^I$ the estimator $\widehat{\varphi}_N$ for φ given by $\widehat{\varphi}_N(y_N, d_N) = \varphi^*(\widehat{\eta}_N(y_N, d_N))$ is weakly asymptotically linear at $P_{\theta,\delta}$ with influence function ψ_θ of the form*

$$\psi_\theta(y, t) = \dot{\varphi}_\beta^* \psi\left(\frac{y - a(t)'\beta}{\sigma(t)}, t\right) \sigma(t).$$

Proof. Lemma 5.4 showed that locally the estimator $\widehat{\varphi}_N$ behaves asymptotically like $\dot{\varphi}_\beta^* \widehat{\eta}_N$. Because $\dot{\varphi}_\beta^* \widehat{\eta}_N$ is an AL-estimator for the linear aspect φ_β, given by $\varphi_\beta(\tilde{\beta}) = \dot{\varphi}_\beta^* A_\mathcal{D} \tilde{\beta} = \dot{\varphi}_\beta \tilde{\beta}$, with influence function $\dot{\varphi}_\beta^* \psi$ the assertion is proved. □

From Theorem 5.3 we get at once the asymptotic normality of the estimators $\widehat{\varphi}_N$ (see also Theorem 5.4).

Theorem 5.6 *For deterministic designs assume $\delta_N(\{t\}) \to \delta(\{t\})$ for $N \to \infty$ for all $t \in \mathcal{T}_0$. If $\widehat{\eta}_N$ is an AL-estimator for $\eta(\beta) = A_{\mathcal{D}}\beta$ at δ with influence function ψ, then under the assumptions of Lemma 5.4 for all $\theta = (\beta', \sigma(\tau_1), ..., \sigma(\tau_I))' \in \mathcal{B} \times (I\!R^+)^I$ the estimator $\widehat{\varphi}_N$ for φ given by $\widehat{\varphi}_N(y_N, d_N) = \varphi^*(\widehat{\eta}_N(y_N, d_N))$ is asymptotically normally distributed for shrinking contamination, i.e.*

$$\begin{aligned}
&\mathcal{L}(\sqrt{N}(\widehat{\varphi}_N - \varphi(\beta))|Q_\theta^N) \\
&\overset{N\to\infty}{\longrightarrow} \quad \mathcal{N}(\dot{\varphi}_\beta^*\, b(\psi\, \sigma, (Q_{N,\delta})_{N\in I\!N}), \\
&\qquad\qquad \dot{\varphi}_\beta^* \int \sigma^2(t)\, \psi(z,t)\, \psi(z,t)'\, P_\delta(dz, dt)\, (\dot{\varphi}_\beta^*)')
\end{aligned}$$

for all $(Q_{N,\delta})_{N\in I\!N} \in \mathcal{U}_{c,\epsilon}^0(P_\delta)$. And the maximum asymptotic bias satisfies

$$\begin{aligned}
&\max\{|\dot{\varphi}_\beta^*\, b(\psi\, \sigma, (Q_{N,\delta})_{N\in I\!N})|;\ (Q_{N,\delta})_{N\in I\!N} \in \mathcal{U}_{c,\epsilon}^0(P_\delta)\} \\
&= \epsilon\ \max_{(z,t)\in I\!R\times \mathrm{supp}(\delta)} |\dot{\varphi}_\beta^*\, \psi(z,t)\, \sigma(t)|.
\end{aligned}$$

If ψ is bounded, then every AL-estimator $\widehat{\eta}_N$ for η with this influence function is asymptotically robust for shrinking contamination (see Definition 5.4). If ψ is bounded, then also $\dot{\varphi}_\beta^*\psi$ is bounded for all $\beta \in \mathcal{B}$ so that also the estimator $\widehat{\varphi}_N = \varphi^*(\widehat{\eta}_N)$ for φ is asymptotically robust for shrinking contamination. Hence, we have the following theorem.

Theorem 5.7 *Let $\widehat{\eta}_N$ be an AL-estimator for $\eta(\beta) = A_{\mathcal{D}}\beta$ at δ with influence function ψ. Then under the assumptions of Theorem 5.6 the estimator $\widehat{\varphi}_N$ for φ given by $\widehat{\varphi}_N = \varphi^*(\widehat{\eta}_N)$ is asymptotically robust for shrinking contamination if and only if ψ is bounded.*

5.4 Robust Estimation in Contaminated Nonlinear Models

M-estimators for linear models can be generalized for nonlinear models in a straightforward way as estimators $\widehat{\beta}_N$ which satisfy

$$\sum\nolimits_{n=1}^{N} \rho(y_{nN} - \mu(t_{nN}, \beta), t_{nN})\, \dot{\mu}(t_{nN}, \beta) \ = \ 0, \tag{5.34}$$

where $\rho : I\!R \times \mathcal{T} \to I\!R$ and $\dot{\mu} : \mathcal{T} \times \Theta \to I\!R^p$ is given by $\dot{\mu}(t, \beta) := (\frac{\partial}{\partial \tilde{\beta}} \mu(t, \tilde{\beta})/_{\tilde{\beta}=\beta})'$ (see Stefanski et al. (1986), Müller (1994b)). Solutions of (5.34) can be calculated by Newton-Raphson iterations. Starting with an initial estimator $\widehat{\beta}_N^0$ the first step of the Newton-Raphson iteration provides

$$\begin{aligned}
\widehat{\beta}_N = \widehat{\beta}_N^0 \ &- \ \left(\tfrac{\partial}{\partial \beta}(\textstyle\sum_{n=1}^{N} \rho(y_{nN} - \mu(t_{nN}, \beta), t_{nN})\, \dot{\mu}(t_{nN}, \beta))/_{\beta=\widehat{\beta}_N^0}\right)^{-1} \\
&\cdot \textstyle\sum_{n=1}^{N} \rho(y_{nN} - \mu(t_{nN}, \widehat{\beta}_N^0), t_{nN})\, \dot{\mu}(t_{nN}, \widehat{\beta}_N^0).
\end{aligned}$$

Now we can also use the result of the first Newton-Raphson iteration as estimate. If $(d_N)_{N\in\mathbb{N}}$ is converging to δ for $N \to \infty$, then under some regularity conditions we have

$$\lim_{N\to\infty} \left(\frac{\partial}{\partial\beta} \left(\frac{1}{N} \sum_{n=1}^{N} \rho(y_{nN} - \mu(t_{nN},\beta), t_{nN})\, \dot{\mu}(t_{nN},\beta) \right) \Big/_{\beta=\widehat{\beta}_N^0} - M(\widehat{\beta}_N^0, \rho) \right) = 0,$$

where

$$M(\beta,\rho) := \int \rho(z,t)\, z\, \dot{\mu}(t,\beta)\, \dot{\mu}(t,\beta)'\, P(dz)\, \delta(dt).$$

Taking into account that the variances $\sigma^2(t)$ may be unknown we now can define one-step M-estimators in nonlinear models as follows:

Definition 5.7 (One-step M-estimator for nonlinear models)
An estimator $\widehat{\beta}_N : \mathbb{R}^N \times T^N \to \mathbb{R}^p$ *is called a one-step-M-estimator for* β *with score function* $\rho : \mathbb{R} \times T \to \mathbb{R}$, *initial estimator* $\widehat{\beta}_N^0 : \mathbb{R}^N \times T^N \to \mathbb{R}^p$ *and variance estimators* $\widehat{\sigma}_N^2(t) : \mathbb{R}^N \times T^N \to \mathbb{R}^+$ *for* $\sigma^2(t)$, $t \in T$, *if*

$$\widehat{\beta}_N = \widehat{\beta}_N^0 + \frac{1}{N} \sum_{n=1}^{N} M(\widehat{\beta}_N^0, \rho)^{-1}\, \dot{\mu}(t_{nN}, \widehat{\beta}_N^0)\, \rho\left(\frac{y_{nN} - \mu(t_{nN}, \widehat{\beta}_N^0)}{\widehat{\sigma}_N(t_{nN})}, t_{nN} \right) \widehat{\sigma}_N(t_{nN}).$$

The weak asymptotic linearity and thus the asymptotic normality of the one-step M-estimators can be derived under the following assumptions as Müller (1994b) has shown:

- $\sqrt{N}(\widehat{\beta}_N^0 - \beta)$ is tight at the ideal model P_θ^N. (5.35)
- $\sqrt{N}(\widehat{\sigma}_N(t) - \sigma(t))$ is tight at the ideal model P_θ^N for all $t \in T_0$. (5.36)
- For every $\beta_0 \in \mathcal{B}$ there exists k_0 so that $\{\beta \in \mathcal{B};\ |\beta - \beta_0| \le k_0\}$ is a compact subset of $\mathbb{R}^p$ (For example $\mathcal{B}$ is an open subset of $\mathbb{R}^p$). (5.37)
- $\mu(t,\cdot)$ has a continuous second order derivative on $\mathcal{B}$ for all $t \in T_0$. (5.38)
- $M(\beta,\rho)$ is regular for all $\beta \in \mathcal{B}$. (5.39)
- $\rho(z,t) = \rho_0(z,t) + c(t)\mathrm{sgn}(z)$, where for all $t \in T_0$ $\rho_0(\cdot,t)$ is antisymmetric, continuous and there exists $l_1(t), ..., l_L(t)$ so that $\rho_0(\cdot,t)$ has bounded and continuous derivatives on $\mathbb{R} \setminus \{l_1(t), ..., l_L(t)\}$. (5.40)

Condition (5.40) can be generalized (see Section 5.2 and Müller (1994e)) but in this form it is easy to check. Moreover it includes all relevant score functions as the signum function $\rho(z,t) = sgn(z)$ for L_1-estimation, $\rho(z,t) = z$ for least squares estimation and redescending score functions. Under some regularity conditions on $\mathcal{B}$ or μ the condition (5.35) is in particular satisfied for the least squares estimator (see for example Jennrich (1969), Läuter (1989) or Section 2.3). Variance estimators which satisfy condition (5.36) are given in Section 5.2. Note that the condition (5.39) is sufficient but not necessary for local identifiability (see Section 1.3).

Lemma 5.6 *If for deterministic designs $\delta_N(\{t\}) \to \delta(\{t\})$ for $N \to \infty$ for all $t \in \mathcal{T}_0 = \{\tau_1, ..., \tau_I\} \supset supp(\delta)$, $\theta = (\beta', \sigma(\tau_1), ..., \sigma(\tau_I))' \in \mathcal{B} \times (\mathbb{R}^+)^I$, then under assumptions (5.35)-(5.40) a one-step M-estimator for β with score function ρ is weakly asymptotically linear at $P_{\theta,\delta}$ with influence function ψ_θ given by*

$$\psi_\theta(y,t) = M(\beta,\rho)^{-1}\,\dot{\mu}(t,\beta)\,\rho\left(\frac{y_{nN} - \mu(\beta,t)}{\sigma(t)}, t\right)\,\sigma(t).$$

Proof. See the proof of Theorem 3.1 in Müller (1994b). □

The asymptotic linearity provides according to Theorem 5.3 that the one-step M-estimators are asymptotically normally distributed for shrinking contamination. For that set

$$\begin{aligned} &C_\theta(\rho,\delta) \\ &:= \; M(\beta,\rho)^{-1} \int \dot{\mu}(t,\beta)\,\dot{\mu}(t,\beta)'\,\rho(z,t)^2\,\sigma(t)^2\,P\otimes\delta(dz,dt) \quad M(\beta,\rho)^{-1} \end{aligned}$$

and

$$\begin{aligned} &b_\theta(\rho\,\sigma,(Q_{N,\delta})_{N\in\mathbb{N}}) \\ &:= \lim_{N\to\infty} \sqrt{N}\,M(\beta,\rho)^{-1} \int \dot{\mu}(t,\beta)\,\rho(z,t)\,\sigma(t)\,Q_{N,\delta}(dz,dt). \end{aligned}$$

Theorem 5.8 *Under the assumptions of Lemma 5.6 a one-step M-estimator for β with score function ρ is asymptotically normally distributed for shrinking contamination, i.e.*

$$\mathcal{L}(\sqrt{N}(\widehat{\beta}_N - \beta)|Q_\theta^N) \overset{N\to\infty}{\longrightarrow} \mathcal{N}(b_\theta(\rho\,\sigma,(Q_{N,\delta})_{N\in\mathbb{N}}), C_\theta(\rho,\delta))$$

for all $(Q_{N,\delta})_{N\in\mathbb{N}} \in \mathcal{U}^0_{c,\epsilon}(P_\delta)$ and $\beta \in \mathcal{B}$, with maximum asymptotic bias

$$\begin{aligned} &\max\{b_\theta(\rho\,\sigma,(Q_{N,\delta})_{N\in\mathbb{N}});\ (Q_{N,\delta})_{N\in\mathbb{N}} \in \mathcal{U}^0_{c,\epsilon}(P_\delta)\} \\ &= \; \epsilon\,\max_{(z,t)\in\mathbb{R}\times\text{supp}(\delta)} |M(\beta,\rho)^{-1}\,\dot{\mu}(t,\beta)\,\rho(z,t)\,\sigma(t)|. \end{aligned}$$

Proof. The assertions follows at once from Theorem 5.3 and Lemma 5.6. See also Müller (1994b). □

Theorem 5.9 *Under the assumptions of Lemma 5.6 a one-step M-estimator for β with score function ρ is asymptotically robust for shrinking contamination if and only if ρ is bounded.*

Proof. The assertion follows at once from the definition of robustness for shrinking contamination (see Definition 5.4) and from Theorem 5.8. □

6

Robustness of Tests

In this chapter some robustness concepts for tests are introduced. In Section 6.1 the concepts of bias and breakdown points are transferred to tests while in Section 6.2 the concept based on shrinking neighbourhoods is transferred to tests.

6.1 Bias and Breakdown Points

If Θ is divided in two disjoined sets Θ_0 and Θ_1 we can test the hypothesis $H_0 : \theta \in \Theta_0$ against the alternative $H_1 : \theta \in \Theta_1$ by a statistical test. The statistical test is based on a test statistic $\tau_N : \mathbb{R}^N \times \mathcal{T}^N \to \mathbb{R}$ and a rejection set $R_N \subset \mathbb{R}$ so that the hypothesis H_0 is rejected if $\tau_N(y_N, d_N) \in R_N$ and H_0 is not rejected if $\tau_N(y_N, d_N) \notin R_N$. The maximum probability of rejecting H_0 although $\theta \in \Theta_0$, i.e. $\max_{\theta \in \theta_0} P_\theta^N(\tau_N \in R_N)$, is called the *error of the first kind* or briefly the *first error* of the test, and the probability of not rejecting H_0 for $\theta \in \Theta_1$, i.e. $P_\theta^N(\tau_N \notin R_N)$, is called the *error of the second kind* of briefly the *second error* of the test for $\theta \in \Theta_1$. If the first error is less than or equal to α, where $\alpha \in (0, 1)$, then the test is called a *level α test* (see for example Lehmann (1959)).

For outlier robust tests the first error and the second errors of the test should not increase too much in the presence of some outliers. I.e. the errors of the test should be stable in some outlier modelling neighbourhood of the ideal distribution $P_{\theta,\delta}$. As for estimators an ideal outlier modelling neighbourhood is the contamination neighbourhood $\mathcal{U}_c(P_{\theta,\delta}, \epsilon)$ (see Section 3.1). We should distinguish between the *bias of the first error* and the *bias of the second error* in this neighbourhood. Compare also with the α and β robustness as defined in Büning (1994a).

Definition 6.1 (Bias of the first and the second error)
a) The bias of the first error of a level α test given by the test statistic $\tau_N : \mathbb{R}^{N\times T} \to \mathbb{R}$ and a rejection set $R_N \subset \mathbb{R}$ is defined as

$$b^1_\epsilon(\tau_N, R_N) := \max\{(Q^N_\theta(\tau_N \in R_N) - \alpha);\\ Q_{N,\theta,\delta} \in \mathcal{U}_c(P_{\theta,\delta}, \epsilon),\ \theta \in \Theta_0\}.$$

b) The bias of the second error of a test given by the test statistic $\tau_N : \mathbb{R}^{N\times T} \to \mathbb{R}$ and a rejection set $R_N \subset \mathbb{R}$ for $\theta \in \Theta_1$ is defined as

$$b^2_\epsilon(\tau_N, R_N, \theta) := \max\{(Q^N_\theta(\tau_N \notin R_N) - P^N_\theta(\tau_N \notin R_N));\\ Q_{N,\theta,\delta} \in \mathcal{U}_c(P_{\theta,\delta}, \epsilon)\}.$$

As for estimators we can define also the breakdown point of a test by basing on the above bias definitions. For other definitions of breakdown points of tests see He, Simpson and Portnoy (1990), He (1991), Coakley and Hettmansperger (1992, 1994) and Zhang (1996).

Definition 6.2 (Breakdown point of the first and the second error)
a) The breakdown point of the first error of a level α test based on τ_N and R_N is defined as

$$\epsilon^1(\tau_N, R_N) := \min\{\epsilon;\ b^1_\epsilon(\tau_N, R_N) = b^1_1(\tau_N, R_N)\}.$$

b) The breakdown point of the second error of a test based on τ_N and R_N at $\theta \in \Theta_1$ is defined as

$$\epsilon^2(\tau_N, R_N, \theta) := \min\{\epsilon;\ b^2_\epsilon(\tau_N, R_N, \theta) = b^2_1(\tau_N, R_N, \theta)\}.$$

A test is called *general robust* if the bias of the first and the second errors are converging to zero for $\epsilon \to 0$.

Definition 6.3 (General robustness of tests)
A level α test based on τ_N and R_N is general robust if $\lim_{\epsilon\to 0} b^1_\epsilon(\tau_N, R_N) = 0$ and $\lim_{\epsilon\to 0} b^2_\epsilon(\tau_N, R_N, \theta) = 0$ for all $\theta \in \Theta_1$.

As for estimators (see Section 4.1) it is clear that for a general robust test the breakdown points of the first and the second errors are greater than zero. Moreover the test statistic τ_N can be regarded as a special estimator so that general robustness of τ_N is defined in Definition 3.8 in Section 3.2. If the distribution of τ_N under $P_{\theta,\delta}$ is absolutely continuous for all $\theta \in \Theta$, then the general robustness of the test statistic τ_N implies the general robustness of the corresponding test.

Lemma 6.1 *If the distribution $(P_\theta^N)^{\tau_N}$ of τ_N under $P_{\theta,\delta}$ is absolutely continuous for all $\theta \in \Theta$ and τ_N is general robust, then every level α test based on τ_N and R_N is general robust.*

Proof. Let be $R_N^\epsilon := \{y \in I\!R;\ |y - \overline{y}| \le \epsilon \text{ for some } \overline{y} \in R_N\}$. Then for every $\theta \in \Theta_0$ and every ϵ_1 there exists ϵ_2 such that

$$(P_\theta^N)^{\tau_N}(R_N^{\epsilon_2}) + \epsilon_2 \le (P_\theta^N)^{\tau_N}(R_N) + \epsilon_1$$

because $(P_\theta^N)^{\tau_N}$ is absolutely continuous. Because τ_N is general robust there exists ϵ_3 with $d_P((P_\theta^N)^{\tau_n}, (Q_\theta^N)^{\tau_N}) < \epsilon_2$ for all $Q_{N,\theta,\delta} \in \mathcal{U}_c(P_{\theta,\delta}, \epsilon_3)$. Hence, the definition of the Prohorov metric d_P (see Section 3.1) provides

$$\begin{aligned} & Q_\theta^N(\tau_N \in R_N) - \alpha \\ \le\ & (Q_\theta^N)^{\tau_N}(R_N) - (P_\theta^N)^{\tau_N}(R_N) \\ \le\ & (P_\theta^N)^{\tau_N}(R_N^{\epsilon_2}) + \epsilon_2 - (P_\theta^N)^{\tau_N}(R_N) \\ \le\ & \epsilon_1 \end{aligned}$$

for all $Q_{N,\theta,\delta} \in \mathcal{U}_c(P_{\theta,\delta}, \epsilon_3)$ such that the assertion is proved for the first error. Similarly the assertion follows for the second errors. □

6.2 Asymptotic Robustness for Shrinking Contamination

As for estimators it is often too difficult to derive the general robustness of tests. Then it will be more tractable to regard the asymptotic behaviour of the test. In particular often the distribution $(P_\theta^N)^{\tau_N}$ of the test statistic is weakly converging to some distribution P_θ^∞ for $N \to \infty$ so that the rejection sets R_N are chosen by $R_N = R$, where $R \subset I\!R$ satisfies $P_\theta^\infty(R) \le \alpha$ for all $\theta \in \Theta_0$. If $(Q_\theta^N)^{\tau_N}$ is also weakly converging to some distribution Q_θ^∞ or at least $Q_\theta^N(\tau_N \in R)$ and $Q_\theta^N(\tau_N \notin R)$ is converging to $Q_\theta^\infty(R)$ and $Q_\theta^\infty(I\!R \setminus R)$, respectively, for $Q_{N,\theta,\delta}$ in some neighbourhood of $P_{\theta,\delta}$, then $Q_\theta^\infty(R)$ and $Q_\theta^\infty(I\!R \setminus R)$ can be compared with $P_\theta^\infty(R)$ and $P_\theta^\infty(I\!R \setminus R)$. Similarly as for the finite sample case and as for estimators we can define the asymptotic bias, the asymptotic breakdown point and the general asymptotic robustness of tests. But even within this approach it is often too difficult to derive the robustness of tests.

Another approach is to define the robustness of the tests via the robustness of their test statistics. As Lemma 6.1 showed the robustness of tests is closely connected with the robustness of the test statistic. Therefore interpreting the test statistics as special estimators all robustness criteria for estimators can be transferred to the test statistics. In particualar, we can regard the influence function and the asymptotic breakdown point of the test statistic. This leads to robustness approaches for tests which

Ronchetti (1982), Hampel et al. (1986), He et al. (1990), Markatou and Hettmansperger (1990), Akritas (1991a) and Markatou and He (1994) used. Finite sample breakdown points for tests can be also defined as for estimation by adding or replacing observations which leads to the notion of resistance of tests (see Ylvisaker (1977), Coakley and Hettmansperger (1992, 1994), Zhang (1996)). But all the approaches which transfer the robustness concepts of estimators to the test statistic have the disadvantage that it is not clear how the robustness of the test statistic is related to the main quantities of a test, namely to the error of the first and the second kind.

But for robustness concepts based on the influence function of the test statistic there exists a relation to the robustness of the error of first and second kind. This was shown for location problems in Ronchetti (1982) and Hampel et al. (1986) and for general parametric models in Heritier and Ronchetti (1994) by assuming Fréchet differentiability of the test statistic. To avoid Fréchet differentiability and to take a planned experiment into account we here use a robustness concept which Müller (1992c) derived. In this robustness concept the test statistics are based on asymptotically linear estimators and their behaviour in shrinking contamination neighbourhoods is investigated. As for estimation this leads to a more model oriented approach and, additionally, to a more test specific approach of robustness of tests than the approach based on the influence function of the test statistic or other approaches applying robustness concepts for estimators on test statistics. Thereby note that originally the robustness concepts for estimators which are based on shrinking neighbourhoods were derived by robustness considerations for simple tests (see Rieder (1978, 1980) and Bednarski (1985)).

To regard the behaviour of tests in shrinking neighbourhoods we assume that $\theta = (\beta', \sigma)' \in \Theta \subset I\!R^r \times I\!R^+$ and that the hypothesis is of the form $H_0 : \varphi(\beta) = l$ such that $\Theta_0 = \{(\beta', \sigma)' \in \Theta;\ \varphi(\beta) = l\}$ and $\Theta_1 = \{(\beta', \sigma)' \in \Theta;\ \varphi(\beta) \neq l\}$. Moreover, we assume that the test statistic is based on a weakly aymptotically linear estimator $\widehat{\varphi}_N$ for $\varphi(\beta)$ with influence function $\psi_\theta = \psi_{\beta,\sigma}$ and that the asymptotic linearity also holds for alternatives of the hypothesis of the form $\theta_N = (\beta_N', \sigma)' = ((\beta + N^{-1/2}\overline{\beta})', \sigma)'$, where $\theta = (\beta', \sigma)' \in \Theta_0$ and $(\overline{\beta}', \sigma)' \in \Theta$ arbitrary for every σ. Because this property in particular holds if $P_{\theta_N}^N$ is contiguous to P_θ^N we call this property *weak asymptotic linearity for contiguous alternatives*. For $P_{\theta,\delta}$ and P_θ^N we also write $P_{\beta,\sigma,\delta}$ and $P_{\beta,\sigma}^N$, respectively.

Definition 6.4 (Weak asymptotic linearity for contiguous alternatives)
An estimator $\widehat{\varphi}_N$ is weakly asymptotically linear for contiguous alternatives of $H_0 : \varphi(\beta) = l$ with influence function $\psi_{\beta,\sigma}$ if $\psi_{\beta,\sigma}$ satisfies conditions (5.11) - (5.13) and

$$\lim_{N\to\infty} P^N_{\beta_N,\sigma}\left(\left\{(y_N, d_N);\ \sqrt{N}\left|\widehat{\varphi}_N(y_N, d_N) - \varphi(\beta_N)\right.\right.\right.$$
$$\left.\left.\left. - \frac{1}{N}\sum\nolimits_{n=1}^{N} \psi_{\beta_N,\sigma}(y_{nN}, t_{nN})\right| > \epsilon\right\}\right) = 0$$

for all $\epsilon > 0$ and $(\beta_N', \sigma)' = ((\beta + N^{-1/2}\overline{\beta})', \sigma)' \in \Theta$ with $\varphi(\beta) = l$.

By replacing the Gauss-Markov estimator by a weakly asymptotically linear estimator for contiguous alternatives we can generalize the classical F-test for testing a hypothesis of the form $H_0 : L\,\beta = l$ against $H_1 : L\,\beta \neq l$ in linear models. Remember that the test statistic of the F-test bases on

$$\tau_N^{LS}(y_N, d_N) = N(\widehat{\varphi}_N^{LS}(y_N, d_N) - l)'\, C_N^{LS}(y_N, d_N)^{-1}(\widehat{\varphi}_N^{LS}(y_N, d_N) - l),$$

where

$$C_N^{LS}(y_N, d_N) = N\, L(A_{d_N}' A_{d_N})^- L'\, \widehat{\sigma}_N^{LS}(y_N, d_N)^2$$

with

$$\widehat{\sigma}_N^{LS}(y_N, d_N)^2 = \frac{y_N'(E_N - A_{d_N}(A_{d_N}' A_{d_N})^- A_{d_N}')y_N}{\mathrm{rk}(E_N - A_{d_N}(A_{d_N}' A_{d_N})^- A_{d_N}')}$$

is converging in probability to the asymptotic covariance matrix of the Gauss-Markov estimator $\widehat{\varphi}_N^{LS}$ (see Section 2.2). Hence in the generalization of the F-test statistic we also substitute C_N^{LS} by any function $C_N : I\!R^N \times \mathcal{T}^N \to I\!R^{s\times s}$ such that $C_N(y_N, d_N)$ is converging in probability to the asymptotic covariance matrix $\int \psi_{\beta,\sigma}\, \psi_{\beta,\sigma}'\, d\,P_{\beta,\sigma,\delta}$ of the weakly asymptotically linear estimator. Hence the generalized F-test statistic for testing $H_0 : \varphi(\beta) = l$ against $H_1 : \varphi(\beta) \neq l$ has the form

$$\tau_N(y_N, d_N) = N(\widehat{\varphi}_N(y_N, d_N) - l)'\, C_N(y_N, d_N)^{-1}\, (\widehat{\varphi}_N(y_N, d_N) - l).$$

The resulting test is sometimes called the *Wald-type test* (see Wald (1943), Markatou et al. (1991), Silvapulle (1992a), Heritier and Ronchetti (1994)) and here it will be called *ALE-test* and its corresponding test statistic is called *ALE-test statistic* because it is based on an asymptotically (A) linear (L) estimator (E). ALE-tests were also regarded in Rieder (1994), p. 153. Moreover, note that these tests behave asymptotically like the score-type tests in Markatou et al. (1991), Markatou and He (1994) and Heritier and Ronchetti (1994) (see Markatou and Manos (1996)).

Definition 6.5 (ALE-test statistic)
A test statistic $\tau_N : \mathbb{R}^N \times \mathcal{T}^N \to \mathbb{R}$ is an ALE-test statistic with influence function $\psi_{\beta,\sigma}$ for testing $H_0 : \varphi(\beta) = l$ against $H_1 : \varphi(\beta) \neq l$ if

$$\tau_N(y_N, d_N) = N(\widehat{\varphi}_N(y_N, d_N) - l)' \, C_N(y_N, d_N)^{-1} \, (\widehat{\varphi}_N(y_N, d_N) - l),$$

where $\widehat{\varphi}_N$ is weakly asymptotically linear for contiguous alternatives of $H_0 : \varphi(\beta) = l$ with influence function $\psi_{\beta,\sigma}$ and

$$\lim_{N\to\infty} C_N = \int \psi_{\beta,\sigma}(y,t) \, \psi_{\beta,\sigma}(y,t)' \, P_{\beta,\sigma,\delta}(dy, dt)$$

in probability for $(P^N_{\beta,\sigma})_{N\in\mathbb{N}}$ with $\varphi(\beta) = l$. A test based on an ALE-test statistic is called an ALE-test.

In the following we will regard mainly homoscedastic linear models and linear aspects $\varphi(\beta) = L\beta$ so that the ALE-test statistic should be based on AL-estimators with influence function ψ (see Section 5.2). Recall that we then have

$$\psi_{\beta,\sigma}(y,t) = \sigma \, \psi\left(\frac{y - a(t)'\beta}{\sigma}, t\right)$$

and $\int \psi_{\beta,\sigma} \, \psi_{\beta,\sigma} \, dP_{\beta,\sigma,\delta} = \sigma^2 \int \psi \, \psi' \, dP_\delta$. Hence, in linear models we define ALE-test statistics for testing $H_0 : \varphi(\beta) = L\beta = l$ against $H_1 : \varphi(\beta) = L\beta \neq l$ as follows.

Definition 6.6 (ALE-test statistic in a linear model)
In a linear model a test statistic $\tau_N : \mathbb{R}^N \times \mathcal{T}^N \to \mathbb{R}$ is an ALE-test statistic with influence function ψ for testing $H_0 : \varphi(\beta) = L\beta = l$ against $H_1 : \varphi(\beta) = L\beta \neq l$ if τ_N is an ALE-test statistic with influence function $\psi_{\beta,\sigma}$ satisfying

$$\psi_{\beta,\sigma}(y,t) = \sigma \, \psi\left(\frac{y - a(t)'\beta}{\sigma}, t\right)$$

for all $(y,t) \in \mathbb{R} \times \mathcal{T}$.

An ALE-test statistic is asymptotically chi-squared distributed in a shrinking neighbourhood which is given by the restricted shrinking contamination neighbourhood around P_δ (compare also with the Theorem 2.2 for the classical F-test statistic). Thereby we regard not only shrinking neighbourhoods around $P_{\beta,\sigma,\delta}$ with $(\beta', \sigma)' \in \Theta_0$ but also around $P_{\beta_N,\sigma,\delta}$, where $\theta_N = (\beta_N', \sigma)' \in \Theta_1$ provides contiguous alternatives. These shrinking neighbourhoods are given by $(Q_{N,\theta_N,\delta})_{N\in\mathbb{N}}$, where $(Q_{N,\delta})_{N\in\mathbb{N}} \in \mathcal{U}^0_{c,\epsilon}(P_\delta)$ and $\theta_N = (\beta_N', \sigma)' = ((\beta + N^{-1/2}\bar{\beta})', \sigma)' \in \Theta$ with $\varphi(\beta_N) = l + N^{-1/2}\gamma$. Note that for linear models, where P is the standard normal distribution $n_{(0,1)}$, for alternatives θ_N as above we really have that $P_{\theta_N,\delta}$ is contiguous to $P_{\beta,\sigma,\delta}$.

Theorem 6.1 *In a linear model let $P_{\theta_N,\delta}$ be contiguous to $P_{\beta,\sigma,\delta}$ for all $\theta_N = (\beta_N', \sigma)' = ((\beta + N^{-1/2}\overline{\beta})', \sigma)' \in \Theta$ with $\varphi(\beta_N) = l + N^{-1/2}\gamma$. If τ_N is an ALE-test statistic for testing $H_0 : \varphi(\beta) = l$ against $H_1 : \varphi(\beta) \neq l$ with influence function ψ, then τ_N has an asymptotic chi-squared distribution, i.e.*

$$\mathcal{L}(\tau_N | Q_{\theta_N}^N) \overset{N\to\infty}{\longrightarrow} \chi^2(s, [\gamma + \sigma\, b(\psi, (Q_{N,\delta})_{N\in\mathbb{N}})]' \\ [\sigma^2\, C(\psi)]^{-1} [\gamma + \sigma\, b(\psi, (Q_{N,\delta})_{N\in\mathbb{N}})])$$

for all $(Q_{N,\delta})_{N\in\mathbb{N}} \in \mathcal{U}_{c,\epsilon}^0(P_\delta)$ and all $\theta_N = (\beta_N', \sigma)' = ((\beta + N^{-1/2}\overline{\beta})', \sigma)' \in \Theta$ with $\varphi(\beta_N) = l + N^{-1/2}\gamma$, where

$$C(\psi) := \int \psi(z,t)\, \psi(z,t)'\, P_\delta(dz, dt)$$

and

$$b(\psi, (Q_{N,\delta})_{N\in\mathbb{N}})) := \sqrt{N} \lim_{N\to\infty} \int \psi(z,t) Q_{N,\delta}(dz, dt).$$

Proof. Consider the following part of τ_N:

$$\sqrt{N}\, C_N^{-1/2} (\widehat{\varphi}_N - l) = \sqrt{N}\, C_N^{-1/2} (\widehat{\varphi}_N - \varphi(\beta_N)) + C_N^{-1/2}\gamma. \quad (6.1)$$

Because $(P_{\theta_N}^N)_{N\in\mathbb{N}}$ is contiguous to $(P_\theta^N)_{N\in\mathbb{N}}$ the covariance estimator C_N also converges to $\sigma^2\, C(\psi)$ in probability $(P_{\theta_N}^N)_{N\in\mathbb{N}}$. Thus because $\widehat{\varphi}_N$ is weakly asymptotically linear for contiguous alternatives the distribution $\mathcal{L}(\sqrt{N}\, C_N^{-1/2} (\widehat{\varphi}_N - \varphi(\beta_N)) | P_{\theta_N}^N)$ behaves asymptotically like

$$\begin{aligned} &\mathcal{L}(N^{-1/2}\, [\sigma^2\, C(\psi)]^{-1/2} \sum_{n=1}^N \psi_{\theta_N}(y_{nN}, t_{nN}) | P_{\theta_N}^N) \\ = \;& \mathcal{L}(N^{-1/2}\, [\sigma^2\, C(\psi)]^{-1/2} \sum_{n=1}^N \sigma\, \psi(z_{nN}, t_{nN}) | P^N) \\ = \;& \mathcal{L}(N^{-1/2} \sum_{n=1}^N C(\psi)^{-1/2}\, \psi(z_{nN}, t_{nN}) | P^N). \end{aligned}$$

The third lemma of LeCam (see Hájek and Šidák, p. 208) provides that $\mathcal{L}(N^{-1/2} \sum_{n=1}^N C(\psi)^{-1/2}\, \psi(z_{nN}, t_{nN}) | Q^N)$ is asymptotically normally distributed with mean

$$\lim_{N\to\infty} \sqrt{N} \int C(\psi)^{-1/2}\, \psi(z,t)\, Q_{N,\delta}(dz, dt) = C(\psi)^{-1/2}\, b(\psi, (Q_{N,\delta})_{N\in\mathbb{N}})$$

and covariance matrix E_s, where E_s is the $(s \times s)$-identity matrix, and in particular that $(Q^N)_{N\in\mathbb{N}}$ is contiguous to $(P^N)_{N\in\mathbb{N}}$ for all $(Q_{N,\delta})_{N\in\mathbb{N}} \in \mathcal{U}^0_{c,\epsilon}(P_\delta)$. Then also $(Q^N_{\theta_N})_{N\in\mathbb{N}}$ is contiguous to $(P^N_{\theta_N})_{N\in\mathbb{N}}$ such that $C_N^{-1/2}\gamma$ converges to $[\sigma^2 C(\psi)]^{-1/2}\gamma$ in probability for $(Q^N_{\theta_N})_{N\in\mathbb{N}}$ and $\mathcal{L}(\sqrt{N}\,[\sigma^2 C(\psi)]^{-1/2}(\widehat{\varphi}_N - \varphi(\beta_N))|Q^N_{\theta_N})$ is asymptotically normally distributed with mean $C(\psi)^{-1/2}\,b(\psi,(Q_{N,\delta})_{N\in\mathbb{N}})$ and covariance matrix E_s. Hence $\mathcal{L}(\sqrt{N}\,[\sigma^2 C(\psi)]^{-1/2}(\widehat{\varphi}_N - l)|Q^N_{\theta_N})$ is asymptotically normally distributed with mean $[\sigma^2 C(\psi)]^{-1/2}[\gamma + \sigma\, b(\psi,(Q_{N,\delta})_{N\in\mathbb{N}})]$ and covariance matrix E_s which provides the assertion. □

Because $b(\psi,(Q_{N,\delta})_{N\in\mathbb{N}}) = 0$ if $(Q_{N,\delta})_{N\in\mathbb{N}} = (P_\delta)_{N\in\mathbb{N}}$ Theorem 6.1 provides the error of the first and the second kind for ideal distributions $P_{\theta,\delta}$. In particular the rejection set R should be chosen as $R := (\chi^2_{1-\alpha,s,0},\infty)$, where $\chi^2_{1-\alpha,s,0}$ is the $(1-\alpha)$-quantile of the chi-squared distribution with s degrees of freedom and noncentrality parameter equal to 0. Then according to Theorem 6.1 the test based on the ALE-test statistic τ_N is an asymptotic level α test for the ideal distributions $P_{\theta,\delta}$, i.e. we have

$$\lim_{N\to\infty} P^N_\theta(\tau_N > \chi^2_{1-\alpha,s,0}) \le \alpha$$

for all $\theta \in \Theta_0 = \{(\beta',\sigma)' \in \Theta;\ \varphi(\beta) = l\}$. Moreover Theorem 6.1 also provides that the asymptotic error of the second kind of this ALE-test for contiguous ideal alternatives based on $P_{\theta_N,\delta}$ with $\theta_N = ((\beta + N^{-1/2}\overline{\beta})',\sigma)' \in \Theta$, where $\varphi(\beta_N) = l + N^{-1/2}\gamma \neq l$, is

$$\lim_{N\to\infty} P^N_{\theta_N}(\tau_N \le \chi^2_{1-\alpha,s,0})) = \mathcal{X}^2_{s,\gamma'[\sigma^2 C(\psi)]^{-1}\gamma}(\chi^2_{1-\alpha,s,0}),$$

where $\mathcal{X}_{s,v}$ is the distribution function of the chi-squared distributions with s degrees of freedom and noncentrality parameter v.

Now the *asymptotic bias of the first error for shrinking contamination* can be defined as the maximum asymptotic bias of the rejection probability in all shrinking contamination neighbourhoods around $P_{\beta,\sigma,\delta}$ with $\varphi(\beta) = l$. And the *asymptotic bias of the second error for shrinking contamination* can be defined as the maximum asymptotic bias of the accepting probability in all shrinking contamination neighbourhoods around $P_{\theta_N,\delta}$ with $\theta_N = ((\beta + N^{-1/2}\overline{\beta})',\sigma)' \in \Theta$, where $\varphi(\beta_N) = l + N^{-1/2}\gamma \neq l$.

Definition 6.7 (Asymptotic bias of first and second error for shrinking contamination)
a) The asymptotic bias of the first error for shrinking contamination of a level α ALE-test with test statistic τ_N and rejection set $R := (\chi^2_{1-\alpha,s,0}, \infty)$ is defined as

$$\begin{aligned} b^1_\epsilon(\tau_N, R) := \max\{ \lim_{N\to\infty} (Q^N_\theta(\tau_N > \chi^2_{1-\alpha,s,0}) - \alpha);\ & \\ (Q_{N,\delta})_{N\in\mathbb{N}} \in \mathcal{U}^0_{c,\epsilon}(P_\delta) \text{ and } & \\ \theta = (\beta', \sigma)' \in \Theta \text{ with } \varphi(\beta) = l\}. & \end{aligned}$$

b) The asymptotic bias of the second error for shrinking contamination of a level α ALE-test with test statistic τ_N, rejection set $R := (\chi^2_{1-\alpha,s,0}, \infty)$ and alternatives given by $\gamma \neq 0$ is defined as

$$\begin{aligned} b^2_\epsilon(\tau_N, R, \gamma) := \max\{ \lim_{N\to\infty} (Q^N_{\theta_N}(\tau_N \le \chi^2_{1-\alpha,s,0}) - P^N_{\theta_N}(\tau_N \le \chi^2_{1-\alpha,s,0}));\ & \\ (Q_{N,\delta})_{N\in\mathbb{N}} \in \mathcal{U}^0_{c,\epsilon}(P_\delta) \text{ and } & \\ \theta_N = (\beta'_N, \sigma)' = ((\beta + N^{-1/2}\overline{\beta})', \sigma)' \in \Theta & \\ \text{with } \varphi(\beta_N) = l + N^{-1/2}\gamma\}. & \end{aligned}$$

Theorem 6.1 provides also a simple characterization of the asymptotic bias of the first error for shrinking contamination. Compare also with Rieder (1994), p. 192-194.

Corollary 6.1 *In a linear model let $P_{\theta_N,\delta}$ be contiguous to $P_{\beta,\sigma,\delta}$ for all $\theta_N = (\beta'_N, \sigma)' = ((\beta + N^{-1/2}\overline{\beta})', \sigma)' \in \Theta$ with $\varphi(\beta_N) = l + N^{-1/2}\gamma$. Then for an ALE-test for testing $H_0 : \varphi(\beta) = l$ against $H_1 : \varphi(\beta) \neq l$ with rejection set $R := (\chi^2_{1-\alpha,s,0}, \infty)$ and test statistic τ_N with influence function ψ the following inequalities are equivalent:*

$$\begin{aligned} a)\ & b^1_\epsilon(\tau_N, R) \le 1 - \mathcal{X}^2_{s,b}(\chi^2_{1-\alpha,s,0}) - \alpha, \\ b)\ & \max\{b(\psi, (Q_{N,\delta})_{N\in\mathbb{N}})' C(\psi)^{-1} b(\psi, (Q_{N,\delta})_{N\in\mathbb{N}}); \\ & \qquad (Q_{N,\delta})_{N\in\mathbb{N}} \in \mathcal{U}^0_{c,\epsilon}(P_\delta)\} \le b, \\ c)\ & \epsilon^2 \max_{(z,t)\in\mathbb{R}\times\text{supp}(\delta)} \psi(z,t)'\, C(\psi)^{-1}\, \psi(z,t) \le b. \end{aligned}$$

Proof. The equivalence of $a)$ and $b)$ follows from Theorem 6.1 and the fact that $\mathcal{X}_{s,b}(k)$ is decreasing in b for all $k > 0$. The equivalence of $b)$ and $c)$ follows from the considerations at the end of Section 5.1 by interpreting the function $C(\psi)^{-1/2}\,\psi$ as influence function. □

Note that Corollary 6.1 in particular shows that the asymptotic bias of the first error does not depend on the variance σ. Moreover, Corollary 6.1 shows that the asymptotic bias of a test based on a M-estimator is bounded if the self-standardized gross-error-sensitivity of the M-estimator is bounded (for the definition of the self-standardized gross-error-sensitivity see Krasker and Welsch (1982), Ronchetti and Rousseeuw (1985), Hampel et al. (1986)). This result corresponds with a result of Heritier and Ronchetti (1994) who derived the robustness of tests by assuming Fréchet differentiability of the test statistic.

As for estimators a test is called *asymptotically robust for shrinking contamination* if the asymptotic bias is bounded so that Corollary 6.1 provides the following definition.

Definition 6.8 (Robustness of tests for shrinking contamination) *In a linear model an ALE-test based on an ALE-test statistic with influence function ψ is asymptotically robust for shrinking contamination if*

$$\max_{(z,t)\in I\!R\times \text{supp}(\delta)} \psi(z,t)'\, C(\psi)^{-1}\, \psi(z,t) < \infty.$$

The following theorem is obvious.

Theorem 6.2 *If an AL-estimator for $\varphi(\beta)$ is asymptotically robust for shrinking contamination, then the ALE-test for testing $\varphi(\beta)$ based on this AL-estimator is also asymptotically robust for shrinking contamination.*

Part III

High Robustness and High Efficiency

7

High Robustness and High Efficiency of Estimation

In this chapter we regard weakly asymptotically linear estimators which are robust for shrinking contamination (see Definition 5.4). In Section 7.1 we characterize "most robust" estimators in linear models, which are weakly asymptotically linear estimators with minimum asymptotic bias for shrinking contamination. We also characterize designs which minimize the asymptotic bias of robust estimation. In Section 7.2 we characterize weakly asymptotically linear estimators in linear models which minimize the trace of the asymptotic covariance matrix within all estimators with an asymptotic bias for shrinking contamination, which is bounded by some bias bound b. Also optimal designs for optimal robust estimation are derived. In Section 7.3 we present efficient robust estimators and designs for estimating a nonlinear aspect in a linear model and in Section 7.4 for estimation in a nonlinear model.

7.1 Estimators and Designs with Minimum Asymptotic Bias

In this section and in Section 7.2 we assume that the ideal model is a homoscedastic linear model with normally distributed errors, i.e. the error Z_{nN} at t_{nN} is distributed according to the normal distribution $n_{(0,\sigma^2)}$ with mean 0 and variance $\sigma^2(t_{nN}) = \sigma^2 \in I\!R^+$ for all $n = 1, ...N$, $N \in I\!N$. In particular, we have $P = n_{(0,1)}$. Then an AL-estimator, a weakly asymptotically linear estimator, $\widehat{\varphi}_N$ for $\varphi(\beta) = L\beta$ with influence function ψ satisfies according to Theorem 5.4 for all $\theta = (\beta', \sigma)' \in I\!R \times I\!R^+$

$$\mathcal{L}\left(\sqrt{N}\,(\widehat{\varphi}_N - \varphi(\beta))\Big|\, Q_\theta^N\right)$$

$$\overset{N\to\infty}{\longrightarrow}\ \mathcal{N}\left(b(\psi\,\sigma, (Q_{N,\delta})_{N\in I\!N}), \sigma^2 \int \psi(z,t)\,\psi(z,t)'\,P(dz)\,\delta(dt)\right)$$

with

$$b(\psi\,\sigma, (Q_{N,\delta})_{N\in I\!N}) = \sigma \lim_{N\to\infty} \sqrt{N} \int \psi(z,t)\,Q_{N,\delta}(dz,dt)$$

for all $(Q_{N,\delta})_{N\in I\!N} \in \mathcal{U}^0_{c,\epsilon}(P_\delta)$. The maximum asymptotic bias for shrinking contamination satisfies

$$b(\psi\,\sigma,\delta) = \epsilon\,\sigma\,||\psi||_\delta,$$

where

$$||\psi||_\delta := \max_{(z,t)\in I\!R\times\mathrm{supp}(\delta)} |\psi(z,t)|.$$

Because the asymptotic covariance matrix and the maximum asymptotic bias $b(\psi\,\sigma,\delta)$ depend on the unknown parameter θ only via σ and σ is only a multiplicator we can set without loss of generality $\sigma = 1$, and for the radius ϵ we also can set $\epsilon = 1$. Then the asymptotic covariance matrix as well as the maximum asymptotic bias depend only on the influence function ψ and the design measure δ so that we can try to minimize the asymptotic covariance matrix and the maximum asymptotic bias with respect to ψ and δ. At first we will regard the problem of minimizing the maximum asymptotic bias with respect to ψ and δ.

For minimizing the maximum asymptotic bias with respect to ψ we should regard only those functions ψ which are really influence functions of some AL-estimator for $\varphi(\beta) = L\beta$ in the linear model. In particular they have to satisfy the conditions (5.30) - (5.32), i.e. they have to be an element of

$$\Psi(\delta, L) := \{\psi : I\!R\times\mathcal{T} \to I\!R^s;\ \psi \text{ satisfies conditions (5.30) - (5.32)}\}.$$

Thereby the set $\Psi(\delta, L)$ is not empty if and only if φ is identifiable at δ (see Kurotschka and Müller (1992), Lemma 1). But it is not clear if every ψ lying in $\Psi(\delta, L)$ is the influence function of some AL-estimator. This means that the set $\Psi(\delta, L)$ may be too large. But if we find an optimal solution ψ_* within $\Psi(\delta, L)$ which satisfies the conditions (5.20) and (5.21), then we know from Lemma 5.3 that a one-step M-estimator with this score function ψ_* is an AL-estimator with influence function ψ_* so that ψ_* is really an influence function. We will see that usually this is the case. Hence we define the *AL-estimator with minimum asymptotic bias* as an AL-estimator with an influence function which minimizes the maximum asymptotic bias for shrinking contamination within $\Psi(\delta, L)$ and the *minimum asymptotic bias at* δ as the minimum value of the maximum asymptotic bias within $\Psi(\delta, L)$.

Definition 7.1 (Minimum asymptotic bias for estimation at δ)
a) $b_0^E(\delta, L)$ is the minimum asymptotic bias for estimating φ at δ if

$$b_0^E(\delta, L) = \min\{||\psi||_\delta;\ \psi \in \Psi(\delta, L)\}.$$

b) An AL-estimator $\widehat{\varphi}_N$ for φ at δ with influence function ψ_0 is an AL-estimator for estimating φ at δ with minimum asymptotic bias if

$$\psi_0 \in \arg\min\{||\psi||_\delta;\ \psi \in \Psi(\delta, L)\}.$$

A design which additionally minimizes the maximum asymptotic bias of AL-estimators with minimum bias within a given set Δ of designs is called a *design with minimum asymptotic bias*, and the minimum value of the maximum asymptotic bias within $\Psi(\delta, L)$ and Δ is called the *minimum asymptotic bias within* Δ.

Definition 7.2 (Minimum asymptotic bias for estimation within Δ)
a) $b_0^E(\Delta, L)$ is the minimum asymptotic bias in Δ for estimating φ if

$$b_0^E(\Delta, L) = \min\{||\psi||_\delta;\ \psi \in \Psi(\delta, L) \text{ and } \delta \in \Delta\}.$$

b) A design δ_0 is a design providing the minimum asymptotic bias in Δ for estimating φ if

$$\delta_0 \in \arg\min\{\min\{||\psi||_\delta;\ \psi \in \Psi(\delta, L)\};\ \delta \in \Delta\}.$$

Rieder (1985, Theorem 3.7(a), (1994), Theorem 7.4.13(c)) gave for estimating the whole parameter vector β, i.e. $L = E_r$, a characterization of the minimum asymptotic bias and of influence functions attaining the minimum asymptotic bias. This can be extended for estimating arbitrary linear aspects as follows (see Müller (1987), Kurotschka and Müller (1992)).

Theorem 7.1 *If φ is identifiable at δ, then:*
a) There exists a matrix $Q_1 \in I\!R^{s\times r}$, so that

$$\begin{aligned} b_0^E(\delta, L) &= \frac{tr\,(Q_1\, L')}{\int |Q_1 a(t)|\delta(dt)} \sqrt{\frac{\pi}{2}} \\ &= \max\left\{\frac{tr\,(Q\, L')}{\int |Q a(t)|\delta(dt)} \sqrt{\frac{\pi}{2}};\ Q \in I\!R^{s\times r} \text{ with } QL' \neq 0\right\}. \end{aligned}$$

b) If Q_1 satisfies a), then the influence function ψ of every AL-estimator for φ with minimum asymptotic bias coincides $P \otimes \delta$-a.e. on $I\!R \times \mathcal{T}_1$ with ψ_1 given by

$$\psi_1(z,t) = \frac{Q_1 a(t)}{|Q_1 a(t)|}\, b_0^E(\delta, L)\, sgn(z) 1_{\mathcal{T}_1}(t),$$

where $\mathcal{T}_1 := \{t \in supp(\delta);\ Q_1 a(t) \neq 0\}$.

Proof. See the proof of Lemma 4.4 in Müller (1987) or of Lemma 2 in Kurotschka and Müller (1992). □

Because in general the set $\mathcal{T}_1$ of Theorem 7.1 may have probability $\delta(\mathcal{T}_1) < 1$, the influence function of an AL-estimator with minimum asymptotic bias may not $P \otimes \delta - a.e.$ determined by the function ψ_1 of Theorem 7.1. The following example demonstrates that really there exists several AL-estimators with minimum asymptotic bias and different influence functions.

Example 7.1 (One-way lay-out, continuation of Example 5.2)
In Example 5.2 for estimating $\varphi(\beta) = (\beta_2 - \beta_1, \beta_3 - \beta_1, \beta_4 - \beta_1)'$ in a one-way lay-out model with three levels (level 2,3,4) and a control level (level 1) two influence functions of AL-estimators at the design $\frac{1}{4}(e_1 + e_2 + e_3 + e_4)$ were given. One of these influence function was ψ_0 given by

$$\psi_0(z,t) = \begin{cases} (-1,-1,-1)' \operatorname{sgn}(z)\, 4\, \sqrt{\frac{\pi}{2}} & \text{for } t = 1, \\ (1_2(t), 1_3(t), 1_4(t))' \operatorname{sgn}(z)\, \frac{\min\{|z|, b\, y\}}{y} & \text{for } t \neq 1, \end{cases}$$

with $b = 4\sqrt{\frac{3\pi}{2}}$ and $y = (2\Phi(b\, y) - 1)\frac{1}{4} \approx 0.2409$. In particular ψ_0 satisfies the conditions (5.30) - (5.32) so that it is an element of $\Psi(\delta, L)$. Hence we have

$$b_0^E(\delta, L) \leq ||\psi_0||_\delta = 4\sqrt{\frac{3\,\pi}{2}}.$$

On the other hand for

$$Q = L \begin{pmatrix} 1 & 0 & 0 & 0 \\ 0 & 0 & 0 & 0 \\ 0 & 0 & 0 & 0 \\ 0 & 0 & 0 & 0 \end{pmatrix} = \begin{pmatrix} -1 & 0 & 0 & 0 \\ -1 & 0 & 0 & 0 \\ -1 & 0 & 0 & 0 \end{pmatrix}$$

we get according to Theorem 7.1 a)

$$\begin{aligned} b_0^E(\delta, L) &\geq \frac{\operatorname{tr}\,(Q\, L')}{\int |Qa(t)|\delta(dt)} \sqrt{\frac{\pi}{2}} \\ &= \operatorname{tr} \begin{pmatrix} 1 & 1 & 1 \\ 1 & 1 & 1 \\ 1 & 1 & 1 \end{pmatrix} \left| \frac{1}{4} \begin{pmatrix} -1 \\ -1 \\ -1 \end{pmatrix} \right|^{-1} \sqrt{\frac{\pi}{2}} = 4\sqrt{\frac{3\,\pi}{2}}. \end{aligned}$$

Hence we have $b_0^E(\delta, L) = 4\sqrt{\frac{3\pi}{2}}$ and ψ_0 is the influence function of an AL-estimator with minimum asymptotic bias. But also the function ψ_{01} given by

$$\psi_{01}(z,t) = \begin{cases} (-1,-1,-1)' \operatorname{sgn}(z)\, 4\, \sqrt{\frac{\pi}{2}} & \text{for } t = 1, \\ (1_2(t), 1_3(t), 1_4(t))' \operatorname{sgn}(z)\, 4\, \sqrt{\frac{\pi}{2}} & \text{for } t \neq 1, \end{cases}$$

satisfies

$$||\psi_{01}||_\delta = b_0^E(\delta, L),$$

and according to Lemma A.2 of the appendix the conditions (5.30) - (5.32) are satisfied. Also condition (5.20) is satisfied. Hence ψ_{01} is also the influence function of an AL-estimator with minimum asymptotic bias, and ψ_0 and ψ_{01} do not coincide $P \otimes \delta$-a.e. □

In the Example 7.1 the design δ has a finite support $\text{supp}(\delta) = \{1, 2, 3, 4\}$ and the regressors $a(1), a(2), a(3), a(4)$ on this support are linearly independent. In general if the support of a design is $\text{supp}(\delta) = \{\tau_1, ..., \tau_I\}$ and the regressors $a(\tau_1), ..., a(\tau_I)$ on this support are linearly independent, then the calculation of the minimum asymptotic bias is easy and two different influence functions of AL-estimators with minimum bias can be given. These influence functions are ψ_0 and ψ_{01} given by

$$\psi_{01}(z,t) = L\mathcal{I}(\delta)^- a(t)\,\text{sgn}(z)\sqrt{\frac{\pi}{2}} \qquad \text{for } t \in \mathcal{T} \tag{7.1}$$

and

$$\psi_0(z,t) = \begin{cases} L\mathcal{I}(\delta)^- a(t)\,\text{sgn}(z)\sqrt{\frac{\pi}{2}}, \\ \qquad \text{for } t \in \text{supp}(\delta) \text{ with} \\ \qquad\qquad |L\mathcal{I}(\delta)^- a(t)|\sqrt{\frac{\pi}{2}} = b_0^E(\delta, L), \\ L\mathcal{I}(\delta)^- a(t)\,\text{sgn}(z)\,\frac{\min\{|z|, b_0^E(\delta,L)\,y_0(t)\}}{|L\mathcal{I}(\delta)^- a(t)|\,y_0(t)}, \\ \qquad \text{for } t \in \text{supp}(\delta) \text{ with} \\ \qquad\qquad 0 < |L\mathcal{I}(\delta)^- a(t)|\sqrt{\frac{\pi}{2}} < b_0^E(\delta, L), \\ 0, \qquad \text{for all other } t \in \mathcal{T}, \end{cases} \tag{7.2}$$

where

$$y_0(t) = \frac{2\Phi(b_0^E(\delta, L)\,y_0(t)) - 1}{|L\mathcal{I}(\delta)^- a(t)|} > 0.$$

Note that ψ_{01} is the influence function of the L_1 estimator (see Section 5.2).

Theorem 7.2 *Let be φ identifiable at $\text{supp}(\delta) = \{\tau_1, ..., \tau_I\}$ and $a(\tau_1), ..., a(\tau_I)$ linearly independent. Then:*

a) The minimum asymptotic bias for estimating φ at δ satisfies

$$b_0^E(\delta, L) = \max\{|L\mathcal{I}(\delta)^- a(t)|\sqrt{\frac{\pi}{2}};\ t \in \text{supp}(\delta)\}.$$

b) ψ_{01} given by (7.1) is the influence function of an AL-estimator for estimating φ at δ with minimum asymptotic bias.

c) ψ_0 given by (7.2) is the influence function of an AL-estimator for estimating φ at δ with minimum asymptotic bias.

Proof. See the proof of Satz 6.10 in Müller (1987), or of Theorem 2 in Kurotschka and Müller (1992). Compare also with Theorem 1 in Müller

(1992a). □

The following theorem shows that at an A-optimal design for estimating φ the influence function of an AL-estimator for φ with minimum asymptotic bias is unique. For A-optimality of designs see Section 2.2.

Theorem 7.3 *If δ_A is A-optimal for φ in $\Delta_{\mathcal{D}}$, then:*

a) $b_0^E(\delta_A, L) = \sqrt{tr\,(L\mathcal{I}(\delta_A)^- L')\frac{\pi}{2}}$.

b) ψ_{01} *given by* $\psi_{01}(z,t) = L\mathcal{I}(\delta_A)^- a(t) sgn(z)\sqrt{\frac{\pi}{2}}$ *is the* $P \otimes \delta_A$*-unique influence function of an AL-estimator for estimating* φ *at* δ_A *with minimum asymptotic bias.*

Proof. The assertion follows from the equivalence theorem for A-optimal designs (see Theorem 2.6) and Theorem 7.1 by setting $Q_1 = L\mathcal{I}(\delta)^-$. See also the proof of Satz 5.6 in Müller (1987) or of Lemma 1 in Müller (1994a). □

If δ is A-optimal in $\Delta(\varphi)$, then it is in particular A-optimal in $\Delta_{\mathcal{D}}$ with $\mathcal{D} = \text{supp}(\delta)$. Hence, Theorem 7.3 holds for all A-optimal designs.

Theorem 7.3 in particular shows that for linearly independent regressors, i.e. under the conditions of Theorem 7.2, the influence functions ψ_0 and ψ_{01} coincide if δ is A-optimal in $\Delta_{\text{supp}(\delta)} = \{\overline{\delta} \in \Delta(\varphi);\ \text{supp}(\overline{\delta}) = \text{supp}(\delta)\}$. If additional δ has minimum support for estimating φ, i.e. for all $\mathcal{D} \subset \text{supp}(\delta)$ with $\mathcal{D} \neq \text{supp}(\delta)$ the aspect φ is not identifiable at $\mathcal{D}$, then conversely the coincidence of ψ_0 and ψ_{01} implies the A-optimality of δ in $\Delta_{\text{supp}(\delta)}$.

Theorem 7.4 *If φ is identifiable at supp(δ) and not identifiable at all $\mathcal{D} \subset$ supp(δ) with $\mathcal{D} \neq$ supp(δ), then the influence function of an AL-estimator for estimating φ at δ with minimum asymptotic bias is $P \otimes \delta$-unique if and only if δ is A-optimal for φ in $\Delta_{\text{supp}(\delta)}$.*

Proof. At first note that because δ has minimum support $\mathcal{D} = \text{supp}(\delta) = \{\tau_1, ..., \tau_I\}$ for estimating φ the regressors $a(\tau_1), ..., a(\tau_I)$ are linearly independent. Then we have setting $A_{\mathcal{D}} = (a(\tau_1), ..., a(\tau_I))'$

$$L\mathcal{I}(\delta)^- a(t)\,\delta(\{t\}) = L\,(A_{\mathcal{D}}' A_{\mathcal{D}})^- a(t) \tag{7.3}$$

(see Lemma 2.2) and according to Theorem 7.2 ψ_0 and ψ_{01} are influence functions of AL-estimators with minimum asymptotic bias. If the influence function of an AL-estimator with minimum asymptotic bias is unique, then in particular ψ_0 and ψ_{01} coincide. This means that for all $t \in \text{supp}(\delta)$ we have

$$|L\mathcal{I}(\delta)^- a(t)|\,\sqrt{\tfrac{\pi}{2}} = b_0^E(\delta, L)$$

or

$$L\mathcal{I}(\delta)^- a(t) = 0.$$

Assume there exists a $t_0 \in \text{supp}(\delta)$ with $L\mathcal{I}(\delta)^- a(t_0) = 0$. Then we have according to (7.3)

$$0 = L\mathcal{I}(\delta)^- a(t_0)\,\delta(\{t\}) = L\,(A'_{\mathcal{D}}A_{\mathcal{D}})^- a(t_0).$$

Moreover, Theorem 2.7 provides that the A-optimal design δ_A in $\Delta_{\mathcal{D}}$ is given by

$$\delta_A(\{t\}) = \frac{|L\,(A'_{\mathcal{D}}A_{\mathcal{D}})^- a(t)|}{\sum_{\tau\in\mathcal{D}} |L\,(A'_{\mathcal{D}}A_{\mathcal{D}})^- a(\tau)|}.$$

This implies the contradiction that φ is identifiable at $\text{supp}(\delta_A) \subset \text{supp}(\delta)\setminus\{t_0\}$. Hence, we have $|L\mathcal{I}(\delta)^- a(t)| = \sqrt{\text{tr}\,(L\mathcal{I}(\delta)^- L')}$ for all $t \in \text{supp}(\delta)$ so that according to Theorem 2.6 the design δ is A-optimal in $\Delta_{\text{supp}(\delta)}$. The converse direction of the assertion follows from Theorem 7.3. □

Theorem 7.4 in particular shows that at a design with minimum support and which is not A-optimal in $\Delta_{\text{supp}(\delta)}$ the influence function of an AL-estimator with minimum bias is not unique. This was the case in Example 7.1. But if the design has not a minimum support, then this may be not anymore true. Then only the sufficiency of the A-optimality for the uniqueness of the influence function remains true as Theorem 7.3 showed. Moreover, we also have uniqueness if the design δ is in a modified form A-optimal in a modified model. For that define for a function $h : \mathcal{T} \to \mathbb{R}^+ \setminus \{0\}$

$$\begin{aligned}
\delta_h(dt) &:= \Big(\int h(t)\delta(dt)\Big)^{-1}\, h(t)\,\delta(dt),\\
a_h(t) &:= h(t)^{-1}a(t),\\
L_h &:= \Big(\int h(t)\delta(dt)\Big)^{-1} L,\\
\mathcal{I}_h(\delta_h) &:= \int a_h(t)a_h(t)'\delta_h(dt).
\end{aligned}$$

Theorem 7.5 *Let φ identifiable at δ. If there exists some function $h : \mathcal{T} \to \mathbb{R}^+ \setminus \{0\}$ so that the modified design δ_h is A-optimal for $\varphi_h(\beta) = L_h\beta$ in $\Delta_{\text{supp}(\delta_h)}$ within the modified model $Y_h(t) = a_h(t)'\beta + Z$, then:*

a) $b_0^E(\delta, L) = \sqrt{tr\,(L_h\,\mathcal{I}_h(\delta_h)^-\,L_h')\,\frac{\pi}{2}}$.

b) ψ_0 given by $\psi_0(z,t) = L_h\mathcal{I}_h(\delta_h)^- a_h(t)sgn(z)\sqrt{\frac{\pi}{2}}$ is the $P\otimes\delta$-unique influence function of an AL-estimator for estimating φ at δ with minimum asymptotic bias.

Proof. The assertion follows from Theorem 7.3 (see also the proof of Theorem 2 in Müller (1992b)). □

A special case of Theorem 7.5 appears when

$$\int \frac{a(t)a(t)^T}{|a(t)|}\delta(dt) = k\, E_r,\ L = E_r,$$

where k is a scalar and E_r the identity matrix. Then setting $h(t) = |a(t)|$ Theorem 7.5 provides

$$b_0^E(\delta, L) = \frac{1}{k}\sqrt{\frac{\pi}{2}} = \frac{r}{\int |a(t)|\delta(dt)}\sqrt{\frac{\pi}{2}},$$

and that

$$\psi_0(z,t) = \frac{a(t)}{|a(t)|}\operatorname{sgn}(z)\, b_0^E(\delta, L)$$

is the $P \otimes \delta$-unique influence function of an AL-estimator with minimum bias at δ. In particular, it provides Theorem 2 of Ronchetti and Rousseeuw (1985) (or Proposition 1(i) in Hampel et al. (1986), p. 318).

We have seen that at A-optimal designs the influence function of an AL-estimator with minimum asymptotic bias is unique. Now we are going to show that A-optimal designs are the designs which provide the minimum asymptotic bias within all designs within $\Delta(\varphi)$.

Theorem 7.6 *If δ_A is A-optimal for φ in $\Delta(\varphi)$, then*

$$b_0^E(\Delta(\varphi), L) = b_0^E(\delta_A, L) = \sqrt{tr(L\mathcal{I}(\delta_A)^- L')}\sqrt{\frac{\pi}{2}},$$

i.e. δ_A provides the minimum asymptotic bias in $\Delta(\varphi)$ for estimating φ.

Proof. For every $\delta \in \Delta(\varphi)$ Theorem 7.1 provides for $Q = L\mathcal{I}(\delta)^-$

$$\begin{aligned} b_0^E(\delta, L) &\geq \frac{\operatorname{tr}(Q\,L')}{\int |Qa(t)|\delta(dt)}\sqrt{\frac{\pi}{2}} \\ &\geq \frac{\operatorname{tr}(Q\,L')}{\sqrt{\int |Qa(t)|^2\delta(dt)}}\sqrt{\frac{\pi}{2}} = \sqrt{\operatorname{tr}(L\mathcal{I}(\delta)^- L')}\sqrt{\frac{\pi}{2}} \\ &\geq \sqrt{\operatorname{tr}(L\mathcal{I}(\delta_A)^- L')}\sqrt{\frac{\pi}{2}} \end{aligned}$$

with equality if and only if δ is A-optimal in $\Delta(\varphi)$. For the equality for A-optimal designs see Theorem 7.3. See also Lemma 1 in Müller (1994a) and its proof. □

Example 7.2 (Continuation of Example 7.1)
For estimating $\varphi(\beta) = (\beta_2 - \beta_1, \beta_3 - \beta_1, \beta_4 - \beta_1)'$ in a one-way lay-out model

with three levels (level 2,3,4) and a control level (level 1) the A-optimal design is

$$\delta_A = \frac{1}{\sqrt{3}+3}(\sqrt{3}e_1 + e_2 + e_3 + e_4)$$

(see Example 2.2) so that according to Theorem 7.6 the design δ_A provides the minimum asymptotic bias within $\Delta(\varphi)$. In particular, we have

$$b_0^E(\Delta(\varphi), L) = b_0^E(\delta_A, L) = (\sqrt{3}+3)\sqrt{\frac{\pi}{2}}$$

and $\psi_{01} = \psi_0$ with

$$\psi_{01}(z,t) = \begin{cases} (-1,-1,-1)' \frac{\sqrt{3}+3}{\sqrt{3}} \operatorname{sgn}(z) \sqrt{\frac{\pi}{2}} & \text{for } t = 1, \\ (1_2(t), 1_3(t), 1_4(t))' (\sqrt{3}+3) \operatorname{sgn}(z) \sqrt{\frac{\pi}{2}} & \text{for } t \neq 1, \end{cases}$$

is according to Theorem 7.3 the unique influence function of an AL-estimator for φ with minimum asymptotic bias at δ_A. I.e. the maximum asymptotic bias of such AL-estimator is $b_0^E(\Delta(\varphi), L) = (\sqrt{3}+3)\sqrt{\frac{\pi}{2}}$. Recall from Example 7.1 that at the design $\delta = \frac{1}{4}(e_1 + e_2 + e_3 + e_4)$ the minimum asymptotic bias is $b_0^E(\delta, L) = 4\sqrt{\frac{3\pi}{2}}$. □

7.2 Optimal Estimators and Designs for a Bias Bound

In Section 7.1 it was shown that at an A-optimal design δ_A the AL-estimator with minimum asymptotic bias for shrinking contamination has a unique influence function ψ_0 of the form

$$\psi_0(z,t) = L\mathcal{I}(\delta_A)^- a(t) \operatorname{sgn}(z) \sqrt{\frac{\pi}{2}}.$$

The equivalence theorem for A-optimality (see Theorem 2.6) provides that the trace of the asymptotic covariance matrix of this AL-estimator is

$$\begin{aligned} \operatorname{tr}\left(\int \psi_0(z,t)\psi_0(z,t)' \, P\otimes\delta_A(dz,dt)\right) &= \int |\psi_0(z,t)|^2 \, P\otimes\delta_A(dz,dt) \\ &= \operatorname{tr}(L\mathcal{I}(\delta_A)^- L')\,\frac{\pi}{2}. \end{aligned}$$

In opposite to the AL-estimator with minimum asymptotic bias, the Gauss-Markov estimator is an AL-estimator with influence function

$$\psi_\infty(z,t) = L\mathcal{I}(\delta_A)^- a(t)\, z$$

(see Section 5.2) so that the trace of its asymptotic covariance matrix is

$$\operatorname{tr}\left(\int \psi_\infty(z,t)\psi_\infty(z,t)'\, P\otimes\delta_A(dz,dt)\right) = \operatorname{tr}\left(L\mathcal{I}(\delta_A)^- L'\right).$$

Hence, the Gauss-Markov estimator is $\frac{\pi}{2}$ times more efficient than the AL-estimator with minimum bias. In general asymptotically the Gauss-Markov estimator is the most efficient AL-estimator if no robustness side condition is used (see also below). But the influence function of the Gauss-Markov estimator is unbounded so that its maximum asymptotic bias is equal to infinity. Hence we have a conflict between efficiency and robustness which is here measured by the maximum asymptotic bias. To combine both criteria we can regard constrained problems. Thereby, as in Section 7.1, we assume a homoscedastic linear model, where the ideal distribution P is the standard normal distribution $n_{(0,1)}$. Then as in Section 7.1, we can formulate optimization problems within the set $\Psi(\delta, L)$ and without loss of generality we can set $\sigma = 1$ and $\epsilon = 1$.

One constrained problem is to minimize the asymptotic covariance matrix under the side condition that the maximum asymptotic bias for shrinking contamination is bounded by some bound b. Thereby for the minimization of the covariance matrix we regard the ordering of matrics, where $C_1 \leq C_2$ if and only if $C_2 - C_1$ is positive-semidefinite. As for designs (see Section 2.2) an AL-estimator which solves this minimization problem is called *asymptotically U-optimal for estimation with bias bound b* because it is universal optimal. But this minimization problem can be solved only for special design situations and only within

$$\Psi^*(\delta, L) := \{\psi \in \Psi(\delta, L);\ \psi(z,t) = M\, a(t)\, \rho(z,t) \text{ for all } (z,t) \in I\!R \times \mathcal{T} \text{ for some } \rho : I\!R \times \mathcal{T} \to I\!R\}$$

(see Theorem 7.10). Therefore, solutions of this general minimization problem are defined within $\Psi^*(\delta, L)$. Note that the restriction to $\Psi^*(\delta, L)$ is not grave because usually the influence functions have the form $\psi(z,t) = M\, a(t)\, \rho(z,t)$, where $\rho : I\!R \times \mathcal{T} \to I\!R$.

A more successful constrained problem for estimation is to minimize the trace of the asymptotic covariance matrix under the side condition that the maximum asymptotic bias for shrinking contamination is bounded by some bound b. As for designs which minimize the trace of the covariance matrix an AL-estimator which solves this problem is called *asymptotically A-optimal for estimation with bias bound b*. A design which additionally minimizes the trace of the asymptotic covariance of the asymptotically A-optimal AL-estimators within a class Δ of designs is called an *asymptotically A-optimal design for estimating with bias bound b*.

Definition 7.3 (Optimal AL-estimators and Designs for a bias bound b)
a) An AL-estimator $\widehat{\varphi}_N$ for φ at δ with influence function $\psi_{b,\delta}$ is asymptotically U-optimal for estimating φ at δ with the bias bound b if

$$\psi_{b,\delta} \in \arg\min\{\int \psi\psi'\, d(P\otimes\delta);\ \psi \in \Psi(\delta, L)\ \text{with}\ \|\psi\|_\delta \leq b\}.$$

b) An AL-estimator $\widehat{\varphi}_N$ for φ at δ with influence function $\psi_{b,\delta}$ is asymptotically A-optimal for estimating φ at δ with the bias bound b if

$$\psi_{b,\delta} \in \arg\min\{\mathrm{tr}\left(\int \psi\psi'\, d(P\otimes\delta)\right);\ \psi \in \Psi(\delta, L)\ \text{with}\ \|\psi\|_\delta \leq b\}. \quad (7.4)$$

c) A design δ_b is asymptotically A-optimal in Δ for estimating φ with the bias bound b if

$$\delta_b \in \arg\min\{\min\{\mathrm{tr}\left(\int \psi\psi'\, d(P\otimes\delta)\right);\ \psi \in \Psi(\delta, L)\ \text{with}\ \|\psi\|_\delta \leq b\};\ \delta \in \Delta\}.$$

Note that $\delta_b = \arg\min\{\mathrm{tr}\left(\int \psi_{b,\delta}\, \psi_{b,\delta}'\, d(P\otimes\delta)\right);\ \delta \in \Delta\}$.

At δ an asymptotically optimal estimator for the bias bound b only can exist if the bias bound b is greater or equal to the minimum asymptotic bias, i.e. $b \geq b_0^E(\delta, L)$. If $b = b_0^E(\delta, L)$ and the influence function of the AL-estimator with minimum asymptotic bias is unique, as for AL-estimators at A-optimal designs and designs satisfying the conditions of Theorem 7.5 (see Section 7.1), then of course the AL-estimator with minimum bias is also U-optimal for the bias bound b. But in Section 7.1 it also was shown that not always the influence function of an AL-estimator with minimum asymptotic bias is unique. Hence also for the bias bound $b = b_0^E(\delta, L)$ we should characterize asymptotically optimal AL-estimators. The following theorem which was shown by Kurotschka and Müller (1992) gives a general characterization of the influence function of asymptotically A-optimal AL-estimators for a bias bound $b \geq b_0^E(\delta, L)$. For the case $b > b_0^E(\delta, L)$ it generalizes the results of Hampel (1978), Krasker (1980), Bickel (1981, 1984), Rieder (1985, 1987, 1994, Theorem 7.4.13) concerning the solution of (7.4) for estimating the whole parameter vector β, i.e. for $L = E_r$. Note that for $b > b_0^E(\delta)$ and $L = E_r$ in the literatur the solution of (7.4) is also called *Hampel-Krasker influence function.*

Theorem 7.7 *Let $\widehat{\varphi}_N$ an AL-estimator for φ at δ with influence function ψ.*

a) If $b > b_0^E(\delta, L)$, then $\widehat{\varphi}_N$ is asymptotically A-optimal for estimating φ at δ with the bias bound b if and only if ψ is of the form

$$\psi(z,t) = Q^* a(t)\, sgn(z) \min\left\{|z|, \frac{b}{|Q^* a(t)|}\right\}, \qquad P \otimes \delta - a.e.,$$

where $Q^* \in I\!R^{s\times r}$ is a solution of $Q^* \int a(t)a(t)'\,[2\Phi(\frac{b}{|Q^* a(t)|}) - 1]\delta(dt) = L$.

b) If $b = b_0^E(\delta, L)$, then $\widehat{\varphi}_N$ is asymptotically A-optimal for estimating φ at δ with the bias bound b if and only if ψ is of the form

$$\begin{aligned}\psi(z,t) \;&=\; \sum\nolimits_{m=1}^{M} Q_m a(t)\, \frac{b}{|Q_m a(t)|}\, sgn(z)\, 1_{I\!R\times \mathcal{T}_m} \\ &\quad + Q_M a(t)\, sgn(z) \min\left\{|z|, \frac{b}{|Q_M a(t)|}\right\} 1_{I\!R\times \mathcal{T}_M},\end{aligned}$$

where for $m = 1, ..., M-1$, $Q_m \in I\!R^{s\times r}$ are solutions of

$$Q_m \sqrt{\frac{2}{\pi}} \int_{\mathcal{T}_m} a(t)a(t)'\, \frac{b}{|Q_m a(t)|}\delta(dt) = L_m \neq 0$$

and $Q_M \in I\!R^{s\times r}$ is a solution of $Q_M \int_{\mathcal{T}_M} a(t)a(t)'\,[2\Phi(\frac{b}{|Q_M a(t)|}) - 1]\delta(dt) = L_M$ with $\mathcal{T}_m := \{t \in supp(\delta) \setminus \bigcup_{k=1}^{m-1} \mathcal{T}_k;\; Q_m a(t) \neq 0\}$, $m = 1, ..., M \geq 2$, and $L = \sum_{m=1}^{M} L_m$.

Proof. The assertion a) follows via the Lagrange principle. The assertion b) follows per induction by using Theorem 7.1 and part a). See the proof of part (ii) and (iii) of Theorem 1 in Kurotschka and Müller (1992). □

Note, if $b = \infty$, then we can set $Q^* = L\mathcal{I}(\delta)^-$ in Theorem 7.7 a) so that the Gauss-Markov estimator is asymptotically A-optimal for estimating φ without bias bound because it has the influence function $\psi(z,t) = L\mathcal{I}(\delta)^- a(t)\, z$.

The following theorem, also given in Kurotschka and Müller (1992), shows that an asymptotically A-optimal AL-estimator for a bias bound $b \geq b_0^E(\delta, L)$ always exists and is unique. In particular, it generalizes the existence result of Bickel (1984) and Rieder (1985, 1994, Theorem 7.4.13) for estimating the whole parameter vector β.

Theorem 7.8 *For every $b \geq b_0^E(\delta, L)$, an asymptotically A-optimal AL-estimator for estimating φ at δ with the bias bound b exists and is unique in the following sense: If ψ and ψ^* are influence functions of two asymptotically A-optimal AL-estimators for estimating φ at δ with the bias bound b, then $\psi = \psi^*$ $P\otimes\delta$-a.e.*

Proof. The existence of solutions of (7.4) follows via compactness arguments, and the uniqueness follows from Theorem 7.7. Then the existence and uniqueness of AL-estimators with influence functions solving (7.4) is a consequence of the special form of the optimal influence functions given by Theorem 7.7 and Theorem 5.3, which shows that one-step M-estimators with the optimal influence functions as score funtions are weakly asymptotically linear with these influence functions. See also the proof of part (i) of Theorem 1 in Kurotschka and Müller (1992). □

From Theorem 7.7 more explicit charcterizations of asymptotically A-optimal estimators can be derived. As for the minimum asymptotic bias the explicit characterizations are possible for linearly independent regresssors and for A-optimal designs. Because the ideal distribution P is the standard normal distribution $n_{(0,1)}$ it is also possible to give an explicit representation of the asymptotic covariance matrix. For that define $W : [0,\infty)^3 \rightarrow I\!R$ as

$$W(b,c,y) := c\,[2\Phi(b\,y) - 1] - y$$

and $w(b,c) > 0$ implicitly by $W(b,c,w(b,c)) = 0$. If we also set

$$g(y) := \int \min\{|z|, y\}^2 n_{(0,1)}(dz),$$

then the asymptotic covariance matrix depends mainly on $v_b : [\frac{\pi}{2b^2}, \infty) \rightarrow I\!R$ given by

$$v_b(c) := \begin{cases} \frac{\pi}{2} & \text{for } c = \frac{\pi}{2b^2}, \\ \frac{c\,g(b\,w(b,\sqrt{c}))}{w(b,\sqrt{c})^2} & \text{for } c > \frac{\pi}{2b^2}. \end{cases}$$

At first we give the more explicit representation of the asymptotically A-optimal AL-estimators for linearly independent regressors, i.e. for the case, where $\text{supp}(\delta) = \{\tau_1, ..., \tau_I\}$ and $a(\tau_1), ..., a(\tau_I)$ are linearly independent. This result was given by Müller (1987) (see also Kurotschka and Müller (1992)). For linearly independent regressors the influence function $\psi_{b,\delta}$ of the asymptotically A-optimal AL-estimator for a bias bound $b \geq b_0^E(\delta, L)$ at δ has the following form:

$$\psi_{b,\delta}(z,t) = \begin{cases} L\,\mathcal{I}(\delta)^- a(t)\,\text{sgn}(z)\,\sqrt{\frac{\pi}{2}}, \\ \qquad\qquad \text{for all } t \in \text{supp}(\delta) \text{ with} \\ \qquad\qquad\qquad |L\,\mathcal{I}(\delta)^- a(t)|\sqrt{\frac{\pi}{2}} = b, \\ L\,\mathcal{I}(\delta)^- a(t)\,\text{sgn}(z)\,\frac{\min\{|z|, b\,y(t)\}}{|L\,\mathcal{I}(\delta)^- a(t)|\,y(t)}, \\ \qquad\qquad \text{for all } t \in \text{supp}(\delta) \text{ with} \\ \qquad\qquad\qquad 0 < |L\,\mathcal{I}(\delta)^- a(t)|\sqrt{\frac{\pi}{2}} < b, \\ 0, \qquad\quad \text{for all other } t \in \mathcal{T}, \end{cases} \tag{7.5}$$

where $y(t) = w(b, \frac{1}{|L\mathcal{I}(\delta)^- a(t)|}) > 0$, i.e. $y(t) = (2\Phi(by(t))-1)\frac{1}{|L\mathcal{I}(\delta)^- a(t)|} > 0$. Note that according to Theorem 7.2 for linearly independent regressors we have

$$b_0^E(\delta, L) = \max\{|L\mathcal{I}(\delta)^- a(t)| \sqrt{\frac{\pi}{2}};\ t \in \text{supp}(\delta)\}$$

so that for $b \geq b_0^E(\delta, L)$ the function $\psi_{b,\delta}$ is not equal to zero for all $(z, t) \in \mathbb{R} \times \mathcal{T}$.

Theorem 7.9 *If φ is identifiable at supp$(\delta) = \{\tau_1, ..., \tau_I\}$, $a(\tau_1), ..., a(\tau_I)$ are linearly independent and $b \geq b_0^E(\delta, L)$, then $\psi_{b,\delta}$ given by (7.5) is the influence function of an asymptotically A-optimal AL-estimator for estimating φ at δ with the bias bound b, and the trace of its asymptotic covariance matrix satisfies*

$$\begin{aligned} &tr \int \psi_{b,\delta}(z,t)\, \psi_{b,\delta}(z,t)'\, P \otimes \delta(dz, dt) \\ &= \sum\nolimits_{t \in \text{supp}(\delta)} |L\mathcal{I}(\delta)^- a(t)|^2\, \delta(\{t\})\, v_b\left(\frac{1}{|L\mathcal{I}(\delta)^- a(t)|^2}\right). \end{aligned}$$

Proof. For $b = b_0^E(\delta, L)$ the assertion follows from Theorem 7.7 b) by setting for $M = 2$

$$\begin{aligned} L_1 &= L\mathcal{I}(\delta)^- \int_{\mathcal{T}_1} a(t)\, a(t)'\, \delta(dt), \\ L_2 &= L\mathcal{I}(\delta)^- \int_{\mathcal{T}_2} a(t)\, a(t)'\, \delta(dt) = L - L_1, \\ Q_1 &= L\left(\int_{\mathcal{T}_1} a(t)\, a(t)'\, \delta(dt)\right)^-, \\ Q_2 &= L\left(\int_{\mathcal{S}_2} a(t)\, a(t)'\, [2\Phi(b\, y(t)) - 1]\, \delta(dt)\right)^-, \end{aligned}$$

where

$$\begin{aligned} \mathcal{T}_1 &= \{t \in \text{supp}(\delta);\ |L\mathcal{I}(\delta)^- a(t)| \sqrt{\frac{\pi}{2}} = b\}, \\ \mathcal{T}_2 &= \text{supp}(\delta) \setminus \mathcal{T}_1 \text{ and} \\ \mathcal{S}_2 &= \mathcal{T}_2 \setminus \{t \in \text{supp}(\delta);\ L\mathcal{I}(\delta)^- a(t) = 0\}. \end{aligned}$$

For $b > b_0^E(\delta, L)$ the set $\mathcal{T}_1$ is empty so that with $Q^* = Q_2$ the assertion follows from Theorem 7.7 a). See the proof of Satz 6.10 in Müller (1987) and of Theorem 2 in Kurotschka and Müller (1992). The representation of the trace of the asymptotic covariance matrix is an immediate consequence of the definition of v_b. □

Within $\Psi^*(\delta, L)$ it can be shown for linearly independent regressors that AL-estimators with influence functions $\psi_{b,\delta}$ are not only asymptotically A-optimal but also asymptotically U-optimal for estimating with bias bound b. To show this we need the following lemma which provides the following property of influence functions in $\Psi^*(\delta, L)$: If the regressors on the support $\{\tau_1, ..., \tau_I\}$ of δ are linearly independent, then every $\psi \in \Psi^*(\delta, L)$ has the form

$$\psi(z,t) = L\, M(\rho)^-\, a(t)\, \rho(z,t),$$

where

$$M(\rho) := \int a(t)\, a(t)'\, \rho(z,t)\, z\, P \otimes \delta(dz, dt)$$

and $\rho : \mathbb{R} \times \mathcal{T} \to \mathbb{R}$. Moreover, the asymptotic covariance matrix and the asymptotic bias have a very simple form. Therefor set

$$\begin{aligned}
A_{\mathcal{D}} &:= (a(\tau_1), ..., a(\tau_I))', \\
Q(\rho) &:= \int a(t)\, a(t)'\, \rho(z,t)^2\, P \otimes \delta(dz, dt), \\
D &:= \operatorname{diag}(\delta(\{\tau_1\}), ..., \delta(\{\tau_I\})), \\
D_1(\rho) &:= \operatorname{diag}(\textstyle\int \rho(z, \tau_1)\, z\, P(dz), ..., \int \rho(z, \tau_I)\, z\, P(dz)), \\
D_2(\rho) &:= \operatorname{diag}(\textstyle\int \rho(z, \tau_1)^2\, P(dz), ..., \int \rho(z, \tau_I)^2\, P(dz)).
\end{aligned}$$

Note also, if $\psi \in \Psi^*(\delta, L)$ exists, then φ is identifiable at δ so that $L = K\, A_{\mathcal{D}}$ for some $K \in \mathbb{R}^{s \times I}$ (see Section 7.1 and Lemma 1.2).

Lemma 7.1 *If $supp(\delta) = \{\tau_1, ..., \tau_I\}$, $a(\tau_1), ..., a(\tau_I)$ are linearly independent and $\psi \in \Psi^*(\delta, L)$, then $\psi(z,t) = L\, M(\rho)^-\, a(t)\, \rho(z,t)$ for all $(z,t) \in \mathbb{R} \times supp(\delta)$, where $\rho : \mathbb{R} \times \mathcal{T} \to \mathbb{R}$, and in particular*

$$\int \psi\, \psi'\, d(P \otimes \delta) = K\, D_1(\rho)^{-1}\, D_2(\rho)\, D_1(\rho)^{-1}\, D^{-1}\, K'$$

and

$$|\psi(z, \tau_i)| = |K\, u_i| \left| \frac{\rho(z, \tau_i)}{\int \rho(y, \tau_i)\, y\, P(dy)} \right| \frac{1}{\delta(\{\tau_i\})}$$

for all $z \in \mathbb{R}$ and $i = 1, ..., I$, where $L = K\, A_{\mathcal{D}}$ and u_i is the ith unit vector in $\mathbb{R}^I$.

Proof. If $\psi \in \Psi^*(\delta, L)$, then $\psi(z,t) = M\, a(t)\, \rho(z,t)$ and ψ satisfies in particular condition (5.31), which implies

$$\begin{aligned}
L &= \int \psi(z,t)\, a(t)'\, z\, P \otimes \delta(dz, dt) \\
&= M \int a(t)\, a(t)'\, \rho(z,t)\, z\, P \otimes \delta(dz, dt) \\
&= M\, M(\rho) = M\, A'_{\mathcal{D}}\, D_1(\rho)\, D\, A_{\mathcal{D}}.
\end{aligned}$$

Then we have with Lemma 2.1

$$\begin{aligned} L\,M(\rho)^-\,A'_{\mathcal{D}} &= M\,A'_{\mathcal{D}}\,D_1(\rho)\,D\,A_{\mathcal{D}}(A'_{\mathcal{D}}\,D_1(\rho)\,D\,A_{\mathcal{D}})^-\,A'_{\mathcal{D}} \\ &= M\,A'_{\mathcal{D}}\,D_1(\rho)\,D\,D_1(\rho)^{-1}\,D^{-1} = M\,A'_{\mathcal{D}} \end{aligned}$$

so that $M\,a(t) = L\,M(\rho)^-\,a(t)$ for all $t \in \text{supp}(\delta)$. Then Lemma 2.1 also provides

$$\begin{aligned} \int \psi\,\psi'\,d(P\otimes\delta) &= L\,M(\rho)^{-1}\int a(t)\,a(t)'\,\rho(z,t)^2\,P\otimes\delta(dz,dt)\,M(\rho)^-\,L' \\ &= K\,A_{\mathcal{D}}\,[A'_{\mathcal{D}}\,D_1(\rho)\,D\,A_{\mathcal{D}}]^-\,A'_{\mathcal{D}}\,D_2(\rho)\,D\,A_{\mathcal{D}}\,[A'_{\mathcal{D}}\,D_1(\rho)\,D\,A_{\mathcal{D}}]^-\,A'_{\mathcal{D}}\,K' \\ &= K\,D_1(\rho)^{-1}\,D^{-1}\,D_2(\rho)\,D\,D_1(\rho)^{-1}\,D^{-1}\,K' \\ &= K\,D_1(\rho)^{-1}\,D_2(\rho)\,D_1(\rho)^{-1}\,D^{-1}\,K' \end{aligned}$$

and

$$\begin{aligned} M\,a(\tau_i) &= L\,M(\rho)^-\,A'_{\mathcal{D}}\,u_i \\ &= K\,A_{\mathcal{D}}(A'_{\mathcal{D}}\,D_1(\rho)\,D\,A_{\mathcal{D}})^-\,A'_{\mathcal{D}}\,u_i = K\,D_1(\rho)^{-1}\,D^{-1}\,u_i \end{aligned}$$

so that the assertions follow. □

Theorem 7.10 *If φ is identifiable at supp$(\delta) = \{\tau_1, ..., \tau_I\}$, $a(\tau_1), ..., a(\tau_I)$ are linearly independent and $b \geq b_0^E(\delta, L)$, then $\psi_{b,\delta}$ given by (7.5) is the influence function of an asymptotically U-optimal AL-estimator for estimating φ at δ with the bias bound b.*

Proof. For any $\psi \in \Psi^*(\delta, L)$ we have according to Lemma 7.1 for some $\rho : I\!R{\times}T \to I\!R$ that $\psi(z,t) = L\,M(\rho)^-\,a(t)\,\rho(z,t)$ for all $(z,t) \in I\!R{\times}\text{supp}(\delta)$. Moreover, Lemma 7.1 and Lemma 2.1 provide

$$\begin{aligned} b &\geq \max\nolimits_{z\in I\!R} |\psi(z,\tau_i)| \\ &= \max\nolimits_{z\in I\!R} |K\,u_i| \left|\frac{\rho(z,\tau_i)}{\int \rho(y,\tau_i)\,y\,P(dy)}\right| \frac{1}{\delta(\{\tau_i\})} \\ &= |L\,\mathcal{I}(\delta)^-\,a(\tau_i)|\,\max\nolimits_{z\in I\!R}\left|\frac{\rho(z,\tau_i)}{\int \rho(y,\tau_i)\,y\,P(dy)}\right| \end{aligned}$$

and

$$\frac{\rho(\cdot,\tau_i)}{\int \rho(z,\tau_i)\,z\,P(dz)} \in \overline{\Psi}(1)$$

for $i = 1, ..., I$, where $\overline{\Psi}(1) := \overline{\Psi}(\delta, 1)$ is defined for the one-dimensional model given by $\overline{Y} = \overline{\beta} + Z$ with $\overline{\beta} \in I\!R$, $\overline{a}(t) = 1$ for all $t \in T$ and $L = 1$. In the definition of $\overline{\Psi}(1)$ we have dropped the design δ because it is not important. Set

$$\rho_{b,\delta}(z,t) = \begin{cases} \operatorname{sgn}(z)\sqrt{\frac{\pi}{2}}, & \\ \qquad \text{for all } t \in \operatorname{supp}(\delta) \text{ with} & \\ \qquad\qquad |L\mathcal{I}(\delta)^{-}a(t)|\sqrt{\frac{\pi}{2}} = b, & \\ \operatorname{sgn}(z)\frac{\min\{|z|, b\, y(t)\}}{|L\mathcal{I}(\delta)^{-}a(t)|\, y(t)}, & \\ \qquad \text{for all } t \in \operatorname{supp}(\delta) \text{ with} & \\ \qquad\qquad 0 < |L\mathcal{I}(\delta)^{-}a(t)|\sqrt{\frac{\pi}{2}} < b, & \\ 0, \qquad \text{for all other } t \in \mathcal{T}. & \end{cases}$$

Because $\int \rho_{b,\delta}(z,\tau_i)\, z\, P(dz) = 1$ for $i = 1, ..., I$ we have $M(\rho_{b,\delta}) = \mathcal{I}(\delta)$ so that $\psi_{b,\delta}(z,t) = L\, M(\rho_{b,\delta})^{-}\, a(t)\, \rho_{b,\delta}(z,t)$ for all $(z,t) \in \mathbb{R} \times \operatorname{supp}(\delta)$. Moreover, Theorem 7.9 applied to the one-dimensional model provides

$$\int \rho_{b,\delta}(z,\tau_i)^2\, P(dz) = \frac{\int \rho_{b,\delta}(z,\tau_i)^2\, P(dz)}{\left(\int \rho_{b,\delta}(z,\tau_i)\, z\, P(dz)\right)^2}$$

$$= \min\left\{\int \psi(z)^2\, P(dz);\ \psi \in \overline{\Psi}(1) \text{ with } |\psi| \leq \frac{b}{|L\mathcal{I}(\delta)^{-}a(\tau_i)|}\right\},$$

and therefore

$$\frac{\int \rho_{b,\delta}(z,\tau_i)^2\, P(dz)}{\left(\int \rho_{b,\delta}(z,\tau_i)\, z\, P(dz)\right)^2} \leq \frac{\int \rho(z,\tau_i)^2\, P(dz)}{\left(\int \rho(z,\tau_i)\, z\, P(dz)\right)^2}$$

for all $i = 1, ..., I$. This implies

$$D_1(\rho_{b,\delta})^{-1}\, D_2(\rho_{b,\delta})\, D_1(\rho_{b,\delta})^{-1} \leq D_1(\rho)^{-1}\, D_2(\rho)\, D_1(\rho)^{-1}$$

so that according to Lemma 7.1 we have

$$\int \psi_{b,\delta}\, \psi_{b,\delta}'\, d(P \otimes \delta) \leq \int \psi\, \psi'\, d(P \otimes \delta)$$

for all $\psi \in \Psi^*(\delta, L)$. □

If δ_A is A-optimal in $\Delta_{\operatorname{supp}(\delta_A)}$, where $\operatorname{supp}(\delta) = \{\tau_1, ..., \tau_I\}$ and $a(\tau_1), ..., a(\tau_I)$ are linearly independent, then the equivalence theorem for A-optimality, Theorem 2.6, provides at once with Theorem 7.9 that an AL-estimator with influence function ψ_{b,δ_A} given by

$$\psi_{b,\delta_A}(z,t) = \begin{cases} L\mathcal{I}(\delta_A)^{-}a(t)\operatorname{sgn}(z)\sqrt{\frac{\pi}{2}}, & \\ \qquad \text{for } \sqrt{\operatorname{tr}(L\mathcal{I}(\delta_A)^{-}L')\frac{\pi}{2}} = b, & \\ L\mathcal{I}(\delta_A)^{-}a(t)\operatorname{sgn}(z)\frac{\min\{|z|, b\, y_b\}}{y_b\sqrt{\operatorname{tr}(L\mathcal{I}(\delta_A)^{-}L')}}, & \\ \qquad \text{for } \sqrt{\operatorname{tr}(L\mathcal{I}(\delta_A)^{-}L')\frac{\pi}{2}} < b, & \end{cases} \tag{7.6}$$

where $y_b = w(b, \frac{1}{\sqrt{\text{tr}\,(L\,\mathcal{I}(\delta_A)^- L')}})$, i.e. $y_b = (2\Phi(b\,y_b) - 1)\frac{1}{\sqrt{\text{tr}\,(L\,\mathcal{I}(\delta_A)^- L')}} > 0$, is asymptotically A-optimal for a bias bound $b \geq b_0^E(\delta_A, L)$. But this holds not only for linearly independent regressors on supp(δ) but also for every A-optimal design in $\Delta_\mathcal{D}$ with $\mathcal{D} \subset \mathcal{T}$, for which the equivalence theorem for A-optimality is satisfied as Müller (1987, 1994a) has shown.

Theorem 7.11 *If φ is identifiable at $\mathcal{D}$, δ_A is A-optimal for φ in $\Delta_\mathcal{D}$ and $b \geq b_0^E(\delta_A, L)$, then ψ_{b,δ_A} given by (7.6) is the influence function of an asymptotically A-optimal AL-estimator for estimating φ at δ with the bias bound b, and the trace of its asymptotic covariance matrix satisfies*

$$\begin{aligned}&\text{tr}\left(\int \psi_{b,\delta_A}(z,t)\,\psi_{b,\delta_A}(z,t)'\,P\otimes\delta_A(dz,dt)\right)\\&= \text{tr}\,(L\,\mathcal{I}(\delta_A)^- L')\,v_b\left(\frac{1}{\text{tr}\,(L\,\mathcal{I}(\delta_A)^- L')}\right).\end{aligned}$$

Proof. The assertion follows from Theorem 2.6 and Theorem 7.7 by setting $Q_1 = L\,\mathcal{I}(\delta)^-$ and $Q_2 = 0_{s\times r}$ with $M = 2$ for $b = b_0^E(\delta_A, L)$ and $Q^* = L\,\mathcal{I}(\delta)^- \frac{1}{y_b\sqrt{\text{tr}\,(L\,\mathcal{I}(\delta)^- L')}}$ for $b > b_0^E(\delta_A, L)$. See Müller (1987, 1994a).
□

Note that for designs δ with linearly independent regressors as well as for A-optimal designs we have

$$\lim_{b\to\infty} \text{tr}\left(\int \psi_{b,\delta}\,\psi_{b,\delta}'\,d(P\otimes\delta)\right) = \text{tr}\,(L\,\mathcal{I}(\delta)^- L')$$

because of

$$\lim_{b\to\infty} v_b(c^2) = 1$$

for all $c \in (0,\infty)$. For A-optimal designs δ_A we also have

$$\lim_{b\downarrow b_0^E(\delta_A,L)} \text{tr}\left(\int \psi_{b,\delta_A}\,\psi_{b,\delta_A}'\,d(P\otimes\delta_A)\right) = \text{tr}\,(L\,\mathcal{I}(\delta_A)^- L')\,\frac{\pi}{2}$$

because of

$$\lim_{b\downarrow\frac{1}{c}\sqrt{\frac{\pi}{2}}} v_b(c^2) = \frac{\pi}{2}$$

for all $c \in (0,\infty)$. See Lemma A.5 in the appendix (or Müller (1994a)).

Moreover $s_b(c) := \frac{1}{c}v_b(c^2)$ is a strictly decreasing and strictly convex function of c (see Lemma A.6 or Müller (1994a)). The monotony and convexity of s_b at once provide that A-optimal designs are also asymptotically A-optimal in $\Delta_\mathcal{D}$ for estimating φ with a bias bound $b \geq b_0^E(\Delta_\mathcal{D}, L)$ if $\mathcal{D} = \{\tau_1, ..., \tau_I\}$ and $a(\tau_1), ..., a(\tau_I)$ are linearly independent. But this results holds also more generally as was shown in Müller (1987, 1994a) and that is shown in the following theorem.

Theorem 7.12 *Let*
a) $\Delta = \{\delta \in \Delta(\varphi);\ supp(\delta)$ *is finite* $\}$ *and* $b \geq b_0^E(\Delta)$*, or*
b) $\Delta = \Delta(\varphi)$*,* $a : \mathcal{D} \to I\!\!R^r$ *continuous,* $b = b_0^E(\Delta)$*, or*
c) $\Delta = \Delta(\varphi)$*,* $a : \mathcal{T} \to I\!\!R^r$ *continuous,* $\mathcal{T}$ *compact and* $b \geq b_0^E(\Delta)$*.*
If Δ *includes an A-optimal design for* φ*, then* δ_A *is A-optimal for* φ *in* Δ *if and only if* δ_A *is asymptotically A-optimal in* Δ *for estimating* φ *with the bias bound* b*.*

Proof. For the proof see also the proofs of Satz 9.9 in Müller (1987) and the main theorem in Müller (1994a).

a) Regard any $\delta \in \Delta(\varphi)$ with $\text{supp}(\delta) = \{\tau_1, \ldots, \tau_I\}$. Let $\psi_{b,\delta}$ the influence function of an asymptotically A-optimal AL-estimator for estimating φ at δ with bias bound b. If $a(\tau_1), \ldots, a(\tau_I)$ are not linearly independent, then we can extend the regressors by $\tilde{a}(t)$ so that $\overline{a}(\tau_1), \ldots, \overline{a}(\tau_I)$ are linearly independent, where $\overline{a}(t) = (a(t)', \tilde{a}(t)')'$. Setting

$$\tilde{L} := \int \psi_{b,\delta}(z,t)\, \tilde{a}(t)' z\, P(dz)\, \delta(dt)$$

we have $\psi_{b,\delta} \in \overline{\Psi}(\delta, \overline{L})$ for $\overline{L} = (L, \tilde{L})$, where $\overline{\Psi}(\delta, \overline{L})$ is defined for the extended model given by $\overline{Y} = \overline{a}(t)'\overline{\beta} + Z$ with $\overline{\beta} = (\beta', \tilde{\beta}')'$. Note that the minimum asymptotic bias for estimating $\overline{\varphi}(\overline{\beta}) = (L, \tilde{L})(\beta', \tilde{\beta}')'$ in the extended model is less or equal to b because $\psi_{b,\delta} \in \overline{\Psi}(\delta, \overline{L})$ and $||\psi_{b,\delta}||_\delta = b$. Hence, according to Theorem 7.9 in the extended model there exists an asymptotically A-optimal AL-estimator for estimating $\overline{\varphi}$ at δ with bias bound b and its influence function $\overline{\psi}_{b,\delta}$ is given by (7.5). Denote by $\overline{\mathcal{I}}(\delta) := \int \overline{a}(t)\overline{a}(t)'\delta(dt)$ the information matrix in the extended model. Because $s_b(c) := \frac{1}{c} v_b(c^2)$ is strictly convex (see Lemma A.6 in the appendix) Theorem 7.9 provides

$$\begin{aligned}
&\text{tr} \int \psi_{b,\delta}(z,t)\, \psi_{b,\delta}(z,t)'\, P \otimes \delta(dz, dt) \\
&\geq \text{tr} \int \overline{\psi}_{b,\delta}(z,t)\, \overline{\psi}_{b,\delta}(z,t)'\, P \otimes \delta(dz, dt) \\
&= \sum\nolimits_{t \in \text{supp}(\delta)} |\overline{L}\,\overline{\mathcal{I}}(\delta)^- \overline{a}(t)|^2\, \delta(\{t\})\, v_b\left(\frac{1}{|\overline{L}\,\overline{\mathcal{I}}(\delta)^- \overline{a}(t)|^2}\right) \\
&= \sum\nolimits_{t \in \text{supp}(\delta)} |\overline{L}\,\overline{\mathcal{I}}(\delta)^- \overline{a}(t)|\, \delta(\{t\})\, s_b\left(\frac{1}{|\overline{L}\,\overline{\mathcal{I}}(\delta)^- \overline{a}(t)|}\right) \\
&= \left(\sum\nolimits_{\tau \in \text{supp}(\delta)} |\overline{L}\,\overline{\mathcal{I}}(\delta)^- \overline{a}(\tau)|\, \delta(\{\tau\})\right) \\
&\quad \cdot \sum\nolimits_{t \in \text{supp}(\delta)} \frac{|\overline{L}\,\overline{\mathcal{I}}(\delta)^- \overline{a}(t)|\, \delta(\{t\})}{\sum_{\tau \in \text{supp}(\delta)} |\overline{L}\,\overline{\mathcal{I}}(\delta)^- \overline{a}(\tau)|\, \delta(\{\tau\})}\, s_b\left(\frac{1}{|\overline{L}\,\overline{\mathcal{I}}(\delta)^- \overline{a}(t)|}\right)
\end{aligned}$$

$$\geq \left(\sum\nolimits_{\tau\in\text{supp}(\delta)} |\overline{L}\overline{\mathcal{I}}(\delta)^- \overline{a}(\tau)|\, \delta(\{\tau\})\right)$$
$$\cdot\, s_b \left(\frac{1}{\sum_{\tau\in\text{supp}(\delta)} |\overline{L}\overline{\mathcal{I}}(\delta)^- \overline{a}(\tau)|\, \delta(\{\tau\})}\right).$$

Let $\overline{\delta}_A$ be the A-optimal design for $\overline{\varphi}$ in the extended model. Then setting $\overline{A}_{\mathcal{D}} := (\overline{a}(\tau_1), \ldots, \overline{a}(\tau_I))'$ Lemma 2.2 and Theorem 2.7 provide

$$\left(\sum\nolimits_{\tau\in\text{supp}(\delta)} |\overline{L}\overline{\mathcal{I}}(\delta)^- \overline{a}(\tau)|\, \delta(\{\tau\})\right)$$
$$\cdot\, s_b \left(\frac{1}{\sum_{\tau\in\text{supp}(\delta)} |\overline{L}\overline{\mathcal{I}}(\delta)^- \overline{a}(\tau)|\, \delta(\{\tau\})}\right)$$
$$= \left(\sum\nolimits_{\tau\in\text{supp}(\delta)} |\overline{L}\,(\overline{A}_{\mathcal{D}}'\overline{A}_{\mathcal{D}})^- \overline{a}(\tau)|\right)$$
$$\cdot\, s_b \left(\frac{1}{\sum_{\tau\in\text{supp}(\delta)} |\overline{L}\,(\overline{A}_{\mathcal{D}}'\overline{A}_{\mathcal{D}})^- \overline{a}(\tau)|}\right)$$
$$= \sqrt{\text{tr}(\overline{L}\overline{\mathcal{I}}(\overline{\delta}_A)^- \overline{L}')} \cdot s_b \left(\frac{1}{\sqrt{\text{tr}(\overline{L}\overline{\mathcal{I}}(\overline{\delta}_A)^- \overline{L}')}}\right).$$

Because s_b is strictly decreasing (see Lemma A.6) and $\overline{L}\overline{\mathcal{I}}(\overline{\delta})^- \overline{L}' \geq L\mathcal{I}(\overline{\delta})^- L'$ for all $\overline{\delta} \in \Delta(\overline{\varphi})$ (see Lemma A.1) we obtain

$$\sqrt{\text{tr}(\overline{L}\overline{\mathcal{I}}(\overline{\delta}_A)^- \overline{L}')} \cdot s_b \left(\frac{1}{\sqrt{\text{tr}(\overline{L}\overline{\mathcal{I}}(\overline{\delta}_A)^- \overline{L}')}}\right)$$
$$\geq \sqrt{\text{tr}(L\mathcal{I}(\overline{\delta}_A)^- L')} \cdot s_b \left(\frac{1}{\sqrt{\text{tr}(L\mathcal{I}(\overline{\delta}_A)^- L')}}\right)$$
$$\geq \sqrt{\text{tr}(L\mathcal{I}(\delta_A)^- L')} \cdot s_b \left(\frac{1}{\sqrt{\text{tr}(L\mathcal{I}(\delta_A)^- L')}}\right)$$
$$= \text{tr}(L\mathcal{I}(\delta_A)^- L') \cdot v_b \left(\frac{1}{\text{tr}(L\mathcal{I}(\delta_A)^- L')}\right), \tag{7.7}$$

where δ_A is the A-optimal design for φ in the original model. According to Theorem 7.11 the lower bound (7.7) is attained by asymptotically A-optimal AL-estimators for estimating φ at the A-optimal design δ_A with the bias bound b. Hence, every A-optimal design for φ is also asymptotically A-optimal for estimating φ with bias bound b. But the above proof also shows that the equality

$$\begin{aligned}\mathrm{tr} \int & \psi_{b,\delta}(z,t)\, \psi_{b,\delta}(z,t)'\, P \otimes \delta(dz,dt) \\ &= \mathrm{tr}(L\mathcal{I}(\delta_A)^- L') \cdot v_b \left(\frac{1}{\mathrm{tr}(L\mathcal{I}(\delta_A)^- L')} \right)\end{aligned}$$

holds if and only if δ is A-optimal for φ.

b) The assertion follows at once from Theorem 7.3 and Theorem 7.6.

c) If the support $\mathrm{supp}(\delta)$ of a given design δ is not finite, then according to the theorem of Carathéodory we can proceed to a design with finite support. Then we can apply the proof of a). See in particular Müller (1994a). □

Theorem 7.12 in particular shows that the asymptotically A-optimal design does not depend on the bias bound b. But the asymptotically A-optimal estimators at the asymptotically A-optimal designs depend on the bias bound b as the asymptotically A-optimal estimators at other designs. Hence we need some rules for choosing the bias bound b. One possibility of choosing b is to choose b such that the maximum asymptotic mean squared error is minimized. The asymptotic mean squared error is given by the bias and the trace of the covariance matrix of the asymptotic normal distribution. According to Theorem 5.4 this is

$$|b(\psi, (Q_{N,\delta})_{N\in\mathbb{N}})|^2 + \mathrm{tr}\left(\int \psi\, \psi'\, d(P \otimes \delta) \right).$$

Then the maximum asymptotic mean squared error is the maximum mean squared error within the shrinking contamination neighbourhood, i.e.

$$\begin{aligned} M(\psi,\delta) := \max\{ |b(\psi, (Q_{N,\delta})_{N\in\mathbb{N}})|^2 &+ \mathrm{tr}\left(\int \psi\, \psi'\, d(P\otimes\delta) \right); \\ &(Q_{N,\delta})_{N\in\mathbb{N}} \in \mathcal{U}^0_{c,\epsilon}(P\otimes\delta)\} \\ = \|\psi\|_\delta^2 + \mathrm{tr}\left(\int \psi\, \psi'\, d(P\otimes\delta)\right). \end{aligned}$$

An AL-estimator which minimizes at δ the maximum asymptotic mean squared error is called a *mean squared error minimizing AL-estimator* or briefly a *MSE minimizing AL-estimator*. A design which additionally minimizes the mean squared error of MSE minimizing AL-estimators within a class Δ of designs is called a *mean squared error minimizing design* or briefly a *MSE minimizing design*. For the definition of MSE minimizing estimators see also Samarov (1985) and Rieder (1994), p. 291.

Definition 7.4 (MSE minimizing estimators and designs)
a) An AL-estimator $\widehat{\varphi}_N$ for φ at δ with influence function ψ_δ is asymptotically mean squared error minimizing (briefly MSE minimizing) at δ if

$$\psi_\delta \in \arg\min\{M(\psi,\delta);\ \psi \in \Psi(\delta)\}.$$

b) A design δ_* is asymptotically mean squared error minimizing (briefly MSE minimizing) for robust estimation of φ in Δ if

$$\delta_* \quad \in \quad \arg\min\{\min\{M(\psi,\delta);\ \psi \in \Psi(\delta)\};\ \delta \in \Delta\}.$$

Note that $\delta_* = \arg\min\{M(\psi_\delta,\delta);\ \delta \in \Delta\}$.

For designs with linearly independent regressors or which are A-optimal the MSE minimizing AL-estimators have an easy characterization. In particular, these AL-estimators have an influence function of the form $\psi_{b_\delta,\delta}$, where $\psi_{b,\delta}$ is defined by (7.5). Moreover the A-optimal designs are also MSE minimizing. See Müller (1987, 1994c,d).

Theorem 7.13
a) Let φ be identifiable at δ and δ be A-optimal in $\Delta_{\text{supp}(\delta)}$ or $supp(\delta) = \{\tau_1, ..., \tau_I\}$, where $a(\tau_1), ..., a(\tau_I)$ are linearly independent. Then an AL-estimator $\widehat{\varphi}_N$ for φ at δ with influence function ψ_δ is MSE minimizing at δ if and only if $\psi_\delta = \psi_{b_\delta,\delta}$ and b_δ is a solution of

$$b_\delta = \arg\min\{b^2 + tr\left(\int \psi_{b,\delta}\, \psi_{b,\delta}'\, d(P\otimes\delta)\right);\ b \geq b_0^E(\delta, L)\}. \tag{7.8}$$

b) Under the conditions of Theorem 7.12 a design δ_A is A-optimal in Δ if and only if δ_A is MSE minimizing in Δ.

Proof. The assertion a) follows from the fact that $||\psi_{b,\delta}||_\delta = b$ and that $\psi_{b,\delta}$ is the unique solution of (7.4) (see Theorem 7.8, Theorem 7.9 and Theorem 7.11). The assertion b) follows from Theorem 7.12. See also Müller (1987, 1994c,d). □

Because $\text{tr}\left(\int \psi_{b,\delta}\, \psi_{b,\delta}'\, d(P\otimes\delta)\right)$ bases on $v_b(c^2)$ (see Theorem 7.9 and Theorem 7.11) and $r_c(b) := v_b(c^2)$ is a convex function in b, where its derivative with respect to b has a simple form (see Lemma A.6 and its proof), a solution of (7.8) can be easily calculated by Newton's method. See Müller (1987, 1994c,d). Moreover, under the assumptions of Theorem 7.13 the solution b_δ of (7.8) always satisfies $b_\delta > b_0^E(\delta, L)$ so that according to (7.5) and (7.6) the influence function ψ_δ is bounded and continuous. Hence, a M-estimator for β with this score function ψ_δ is not only MSE minimizing but also strongly asymptotically robust, because it is Fréchet differentiable (see Theorem 3.2 and the definition of strong asymptotic robustness in Section 3.2).

Corollary 7.1 *Under the assumptions of Theorem 7.13 any M-estimator for β at δ with score function $\psi_\delta = \psi_{b_\delta,\delta}$ is MSE minimizing and strongly asymptotically robust.*

Example 7.3 (One-way lay-out, continuation of Example 7.2)
Example 7.1 provided that for estimating $\varphi(\beta) = (\beta_2 - \beta_1, \beta_3 - \beta_1, \beta_4 - \beta_1)'$ at $\delta = \frac{1}{4}(e_1 + e_2 + e_3 + e_4)$ in a one-way lay-out model with 3 levels (levels 2,3,4) and a control level (level 1) the minimum asymptotic bias is $b_0^E(\delta, L) = 4\sqrt{\frac{3\pi}{2}}$. The influence function of an AL-estimator, which is asymptotically A-optimal as well as asymptotically U-optimal for estimating φ at δ with bias bound b, is given by Theorem 7.9 and Theorem 7.10. For $b = 4\sqrt{\frac{3\pi}{2}} \approx 8.6832$ the influence function coincides with ψ_0 given in Example 7.1, i.e.

$$\begin{aligned}\psi_{b,\delta}(z,t) &= \psi_0(z,t)\\ &= \begin{cases} (-1,-1,-1)'\, 4\,\mathrm{sgn}(z)\, \sqrt{\frac{\pi}{2}}, & \text{for } t = 1,\\ (1_{\{2\}}(t), 1_{\{3\}}(t), 1_{\{4\}}(t))'\,\mathrm{sgn}(z)\, \frac{\min\{|z|, b\,y\}}{y}, & \text{for } t \neq 1,\end{cases}\end{aligned}$$

with $y = (2\Phi(b\,y) - 1)\frac{1}{4} \approx 0.2409$, and for $b > 4\sqrt{\frac{3\pi}{2}} \approx 8.6832$ it has the form

$$\psi_{b,\delta}(z,t) = \begin{cases} (-1,-1,-1)'\,\mathrm{sgn}(z)\, \frac{\min\{|z|, b\,w_b\}}{w_b\sqrt{3}}, & \text{for } t = 1,\\ (1_{\{2\}}(t), 1_{\{3\}}(t), 1_{\{4\}}(t))'\,\mathrm{sgn}(z)\, \frac{\min\{|z|, b\,v_b\}}{v_b}, & \text{for } t \neq 1,\end{cases}$$

with $w_b = (2\Phi(b\,w_b) - 1)\frac{1}{4\sqrt{3}} > 0$ and $v_b = (2\Phi(b\,v_b) - 1)\frac{1}{4} > 0$. The MSE minimizing AL-estimator is an asymptotically A-optimal AL-estimator for the bias bound $b_\delta \approx 8.7213$. For this bias bound we have $w_{b_\delta} \approx 0.0186$ and $v_{b_\delta} \approx 0.2411$, and the maximum asymptotic mean squared error is $M(\psi_{b_\delta,\delta}, \delta) \approx 105.53$ (see Müller (1994c)).

The A-optimal design for estimating φ is $\delta_A = \frac{1}{\sqrt{3}+3}(\sqrt{3}e_1 + e_2 + e_3 + e_4)$ (see Example 2.2). In Example 7.2 it was shown that the minimum asymptotic bias in $\Delta(\varphi)$ is $b_0^E(\Delta(\varphi), L) = b_0^E(\delta_A, L) = (\sqrt{3}+3)\sqrt{\frac{\pi}{2}}$. Then according to Theorem 7.12 for every $b \geq b_0^E(\Delta(\varphi))$ the A-optimal design δ_A is asymptotically A-optimal in $\Delta(\varphi)$ for estimating φ with bias bound b. The influence function of an asymptotically A-optimal AL-estimator for estimating φ at δ_A with bias bound b, which is also asymptotically U-optimal according to Theorem 7.10, is given by Theorem 7.11 or Theorem 7.9. As in Example 7.2 already was shown for $b = b_0^E(\delta_A, L) = (\sqrt{3} + 3)\sqrt{\frac{\pi}{2}} \approx 5.9307$ the influence function has the form

$$\begin{aligned}\psi_{b,\delta_A}(z,t) &= \psi_{01}(z,t)\\ &= \begin{cases} (-1,-1,-1)'\,\mathrm{sgn}(z)\, \frac{\sqrt{3}+3}{\sqrt{3}}\, \sqrt{\frac{\pi}{2}}, & \text{for } t = 1,\\ (1_{\{2\}}(t), 1_{\{3\}}(t), 1_{\{4\}}(t))'\,\mathrm{sgn}(z)\, (\sqrt{3}+3)\, \sqrt{\frac{\pi}{2}}, & \text{for } t \neq 1.\end{cases}\end{aligned}$$

For $b > b_0^E(\delta_A, L) = (\sqrt{3}+3)\sqrt{\frac{\pi}{2}} \approx 5.9307$ it has the form

$$\psi_{b,\delta_A}(z,t) = \begin{cases} (-1,-1,-1)' \operatorname{sgn}(z) \frac{1}{\sqrt{3}} \frac{\min\{|z|, b\, v_b\}}{v_b}, & \text{for } t=1, \\ (1_{\{2\}}(t), 1_{\{3\}}(t), 1_{\{4\}}(t))' \operatorname{sgn}(z) \frac{\min\{|z|, b\, v_b\}}{v_b}, & \text{for } t \neq 1, \end{cases}$$

with $v_b = (2\Phi(b\, v_b) - 1)\frac{1}{\sqrt{3}+3} > 0$. The MSE minimizing AL-estimator is an asymptotically A-optimal AL-estimator for the bias bound $b_* := b_{\delta_A} \approx 6.1195$, where $v_{b_*} \approx 0.0713$. The maximum asymptotic mean squared error of this estimator is $M(\psi_{b_*,\delta_A}, \delta_A) \approx 66.367$. □

For other examples see Müller (1987, 1993a, 1994c,d) and Kurotschka and Müller (1992). In Müller (1993a) it also was shown by a simulation study that for finite sample sizes about 30-50 the approximation of the asymptotic behaviour is very good.

7.3 Robust and Efficient Estimation of Nonlinear Aspects

For robust estimation of a nonlinear aspect $\varphi(\beta)$ of a linear model in Section 5.3 we regarded estimators $\widehat{\varphi}_N$ for φ which has the form $\widehat{\varphi}_N(y_N, d_N) = \varphi^*(\widehat{\eta}_N(y_N, d_N))$, where $\widehat{\eta}_N$ is an AL-estimator for $\eta(\beta) = A_{\mathcal{D}}\beta$ and φ is identifiable at $\operatorname{supp}(\delta) = \mathcal{D}$ with $\varphi(\beta) = \varphi^*(A_{\mathcal{D}}\beta)$ for all $\beta \in \mathcal{B}$. If we have a homoscedastic linear model, i.e. $\sigma(t) = \sigma$ for all $t \in \mathcal{T}$, and ψ is the influence function of the AL-estimator $\widehat{\eta}_N$, then Theorem 5.6 provides that the estimator $\widehat{\varphi}_N$ is asymptotically normally distributed for shrinking contamination, i.e.

$$\mathcal{L}(\sqrt{N}(\widehat{\varphi}_N - \varphi(\beta))|Q_\theta^N) \xrightarrow{N\to\infty} \mathcal{N}(\sigma\, \dot{\varphi}_\beta^*\, b(\psi, (Q_{N,\delta})_{N\in\mathbb{N}}),\, \dot{\varphi}_\beta^*\, \sigma^2 \int \psi\psi'\, d(P\otimes\delta)\, (\dot{\varphi}^*)')$$

for all $(Q_{N,\delta})_{N\in\mathbb{N}} \in \mathcal{U}_{c,\epsilon}^0(P\otimes\delta)$. And the maximum asymptotic bias satisfies

$$\max\{|\sigma\, \dot{\varphi}_\beta^*\, b(\psi, (Q_{N,\delta})_{N\in\mathbb{N}})|;\ (Q_{N,\delta})_{N\in\mathbb{N}} \in \mathcal{U}_{c,\epsilon}^0(P\otimes\delta)\} = \epsilon\, \sigma\, \|\dot{\varphi}_\beta^*\, \psi\|_\delta.$$

For robust and efficient estimation of φ the maximum asymptotic bias should be bounded and the asymptotic covariance matrix, and in particular its trace, should be small. If there is no bias bound for the maximum asymptotic bias, then the matrix $\int \psi\, \psi'\, d(P\otimes\delta)$ is minimized uniformly by $\psi_\infty(z,t) := A_{\mathcal{D}}\, \mathcal{I}(\delta)^{-1} a(t) z$ within all influence functions of AL estimators for $\eta(\beta) = A_{\mathcal{D}}\beta$. Because ψ_∞ is the influence function of the Gauss Markov estimator (see Section 5.2) this means that the estimator $\widehat{\varphi}_N = \varphi^*(\widehat{\eta}_N^{LS})$ based on the Gauss Markov estimator for η has minimum asymptotic covariance matrix. But if the maximum asymptotic bias should be bounded by some bias bound b, in principle it is impossible to minimize $\int \psi\, \psi'\, d(P\otimes\delta)$ uniformly so that an optimal robust estimator cannot be found independently of β. Also optimal designs δ cannot be found independently of β, and this problem already occurs for classical nonrobust estimation. All classical nonlinear optimal experimental design theory suffer on this β-dependence so that strategies were invented to overcome this problem. See Section 2.3 or Ford et al. (1989).

One basic strategy is to compare a given estimator at a given design with the locally optimal estimator at the locally optimal design, i.e. with the optimal estimator at the optimal design for a given value of β. For robust estimation this strategy was used in Kitsos and Müller (1995b), and here this strategy will be also used.

Because according to Lemma 5.5 at $P_{\beta,\delta}$ the estimator $\widehat{\varphi}_N = \varphi^*(\widehat{\eta}_N)$ is an AL-estimator for the linear aspect φ_β given by $\varphi_\beta(\tilde{\beta}) = \dot{\varphi}_\beta\tilde{\beta}$ the locally optimal estimator at β for the bias bound b at δ is an optimal AL-estimator of the linear aspect φ_β for the bias bound b at δ as defined in Section 7.2. Also the *locally optimal design at β for robust estimation with bias bound b* is an optimal design for robust estimation of φ_β with bias bound b. Thereby again, we assume that the ideal distribution P is the standard normal distribution, i.e. $P = n_{(0,1)}$. Without loss of generality we also can set $\epsilon = 1 = \sigma$.

Definition 7.5 (Locally A-optimal estimator and design for a bias bound b)
a) The AL-estimator $\widehat{\varphi}_{N,\beta}$ with influence function $\psi_{\beta,b,\delta}$ is locally A-optimal at β and δ for estimating φ with the bias bound b if $\widehat{\varphi}_{N,\beta}$ is an asymptotically A-optimal AL-estimator for estimating φ_β at δ with the bias bound b, i.e.

$$\psi_{\beta,b,\delta} \in \arg\min\left\{\operatorname{tr}\ \left(\int \psi\, \psi'\, d(P\otimes\delta)\right)\ ;\ \psi \in \Psi(\delta, \dot{\varphi}_\beta)\ \text{with}\ \|\psi\|_\delta \le b\right\}.$$

b) A design $\delta_{\beta,b}$ is locally A-optimal in Δ for estimating φ with the bias bound b if it is asymptotically A-optimal in $\Delta \cap \Delta(\varphi_\beta)$ for estimating φ_β with the bias bound b, i.e.

$$\delta_{\beta,b} \in \arg\min\left\{\operatorname{tr}\ \left(\int \psi_{\beta,b,\delta}\, \psi'_{\beta,b,\delta}\, d(P\otimes\delta)\right)\ ;\ \delta \in \Delta \cap \Delta(\varphi_\beta)\right\}.$$

From Section 7.2 characterizations of the locally A-optimal estimators can be obtained. In particular Theorem 7.12 provides that the locally A-optimal designs are independent of the bias bound and coincide with the locally A-optimal designs $\delta_{\beta,A}$ defined in Section 2.3. Moreover, Theorem 7.11 provides that the locally A-optimal estimators at the locally A-optimal designs are easy to obtain. For simplicity we write for their influence function $\psi_{\beta,b}$, i.e. we define

$$\psi_{\beta,b} := \psi_{\beta,b,\delta_{\beta,A}} = \psi_{\beta,b,\delta_{\beta,b}}.$$

If φ is identifiable at $\text{supp}(\delta) = \{\tau_1, ..., \tau_I\}$, where $a(\tau_1), ..., a(\tau_I)$ are linearly independent, then according to Theorem 7.9 the asymptotically A-optimal AL-estimator for estimating φ_β at δ with the bias bound b has an influence function of the form $\psi_{\beta,b,\delta} = \dot{\varphi}_\beta^* \psi_{\beta,b,\delta}^*$, where $\psi_{\beta,b,\delta}^*$ is the influence function of an AL-estimator for $\eta(\beta) = A_\mathcal{D}\beta$. Hence, there exists an estimator $\widehat{\varphi}_N = \varphi^*(\widehat{\eta}_N)$ for φ which at $P_{\beta,\delta}$ behaves asymptotically like the locally A-optimal robust estimator. But in general this is not the case. Often there is no estimator $\widehat{\varphi}_N = \varphi^*(\widehat{\eta}_N)$ for φ which at $P_{\beta,\delta}$ behaves asymptotically like the locally A-optimal robust estimator at δ. Moreover, usually the nonlinear aspect φ is not identifiable at all locally A-optimal designs. But the locally A-optimal estimators and designs are useful for efficiency comparisons.

If a robust estimator $\widehat{\varphi}_N = \varphi^*(\widehat{\eta}_N)$ for φ is given, then the trace of its asymptotic covariance matrix can be compared with the trace of the asymptotic covariance matrix of the locally A-optimal robust estimator at the locally A-optimal design. To make fair comparisons the bound for the asymptotic bias for the locally A-optimal estimators should be the same as the maximum asymptotic bias of the given estimator. Because according to Theorem 5.6 the maximum asymptotic bias depends on β we have only *local bias bounds*.

Definition 7.6 (Local bias bound)
$b(\beta)$ is the local bias bound at β of an estimator $\widehat{\varphi}_N = \varphi^*(\widehat{\eta}_N)$ for φ at δ, where $\widehat{\eta}_N$ is an AL-estimator for η with influence function ψ, if

$$b(\beta) = ||\dot{\varphi}_\beta^* \psi||_\delta.$$

Because $\dot{\varphi}_\beta^* \psi \in \Psi(\delta, \dot{\varphi}_\beta)$ we have according to Theorem 7.6 $b(\beta) \geq b_0^E(\delta, \dot{\varphi}_\beta) \geq b_0^E(\delta_{\beta,A}, \dot{\varphi}_\beta)$. Hence, for the bias bound $b(\beta)$ always a locally A-optimal estimator at the locally A-optimal design exists so that the *relative asymptotic efficiency* can be defined as follows.

Definition 7.7 (Relative asymptotic efficiency)
$E(\psi, \delta, \beta)$ is the relative asymptotic efficiency of an estimator $\widehat{\varphi}_N = \varphi^*(\widehat{\eta}_N)$ for φ at δ and β, where $\widehat{\eta}_N$ is an AL-estimator for η with influence function ψ, if

$$E(\psi, \delta, \beta) = \frac{tr\left(\int \psi_{\beta,b(\beta)}\, \psi'_{\beta,b(\beta)}\; d(P \otimes \delta_{\beta,A})\right)}{tr\left(\dot{\varphi}_\beta^* \int \psi\, \psi'\; d(P \otimes \delta)\, (\dot{\varphi}_\beta^*)'\right)} \in [0, 1].$$

Note that if $\widehat{\varphi}_N$ is based on the Gauss Markov estimator $\widehat{\eta}_N^{LS}$ for η, which has ψ_∞, given by $\psi_\infty(z, t) = A_\mathcal{D}\, \mathcal{I}(\delta)^- a(t)\, z$ (see Section 5.2), as influence function, then its relative efficiency satisfies

$$E(\psi_\infty, \delta, \beta) = \frac{\text{tr}\,(\dot{\varphi}_\beta\, \mathcal{I}(\delta_{\beta,A})^- \dot{\varphi}_\beta')}{\text{tr}\,(\dot{\varphi}_\beta\, \mathcal{I}(\delta)^- \dot{\varphi}_\beta')} = E(\delta, \beta),$$

where $E(\delta,\beta)$ is the relative efficiency of the design δ in classical nonrobust estimation (see Section 2.3 and Kitsos et al. (1988)).

Moreover, if $\widehat{\eta}_N^*$ is an optimal AL-estimator for estimating η at an A-optimal design δ_A with the bias bound b, then its influence function ψ_{b,δ_A} is given by Theorem 7.11 as

$$\psi_{b,\delta_A}(z,t) = \begin{cases} A_{\mathcal{D}}\, \mathcal{I}(\delta_A)^{-1} a(t)\, \mathrm{sgn}(z)\, \sqrt{\frac{\pi}{2}}, \\ \qquad \text{for } b = \sqrt{\mathrm{tr}\,(A_{\mathcal{D}}\, \mathcal{I}(\delta_A)^{-1} A'_{\mathcal{D}})\, \frac{\pi}{2}}, \\ A_{\mathcal{D}}\, \mathcal{I}(\delta_A)^{-1} a(t)\, \mathrm{sgn}(z) \frac{\min\{|z|, b\, y_b\}}{y_b \sqrt{\mathrm{tr}\,(A_{\mathcal{D}}\, \mathcal{I}(\delta_A)^{-1} A'_{\mathcal{D}})}}, \\ \qquad \text{for } b > \sqrt{\mathrm{tr}\,(A_{\mathcal{D}}\, \mathcal{I}(\delta_A)^{-1} A'_{\mathcal{D}})\, \frac{\pi}{2}}, \end{cases}$$

where y_b is the positive solution of $y\sqrt{\mathrm{tr}\,(A_{\mathcal{D}}\, \mathcal{I}(\delta_A)^{-1} A'_{\mathcal{D}})} = 2\,\Phi(b\,y) - 1 > 0$. Hence the asymptotic efficiency of the estimator $\varphi^*(\widehat{\eta}_N^*)$ for φ equals

$$E(\psi,\delta,\beta) = \frac{\mathrm{tr}\,(\dot{\varphi}_\beta\, \mathcal{I}(\delta_{\beta,A})^- \dot{\varphi}'_\beta)\; v_{b(\beta)}\left(\frac{1}{\mathrm{tr}\,(\dot{\varphi}_\beta\, \mathcal{I}(\delta_{\beta,A})^- \dot{\varphi}'_\beta)}\right)}{\mathrm{tr}\,(\dot{\varphi}_\beta\, \mathcal{I}(\delta_A)^- \dot{\varphi}'_\beta)\; v_b\left(\frac{1}{\mathrm{tr}\,(A_{\mathcal{D}}\, \mathcal{I}(\delta_A)^- A'_{\mathcal{D}})}\right)}. \tag{7.9}$$

Because $b(\beta) \to \infty$ if $b \to \infty$ this efficiency tends to $E(\delta_A,\beta)$ for $b \to \infty$, i.e. to the design efficiency of δ_A relative to $\delta_{\beta,A}$. Despite this convergence and the similar form of $E(\psi,\delta_A,\beta)$ and $E(\delta_A,\beta)$ the behaviour of $E(\psi,\delta_A,\beta)$ and $E(\delta_A,\beta)$ can be different. In particular $E(\psi,\delta_A,\beta)$ can depend on some components of β on which $E(\delta_A,\beta)$ does not depend. This shows the example of a rythmometry problem in Kitsos and Müller (1995b).

Nevertheless there are some similarities. In particular a similarity appears if we regard only estimators $\widehat{\varphi}_N$ for φ, which are of the form $\widehat{\varphi}_N(y_N,d_N) = \varphi^*(\widehat{\eta}_N(y_N,d_N))$, where $\widehat{\eta}_N$ is an optimal AL-estimator of $\eta(\beta) = A_{\mathcal{D}}\beta$ for the bias bound b at $\delta \in \Delta \cap \Delta(\varphi)$. Then for some classes Δ the maximin efficient designs for robust estimation with bias bound b are the same as for nonrobust estimation. For that define the *maximin efficient design for the bias bound* b as follows. Thereby $\psi_{b,\delta}$ is the influence function of the optimal AL-estimator $\widehat{\eta}_N$ of η for the bias bound b at δ.

Definition 7.8 (Maximin efficient design for the bias bound b)
$\delta_{M,b}$ is maximin efficient in Δ for estimating φ with the bias bound b if

$$\delta_{M,b} \in \arg\max\{\min_{\beta\in\mathcal{B}} E(\psi_{b,\delta},\delta,\beta);\ \delta \in \Delta \cap \Delta(\varphi)\}.$$

If the class Δ of the designs is given by $\Delta_{\mathcal{D}}^* = \{\delta \in \Delta_0;\ \mathrm{supp}(\delta) \subset \mathcal{D}\}$, where the regressors on $\mathcal{D}$ are linearly independent, then the locally A-optimal design at β is the A-optimal design $\delta_{\beta,A}$ for estimating φ_β, which is given according to Theorem 2.7 as

$$\delta_{\beta,A}(\{t\}) = \frac{|\dot{\varphi}_\beta\,(A'_{\mathcal{D}} A_{\mathcal{D}})^- a(t)|}{\sum_{\tau\in\mathcal{D}} |\dot{\varphi}_\beta\,(A'_{\mathcal{D}} A_{\mathcal{D}})^- a(\tau)|}.$$

If $\max\{\delta_{\beta,A}(\{t\}); \beta \in \mathcal{B}\} = 1$ for all $t \in \mathcal{D}$, then the maximin efficient designs for a bias bound b and the maximin efficient design for no bias bound coincide. Recall from Theorem 2.10 that the maximin efficient design for no bias bound in $\Delta_{\mathcal{D}}^*$ is $\delta_M = \frac{1}{I}\sum_{t\in\mathcal{D}} e_t$.

Theorem 7.14 *Let φ be identifiable at $\mathcal{D} = \{\tau_1, ..., \tau_I\}$, $a(\tau_1), ..., a(\tau_I)$ be linearly independent,* $\max\{\delta_{\beta,A}(\{t\}); \beta \in \mathcal{B}\} = 1$ *for all $t \in \mathcal{D}$ and $b \geq I\sqrt{\frac{\pi}{2}}$. Then $\delta_{M,b}$ is maximin efficient in $\Delta_{\mathcal{D}}^*$ for estimating φ with the bias bound b if and only if $\delta_{M,b} = \delta_M = \frac{1}{I}\sum_{t\in\mathcal{D}} e_t$, and the maximin efficiency is equal to*

$$\frac{1}{I}\,\frac{v_b(1)}{v_b(\frac{1}{I^2})}.$$

Proof. According to Lemma 2.2 we have $|A_{\mathcal{D}}(A_{\mathcal{D}}'A_{\mathcal{D}})^- a(t)| = 1$ for all $t \in \mathcal{D}$, and according Theorem 2.7 we have that the A-optimal design in $\Delta_{\mathcal{D}}^*$ for estimating $\eta(\beta) = A_{\mathcal{D}}\beta$ is $\delta_A = \frac{1}{I}\sum_{t\in\mathcal{D}} e_t$ with

$$\operatorname{tr}\,(A_{\mathcal{D}}\mathcal{I}(\delta_A)^- A_{\mathcal{D}}') = \left(\sum\nolimits_{t\in\mathcal{D}} |A_{\mathcal{D}}(A_{\mathcal{D}}'A_{\mathcal{D}})^- a(t)|\right)^2 = I^2.$$

Then according to Theorem 7.3 we have $b_0^E(\delta_A, A_{\mathcal{D}}) = I\sqrt{\frac{\pi}{2}} \leq b$. Hence an asymptotically A-optimal AL-estimator for estimating η at δ_A with the bias bound b exists.

Now let $\psi_{b,\delta}$ be the influence function of any A-optimal AL-estimator for η at $\delta \in \Delta_{\mathcal{D}}^*$. Then, according to Lemma 2.2 we have $|A_{\mathcal{D}}\mathcal{I}(\delta)^- a(t)| = \frac{1}{\delta(\{t\})}$ for all $t \in \operatorname{supp}(\delta)$ so that according to Theorem 7.9 and Theorem 7.8 the influence function $\psi_{b,\delta}$ has the form

$$\psi_{b,\delta}(z,t) = \begin{cases} A_{\mathcal{D}}\mathcal{I}(\delta)^- a(t)\operatorname{sgn}(z)\sqrt{\frac{\pi}{2}}, & \\ \qquad \text{for all } t \in \operatorname{supp}(\delta) \text{ with } \frac{1}{\delta(\{t\})}\sqrt{\frac{\pi}{2}} = b, & \\ A_{\mathcal{D}}\mathcal{I}(\delta)^- a(t)\operatorname{sgn}(z)\,\delta(\{t\})\,\frac{\min\{|z|, b\,y(t)\}}{y(t)}, & \\ \qquad \text{for all } t \in \operatorname{supp}(\delta) \text{ with } 0 < \frac{1}{\delta(\{t\})}\sqrt{\frac{\pi}{2}} < b, & \\ 0, \qquad \text{for all other } t \in \mathcal{T}, & \end{cases}$$

where $y(t) = (2\Phi(by(t)) - 1)\delta(\{t\}) > 0$. Hence with Lemma 2.2 we get

$$\begin{aligned} &\operatorname{tr}\left(\dot{\varphi}_\beta^* \int \psi_{b,\delta}\psi_{b,\delta}' d(P\otimes\delta)\,(\dot{\varphi}_\beta^*)'\right) \\ &= \sum\nolimits_{\tau\in\mathcal{D}} |\dot{\varphi}_\beta\,\mathcal{I}(\delta)^- a(\tau)|^2\,\delta(\{\tau\})\,v_b(\delta(\{\tau\})^2) \\ &= \sum\nolimits_{\tau\in\mathcal{D}} |\dot{\varphi}_\beta\,(A_{\mathcal{D}}'A_{\mathcal{D}})^- a(\tau)|^2\,\frac{1}{\delta(\{\tau\})}\,v_b(\delta(\{\tau\})^2) \end{aligned}$$

and

$$b(\beta) = ||\dot{\varphi}_\beta^*\psi_{b,\delta}||_\delta$$

$$
\begin{aligned}
&= \max_{t\in\mathcal{D}} \delta(\{t\})\, |\dot\varphi_\beta\, \mathcal{I}(\delta)^- a(t)|\, b \\
&= b\, \max_{t\in\mathcal{D}} |\dot\varphi_\beta\, (A_\mathcal{D}' A_\mathcal{D})^- a(t)|
\end{aligned}
$$

The locally A-optimal designs $\delta_{\beta,A}$ are also given by Theorem 2.7, and in particular, they satisfy

$$
\operatorname{tr}(\dot\varphi_\beta\, \mathcal{I}(\delta_{\beta,A})^-\, \dot\varphi_\beta') = \left(\sum\nolimits_{\tau\in\mathcal{D}} |\dot\varphi_\beta\, (A_\mathcal{D}' A_\mathcal{D})^- a(\tau)|\right)^2 .
$$

Hence with Theorem 7.11 and the property $v_{b\cdot c}(a^2) = v_b(a^2 c^2)$ (see Lemma A.7) the asymptotic covariance matrix of the locally A-optimal estimators for the local bias bound $b(\beta)$ at the locally A-optimal design satisfies

$$
\begin{aligned}
&\operatorname{tr}\left(\int \psi_{\beta,b(\beta)}\, \psi_{\beta,b(\beta)}'\, d(P\otimes\delta_{\beta,A})\right) \\
&= \operatorname{tr}(\dot\varphi_\beta\, \mathcal{I}(\delta_{\beta,A})^-\, \dot\varphi_\beta')\, v_{b(\beta)}\left(\frac{1}{\operatorname{tr}(\dot\varphi_\beta\, \mathcal{I}(\delta_{\beta,A})^-\, \dot\varphi_\beta')}\right) \\
&= \left(\sum\nolimits_{\tau\in\mathcal{D}} |\dot\varphi_\beta\, (A_\mathcal{D}' A_\mathcal{D})^- a(\tau)|\right)^2 v_{b(\beta)}\left(\frac{1}{(\sum_{\tau\in\mathcal{D}} |\dot\varphi_\beta\, (A_\mathcal{D}' A_\mathcal{D})^- a(\tau)|)^2}\right) \\
&= \left(\sum\nolimits_{\tau\in\mathcal{D}} |\dot\varphi_\beta\, (A_\mathcal{D}' A_\mathcal{D})^- a(\tau)|\right)^2 \\
&\qquad \cdot\, v_b\left(\max_{t\in\mathcal{D}} \frac{|\dot\varphi_\beta\, (A_\mathcal{D}' A_\mathcal{D})^- a(t)|^2}{(\sum_{\tau\in\mathcal{D}} |\dot\varphi_\beta\, (A_\mathcal{D}' A_\mathcal{D})^- a(\tau)|)^2}\right) \\
&= \left(\sum\nolimits_{\tau\in\mathcal{D}} |\dot\varphi_\beta\, (A_\mathcal{D}' A_\mathcal{D})^- a(\tau)|\right)^2 v_b\left(\max_{t\in\mathcal{D}} \delta_{\beta,A}(\{t\})^2\right).
\end{aligned}
$$

Then, because $s_b(c) := \frac{1}{c} v_b(c^2)$ is a decreasing function of c (see Lemma A.6), for the inverse of the efficiency we have

$$
\begin{aligned}
&E(\psi_{b,\delta}, \delta, \beta)^{-1} \\
&= \frac{\operatorname{tr}(\dot\varphi_\beta^* \int \psi_{b,\delta}\, \psi_{b,\delta}'\, d(P\otimes\delta)\, (\dot\varphi_\beta^*)')}{\operatorname{tr}(\int \psi_{\beta,b(\beta)}\, \psi_{\beta,b(\beta)}'\, d(P\otimes\delta_{\beta,A}))} \\
&= \sum\nolimits_{\tau\in\mathcal{D}} \frac{|\dot\varphi_\beta (A_\mathcal{D}' A_\mathcal{D})^- a(\tau)|^2}{(\sum_{t\in\mathcal{D}} |\dot\varphi_\beta (A_\mathcal{D}' A_\mathcal{D})^- a(t)|)^2}\, \frac{1}{\delta(\{\tau\})}\, \frac{v_b\left(\delta(\{\tau\})^2\right)}{v_b\left(\max_{t\in\mathcal{D}} \delta_{\beta,A}(\{t\})^2\right)} \\
&\le \sum\nolimits_{\tau\in\mathcal{D}} \delta_{\beta,A}(\{\tau\})\, (\max_{t\in\mathcal{D}} \delta_{\beta,A}(\{t\}))\, \frac{1}{\delta(\{\tau\})} \\
&\qquad \cdot\, \frac{v_b\left(\delta(\{\tau\})^2\right)}{v_b\left((\max_{t\in\mathcal{D}} \delta_{\beta,A}(\{t\}))^2\right)} \\
&= \sum\nolimits_{\tau\in\mathcal{D}} \frac{\delta_{\beta,A}(\{\tau\})}{\delta(\{\tau\})}\, \frac{v_b\left(\delta(\{\tau\})^2\right)}{s_b\left(\max_{t\in\mathcal{D}} \delta_{\beta,A}(\{t\})\right)} \\
&\le \sum\nolimits_{\tau\in\mathcal{D}} \frac{1}{\delta(\{\tau\})}\, \frac{v_b\left(\delta(\{\tau\})^2\right)}{s_b(1)}\, \delta_{\beta,A}(\{\tau\})
\end{aligned}
$$

$$= \sum_{\tau\in\mathcal{D}} \frac{1}{\delta(\{\tau\})} \frac{v_b\left(\delta(\{\tau\})^2\right)}{v_b(1)}\, \delta_{\beta,A}(\{\tau\})$$
$$\leq \max\left\{\frac{1}{\delta(\{\tau\})} \frac{v_b\left(\delta(\{\tau\})^2\right)}{v_b\,(1)};\ \tau\in\mathcal{D}\right\} \qquad (7.10)$$

for all $\beta\in\mathcal{B}$. Because of $\max\{\delta_{\beta,A}(\{t\});\beta\in\mathcal{B}\}=1$ for all $t\in\mathcal{D}$ the upper bound (7.10) is also attained. Hence we have

$$\min_{\beta\in\mathcal{B}} E(\psi_{b,\delta},\delta,\beta) = \min\left\{\delta(\{\tau\})\,\frac{v_b(1)}{v_b(\delta(\{\tau\})^2)};\ \tau\in\mathcal{D}\right\}$$

which is only maximized by $\delta_{M,b}=\frac{1}{I}\sum_{t\in\mathcal{D}} e_t$. □

Example 7.4 (Linear calibration, continuation of Example 2.3) For estimating $\varphi(\beta)=\frac{y-\beta_0}{\beta_1}$ in a simple linear regression model in Example 2.3 it was shown that all locally A-optimal designs in $\Delta^*_{\mathcal{T}}=\{\delta;\ \text{supp}(\delta)\subset \mathcal{T}\}$ with $\mathcal{T}=[-1,1]$ has a support included in $\{-1,1\}$ and that the maximin design for nonrobust estimation in $\Delta_{\mathcal{D}}$ with $\mathcal{D}=\{-1,1\}$ is $\delta_M=\frac{1}{2}(e_{-1}+e_1)$. Hence according to Theorem 7.12 also all locally A-optimal designs for robust estimation with bias bound b in $\Delta^*_{\mathcal{T}}$ have a support that is included in $\{-1,1\}$. Moreover, in Example 2.3 it was shown that the conditions of Theorem 2.10 and therefore also of Theorem 7.14 are satisfied. Hence Theorem 7.14 provides that $\delta_M=\frac{1}{2}(e_{-1}+e_1)$ is also maximin efficient for robust estimation with bias bound $b\geq I\sqrt{\frac{\pi}{2}}$ in $\Delta_{\mathcal{D}}$ and that the maximin efficiency is equal to $\frac{1}{2}\frac{v_b(1)}{v_b(\frac{1}{4})}$. This result was already shown in Kitsos and Müller (1995a) by straightforward calculations for the calibration problem. □

7.4 Robust and Efficient Estimation in Nonlinear Models

For robust estimation of β in a nonlinear model we proposed in Section 5.4 one-step M-estimators. For homoscedastic nonlinear model, i.e. $\sigma(t)=\sigma$ for all $t\in\mathcal{T}$, Theorem 5.8 provides for these one-step M-estimators their asymptotic normality for shrinking contamination, where

$$C_\theta(\rho,\delta) := \sigma^2\, M(\beta,\rho)^{-1}\int \dot\mu(t,\beta)\,\dot\mu(t,\beta)'\,\rho(z,t)^2\,P\otimes\delta(dz,dt)\; M(\beta,\rho)^{-1}$$

is their asymptotic covariance matrix and

$$\epsilon\,\sigma\,\max_{(z,t)\in\mathbb{R}\times\text{supp}(\delta)} |M(\beta,\rho)^{-1}\,\dot\mu(t,\beta)\,\rho(z,t)|$$

is their maximum asymptotic bias for shrinking contamination. This shows that in nonlinear models the asymptotic covariance matrix as well as the maximum asymptotic bias for shrinking contamination depend on the unknown parameter β. Then as for estimating a nonlinear aspect in a linear model we only can define *locally A-optimal estimators* and *locally A-optimal designs*. For that we again assume that the ideal error distribution P is the standard normal distribution, i.e. $P = n_{(0,1)}$, and without loss of generality we also can assume $\sigma = 1 = \epsilon$. Then we also will write $C_\beta(\rho, \delta)$ instead of $C_\theta(\rho, \delta)$. Moreover, we will write $M_\beta(\rho, \delta)$ instead of $M(\beta, \rho)$ to express here the dependence on the design δ.

In linear models the score function of a weakly asymptotically linear one-step M-estimator is also the influence function of the one-step M-estimator. Therefore by defining optimal estimators via the score/influence function we defined the optimal estimators within the more general class of AL-estimators. But in nonlinear models according to Lemma 5.6 the influence function of a weakly asymptotically linear one-step M-estimator with score function ρ is $\psi_\beta = M_\beta(\rho, \delta)^{-1}\, \dot{\mu}(\cdot, \beta)\, \rho$ which has a very special form. Hence for nonlinear models it is less complicated to define only locally A-optimal one-step M-estimators and for them locally A-optimal designs.

Definition 7.9 (Locally optimal estimators and designs in nonlinear models)
a) An one-step M-estimator $\widehat{\beta}_N$ for β with score function $\rho_{\beta,b,\delta}$ is locally A-optimal for estimating β at δ with the bias bound b if

$$\begin{aligned}\rho_{\beta,b,\delta} \in \arg\min\{&\operatorname{tr}(C_\beta(\rho, \delta));\ \rho \text{ satisfies condition (5.40) and} \\ &\max_{(z,t)\in \mathbb{R}\times \operatorname{supp}(\delta)} |M_\beta(\rho, \delta)^{-1}\, \dot{\mu}(\beta, t)\, \rho(z, t)| \le b\}.\end{aligned} \tag{7.11}$$

b) A design $\delta_{\beta,b}$ is locally A-optimal in Δ for estimating β with the bias bound b if

$$\delta_{\beta,b} \in \arg\min\{\operatorname{tr}(C_\beta(\rho_{\beta,b,\delta}, \delta));\ \delta \in \Delta\}.$$

Using Theorem 7.11 and Theorem 7.12 for linear models given by the response function $\tilde{\mu}(t, \tilde{\beta}) = \dot{\mu}(t, \beta)'\tilde{\beta}$ we can see that the locally A-optimal design $\delta_{\beta,b}$ for robust estimation with bias bound b is the A-optimal design for estimating $\tilde{\beta}$ in the linear model given by the response function $\tilde{\mu}(t, \tilde{\beta}) = \dot{\mu}(t, \beta)'\tilde{\beta}$, i.e. the locally A-optimal design $\delta_{\beta,A}$ for nonlinear models as defined in Definition 2.12. The score functions of locally A-optimal estimators at the locally A-optimal designs have a very simple form, namely

$$\begin{aligned}\rho_{\beta,b}(z,t) &:= \rho_{\beta,b,\delta_\beta}(z,t) \\ &= \begin{cases} \operatorname{sgn}(z)\sqrt{\frac{\pi}{2}} & \text{for } b = \sqrt{\operatorname{tr}\mathcal{I}_\beta(\delta_{\beta,A})^{-1}\,\frac{\pi}{2}}, \\ \operatorname{sgn}(z)\,\frac{\min\{|z|,\, b\, y_b\}}{y_b\sqrt{\operatorname{tr}\mathcal{I}_\beta(\delta_{\beta,A})^{-1}}} & \text{for } b > \sqrt{\operatorname{tr}\mathcal{I}_\beta(\delta_{\beta,A})^{-1}\,\frac{\pi}{2}}, \end{cases}\end{aligned}$$

where y_b is a positive solution of $y\sqrt{\operatorname{tr}\mathcal{I}_\beta(\delta_{\beta,A})^{-1}} = (2\Phi(b\,y) - 1)$ and

$$\mathcal{I}_\beta(\delta) := \int \dot\mu(\beta,t)\,\dot\mu(\beta,t)'\,\delta(dt)$$

is the information matrix of δ in the linear model given by $\tilde\mu(t,\tilde\beta) = \dot\mu(t,\beta)'\tilde\beta$, i.e. the local information matrix. This shows the following theorem.

Theorem 7.15 *If the conditions of Theorem 7.12 for Δ and b are satisfied for the linear model given by the response function $\tilde\mu(t,\tilde\beta) = \dot\mu(t,\beta)'\tilde\beta$, then the locally A-optimal design $\delta_{\beta,A}$ for nonlinear models is locally A-optimal in Δ for estimating β with the bias bound b and a one-step M-estimator for β with score function $\rho_{\beta,b}$ is locally A-optimal for estimating β at $\delta_{\beta,A}$ with the bias bound b.*

Proof. Because the influence function of a weakly asymptotically linear one-step M-estimator with score function ρ is $\psi_\beta = M_\beta(\rho,\delta)^{-1}\,\dot\mu(\cdot,\beta)\,\rho$ the special form of $M_\beta(\rho,\delta)$ provides

$$\int \psi_\beta(z,t)\,\dot\mu(t,\beta)'\,z\,P{\otimes}\delta(dz,dt) = E_r,$$

so that ψ_β satisfies the conditions (5.30) - (5.32) for a influence function of an AL-estimator for $\tilde\beta$ in a linear model given by the reponse function $\tilde\mu(t,\tilde\beta) = \dot\mu(t,\beta)'\tilde\beta$. Generalize $\Psi(\delta,E_r)$ to

$$\begin{aligned}\Psi_\beta(\delta,E_r) \;:=\; \{\psi : I\!R\times\mathcal{T}\to I\!R^r;\ &\psi \text{ satisfies conditions}\\ &(5.30)\text{ - }(5.32) \text{ for the linear model given by the}\\ &\text{response function } \tilde\mu(t,\tilde\beta) = \dot\mu(t,\beta)'\tilde\beta\}.\end{aligned}$$

Then we have

$$\begin{aligned}&\{\psi : I\!R\times\mathcal{T}\to I\!R^r;\ \psi = M_\beta(\rho,\delta)^{-1}\,\dot\mu(\cdot,\beta)\,\rho, \text{ where}\\ &\qquad\qquad \rho \text{ satisfies condition (5.40)}\}\\ &\subset\ \Psi_\beta(\delta,E_r).\end{aligned} \tag{7.12}$$

If $\psi_{\beta,b,\delta}$ is a solution of

$$\begin{aligned}\psi_{\beta,b,\delta} \in \arg\min\{\operatorname{tr}\left(\int \psi\,\psi'\,d(P{\otimes}\delta)\right);\\ \psi\in\Psi_\beta(\delta,E_r) \text{ with } \|\psi\|_\delta \le b\}\end{aligned} \tag{7.13}$$

and $\psi_{\beta,b,\delta} = M_\beta(\rho_{\beta,b,\delta},\delta)^{-1}\,\dot\mu(\cdot,\beta)\,\rho_{\beta,b,\delta}$ for some $\rho_{\beta,b,\delta}$ satisfying condition (5.40), then $\rho_{\beta,b,\delta}$ is a solution of (7.11). Theorem 7.12 provides that the locally A-optimal design $\delta_{\beta,A}$ satisfies

$$\delta_{\beta,A} \in \arg\min\{\operatorname{tr}\left(\int \psi_{\beta,b,\delta}\,\psi_{\beta,b,\delta}\,d(P{\otimes}\delta)\right);\ \delta\in\Delta\}.$$

Moreover, Theorem 7.11 provides that $\psi_{\beta,b,\delta_{\beta,A}} := \mathcal{I}_\beta(\delta_{\beta,A})^{-1}\dot\mu(t,\beta)\,\rho_{\beta,b}$ is a solution of (7.13) with $\delta = \delta_{\beta,A}$. Thereby we have $M_\beta(\rho_{\beta,b},\delta_{\beta,A}) = \mathcal{I}_\beta(\delta_{\beta,A})$ so that $\psi_{\beta,b,\delta_{\beta,A}}$ satisfies $\psi_{\beta,b,\delta_{\beta,A}} = M_\beta(\rho_{\beta,b},\delta_{\beta,A})^{-1}\,\dot\mu(\cdot,\beta)\,\rho_{\beta,b}$. Hence, a one-step M-estimator with score function $\rho_{\beta,b}$ is locally A-optimal for the bias bound b at $\delta_{\beta,A}$. Because of (7.12) this implies that $\delta_{\beta,A}$ is locally optimal for the bias bound b. See also Müller (1994b). □

Example 7.5 (Generalized linear model, continuation of Example 2.4)
If we have a generalized linear model, where the observation at t_{nN} is given by

$$Y_{nN} = e^{\beta_0+\beta_1 t_{nN}} + Z_{nN},$$

then in Example 2.4 it was shown that the locally A-optimal designs in $\Delta^*_{\mathcal{D}}$ with $\mathcal{D} = \{0,1\}$ are given by

$$\delta_{\beta,A} = \frac{1}{\sqrt{2}+e^{-\beta_1}}\,(\sqrt{2}e_0 + e^{-\beta_1}\,e_1).$$

Theorem 7.15 provides that these designs are also locally A-optimal for robust estimation. □

The locally A-optimal robust estimators at the locally A-optimal designs can be used for efficiency comparisons as Ford et al. (1989) described for designing experiments for classical nonrobust estimation in nonlinear models and as was described above for estimating a nonlinear aspect in a linear model. Given a one-step M-estimator with score function ρ at a design δ its relative efficiency is given by

$$E(\rho,\delta,\beta) := \frac{\mathrm{tr}\,(C_\beta(\rho_{\beta,b(\beta)},\delta_{\beta,A}))}{\mathrm{tr}\,(C_\beta(\rho,\delta))} \in [0,1],$$

where the local bias bound

$$b(\beta) := b(\beta,\rho,\delta) := \max_{(z,t)\in\mathbb{R}\times\mathrm{supp}(\delta)} |M_\beta(\rho,\delta)^{-1}\,\dot\mu(t,\beta)\,\rho(z,t)|$$

should be used to make fair comparisons.

8

High Robustness and High Efficiency of Tests

In this chapter we regard tests based on ALE-test statistics as defined in Section 6.2. In Section 8.1 we characterize "most robust" ALE-tests in linear models, which are ALE-tests with minimum asymptotic bias of the first error for shrinking contamination. We also characterize designs which minimize the asymptotic bias of the first error. In Section 8.2 we characterize admissible ALE-tests and ALE-tests which minimize the determinant of the asymptotic covariance matrix within all ALE-tests with an asymptotic bias of the first error bounded by some bias bound b. Also optimal designs for optimal robust testing are derived. Thereby, in both sections we assume that the ideal model is a homoscedastic linear model with normally distributed errors, i.e. the error Z_{nN} at t_{nN} is distributed according to the normal distribution $n_{(0,\sigma^2)}$ with mean 0 and variance $\sigma(t_{nN})^2 = \sigma^2 \in \mathbb{R}^+$ for all $n = 1, ...N$, $N \in \mathbb{N}$. In particular, we have $P = n_{(0,1)}$.

8.1 Tests and Designs with Minimum Asymptotic Bias

In Corollary 6.1 it was shown that in a homoscedastic linear model the asymptotic bias of the first error for shrinking contamination of an α level ALE-test for testing $H_0 : \varphi(\beta) = l$ is bounded by $1 - \mathcal{X}^2_{s,b}(\chi^2_{1-\alpha,s,0})$ if and only if

$$\epsilon^2 \, \|\psi' \, C(\psi)^{-1} \, \psi\|_\delta \leq b,$$

where

$$\|\psi' \, C(\psi)^{-1} \, \psi\|_\delta \tag{8.1}$$
$$:= \max_{(z,t)\in \mathbb{R}\times \mathrm{supp}(\delta)} \psi(z,t)' \left(\int \psi\,\psi \, d(P\otimes\delta)\right)^{-1} \psi(z,t),$$

and ψ is the influence function of the ALE-test statistic. An ALE-test will be called "most robust" or with minimum asymptotic bias if it minimizes the asymptotic bias of the first error within all ALE-tests, or equivalently, if its influence function minimizes (8.1) within all influence functions of ALE-tests. In Chapter 5 and Chapter 7 we have seen that in linear models usually the influence functions of AL-estimators have the form $\psi(z,t) = M\,a(t)\,\rho(z,t)$, where $\rho : \mathbb{R} \times \mathcal{T} \to \mathbb{R}$. Because ALE-test statistics are based on AL-estimators, here we will regard only those influence function, and we will try to minimize (8.1) within the following set of influence functions:

$$\begin{aligned}\Psi^*(\delta, L) := \{\psi : \mathbb{R} \times \mathcal{T} \to \mathbb{R}^s;\ &\psi \text{ satisfies conditions (5.30) - (5.32)}\\ &\text{and } \psi(z,t) = M\,a(t)\,\rho(z,t) \text{ for all } (z,t) \in \mathbb{R} \times \mathcal{T} \text{ for}\\ &\text{some } M \in \mathbb{R}^{s \times r} \text{ and } \rho : \mathbb{R} \times \mathcal{T} \to \mathbb{R}\} \subset \Psi(\delta, L).\end{aligned}$$

Thereby, as for robust estimation (see Chapter 7), it is not clear if every ψ lying in $\Psi^*(\delta, L)$ is the influence function of an ALE-test statistic. But if we find an optimal solution ψ_* within $\Psi^*(\delta, L)$ which satisfies the conditions (5.20) and (5.21), then we know from Lemma 5.3 that a one-step M-estimator with this score function ψ_* is an AL-estimator with influence function ψ_* so that ψ_* is also an influence function of an ALE-test statistic. We will see that as for robust estimation also for robust testing this is usually the case. Hence we define the *ALE-test statistic with minimum asymptotic bias* as an ALE-test statistic with an influence function which minimizes (8.1) within $\Psi^*(\delta, L)$ and the *minimum asymptotic bias for testing at* δ as the minimum value of (8.1) within $\Psi^*(\delta, L)$. Compare also with the definitions of minimum asymptotic bias for estimation in Section 7.1.

Definition 8.1 (Minimum asymptotic bias for testing at δ)
a) $b_0^T(\delta, L)$ is the minimum asymptotic bias for testing φ at δ if

$$b_0^T(\delta, L) = \min\{\|\psi'\,C(\psi)^{-1}\,\psi\|_\delta;\ \psi \in \Psi^*(\delta, L)\}.$$

b) An ALE-test statistic $\widehat{\tau}_N$ for testing $H_0 : \varphi(\beta) = l$ at δ with influence function ψ_0 is an ALE-test statistic for testing φ at δ with minimum asymptotic bias if

$$\psi_0 \in \arg\min\{\|\psi'\,C(\psi)^{-1}\,\psi\|_\delta;\ \psi \in \Psi^*(\delta, L)\}.$$

A design which additionally minimizes the asymptotic bias of ALE-test statistics with minimum bias within a given set Δ of designs is called a *design with minimum asymptotic bias*, and the minimum value of the asymptotic bias within $\Psi^*(\delta, L)$ and Δ is called the *minimum asymptotic bias within* Δ.

Definition 8.2 (Minimum asymptotic bias for testing within Δ)
a) $b_0^T(\Delta, L)$ *is the minimum asymptotic bias in* Δ *for testing* φ *if*

$$b_0^T(\Delta, L) = \min\{\|\psi' C(\psi)^{-1} \psi\|_\delta;\ \psi \in \Psi^*(\delta, L) \text{ and } \delta \in \Delta\}.$$

b) A design δ_0 *is a design providing the minimum asymptotic bias in* Δ *for testing* φ *if*

$$\delta_0 \in \arg\min\{\min\{\|\psi' C(\psi)^{-1} \psi\|_\delta;\ \psi \in \Psi^*(\delta, L)\};\ \delta \in \Delta\}.$$

The problem of minimizing (8.1) appears also for deriving "most robust" estimators when the robustness of estimators is defined via the (self)-standardized gross-error-sensitivity (see Ronchetti and Rousseeuw (1985) and Hampel et al. (1986), Section 6.3). For the case of estimating the whole parameter vector β, i.e. for $L = E_r$, they derive a lower bound for (8.1) (see Theorem 2' in Ronchetti and Rousseeuw (1985) or Proposition 1 (ii) in Hampel et al. (1986), p. 318). This result can be generalized for arbitrary $L \in \mathbb{R}^{s \times r}$ of full rank s, i.e. for testing arbitrary linear hypotheses $H_0 : \varphi(\beta) = l$.

Lemma 8.1 *For all* $\delta \in \Delta_0$ *we have* $b_0^T(\delta, L) \geq s$.

Proof. Because $C(\psi) = \int \psi\psi'\, d(P \otimes \delta)$ we have

$$\begin{aligned} s &= \operatorname{tr}(E_s) = \operatorname{tr}\left(C(\psi)^{-1} \int \psi\, \psi'\, d(P \otimes \delta)\right) \\ &= \int \operatorname{tr}(C(\psi)^{-1}\, \psi\, \psi')\, d(P \otimes \delta) = \int \psi'\, C(\psi)^{-1}\, \psi\, d(P \otimes \delta) \\ &\leq \|\psi'\, C(\psi)^{-1}\, \psi\|_\delta. \end{aligned}$$

Compare also with the proofs of Theorem 2' in Ronchetti and Rousseeuw (1985) and Proposition 1 (ii) in Hampel et al. (1986), p. 318. □

The following theorem shows that the lower bound for the asymptotic bias given in Lemma 8.1 is attained by an ALE-test statistics based on the L_1 estimator at a D-optimal design. This result is analogous to Theorem 7.3 which shows that the minimum asymptotic bias for estimation is attained by L_1 estimators at A-optimal designs. Recall from Section 5.2 that the influence function of the L_1 estimator is given by $\psi_{01}(z,t) := L\,\mathcal{I}(\delta)^-\, a(t)\, \operatorname{sgn}(z)\, \sqrt{\frac{\pi}{2}}$. Recall also that $\Delta_\mathcal{D} = \{\delta \in \Delta_0;\ \operatorname{supp}(\delta) = \mathcal{D}\}$.

Theorem 8.1 *If δ_D is D-optimal for φ in $\Delta_{\mathcal{D}}$, then*

a) $b_0^T(\Delta_{\mathcal{D}}, L) = b_0^T(\delta_D, L) = s.$

b) ψ_{01} *given by* $\psi_{01}(z,t) = L\mathcal{I}(\delta_D)^- a(t)\, sgn(z)\sqrt{\frac{\pi}{2}}$ *is the influence function of an ALE-test statistic for testing φ at δ_D with minimum asymptotic bias.*

Proof. Because of $\int \psi_{01}\,\psi_{01}'\, d(P\otimes\delta_D) = L\mathcal{I}(\delta_D)^-\, L'\,\frac{\pi}{2}$ the equivalence theorem for D-optimality (Theorem 2.5) provides

$$\psi_{01}(z,t)' \left(\int \psi_{01}\,\psi_{01}'\, d(P\otimes\delta_D)\right)^{-1} \psi_{01}(z,t)$$
$$= \; a(t)'\mathcal{I}(\delta_D)^-\, L'\,(L\mathcal{I}(\delta_D)^-\, L')^{-1}\, L\mathcal{I}(\delta_D)^-\, a(t) = s$$

for all $t\in\mathcal{D} = \text{supp}(\delta_D)$. Hence we have $b_0^T(\delta_D, L) \le s$. Because Lemma 8.1 yields $b_0^T(\delta_D, L) \ge b_0^T(\Delta_{\mathcal{D}}, L) = s$ the assertion follows. □

Theorem 8.1 provides that the minimum asymptotic bias within Δ_D is independent of the support $\mathcal{D}$. Hence the minimum asymptotic bias will be not improved if we regard the D-optimal design within all designs at which φ is identifiable, i.e. within $\Delta(\varphi)$. In particular, for a D-optimal design in $\Delta(\varphi)$ we get the same result as in Theorem 8.1 because a D-optimal design δ_D in $\Delta(\varphi)$ is in particular D-optimal in $\Delta_{\mathcal{D}}$ with $\mathcal{D} = \text{supp}(\delta_D)$. Hence we have the following corollary which is analogous to Theorem 7.6.

Corollary 8.1 *If δ_D is D-optimal for φ in $\Delta(\varphi)$, then*

$$b_0^T(\delta_D, L) = b_0^T(\Delta(\varphi), L) = s,$$

i.e. δ_D provides the minimum asymptotic bias in $\Delta(\varphi)$ for testing φ.

As for estimation with minimum asymptotic bias Theorem 8.1 and Corollary 8.1 hold also for D-optimal designs in a modified model (compare with Theorem 7.5). For that define for a function $h:\mathcal{T}\to I\!R^+\setminus\{0\}$

$$a_h(t) := h(t)^{-1}a(t),$$
$$L_h := L\,J_h^-\,\mathcal{I}_h(\delta),$$

where

$$J_h := \int a_h(t)\,a(t)'\,\delta(dt),$$
$$\mathcal{I}_h(\delta) := \int a_h(t)a_h(t)'\delta(dt).$$

Theorem 8.2 *If there exists some function $h : \mathcal{T} \to I\!R^+ \setminus \{0\}$ so that δ is D-optimal in $\Delta_{\text{supp}(\delta)}$ for testing $\varphi_h(\beta) = L_h\beta$ in the modified model $Y_h(t) = a_h(t)'\beta + Z$, then:*

a) $b_0^T(\delta, L) = b_0^T(\Delta_{\text{supp}(\delta)}) = s.$

b) ψ_0 *given by* $\psi_0(z,t) = L_h\mathcal{I}_h(\delta)^- a_h(t) sgn(z)\sqrt{\frac{\pi}{2}}$ *is the influence function of an ALE-test statistic for testing* φ *at* δ *with minimum asymptotic bias.*

Proof. Because $h(t) > 0$ for all $t \in \mathcal{T}$ and φ is identifiable at $\mathcal{D} = \text{supp}(\delta) = \{\tau_1, ..., \tau_I\}$ the aspect φ is also identifiable at δ_h given by

$$\delta_h(\{t\}) := \left(\int \frac{1}{h(t)}\,\delta(dt)\right)^{-1} \frac{1}{h(t)}\delta(\{t\})$$

so that there exists $K_1 \in I\!R^{s\times r}$ such that

$$L = \left(\int \frac{1}{h(t)}\,\delta(dt)\right)\, K_1\,\mathcal{I}(\delta_h) = K_1\,J_h.$$

Setting

$$\begin{aligned}
A_{\mathcal{D}} &:= (a(\tau_1), ..., a(\tau_I))', \\
D_{h1} &:= \text{diag}\left(\frac{1}{h(\tau_1)}\delta(\{\tau_1\}), ..., \frac{1}{h(\tau_1)}\delta(\{\tau_I\})\right), \\
D_{h1}^{1/2} &:= \text{diag}\left(\sqrt{\frac{1}{h(\tau_1)}\delta(\{\tau_1\})}, ..., \sqrt{\frac{1}{h(\tau_1)}\delta(\{\tau_I\})}\right) \\
D_{h1}^{-1/2} &:= (D_{h1}^{1/2})^{-1}
\end{aligned}$$

and

$$\begin{aligned}
D_{h2} &:= \text{diag}\left(\frac{1}{h(\tau_1)^2}\delta(\{\tau_1\}), ..., \frac{1}{h(\tau_1)^2}\delta(\{\tau_I\})\right) \\
D_{h2}^{1/2} &:= \text{diag}\left(\frac{1}{h(\tau_1)}\sqrt{\delta(\{\tau_1\})}, ..., \frac{1}{h(\tau_1)}\sqrt{\delta(\{\tau_I\})}\right) \\
D_{h2}^{-1/2} &:= (D_{h2}^{1/2})^{-1}
\end{aligned}$$

the properties of generalized inverses provide

$$\begin{aligned}
&\int \psi_0(z,t)\,a(t)'\,z\,P\otimes\delta(dz,dt) = L_h\,\mathcal{I}_h(\delta)^- \int a_h(t)\,a(t)'\,\delta(dt) \\
&= L\,J_h^-\,\mathcal{I}_h(\delta)\,\mathcal{I}_h(\delta)^-\,J_h \\
&= K_1\,J_h\,[A'_{\mathcal{D}}D_{h1}\,A_{\mathcal{D}}]^-\,A'_{\mathcal{D}}D_{h2}\,A_{\mathcal{D}}\,[A'_{\mathcal{D}}D_{h2}\,A_{\mathcal{D}}]^-\,A'_{\mathcal{D}}D_{h1}\,A_{\mathcal{D}}
\end{aligned}$$

$$
\begin{aligned}
&= K_1\, A_{\mathcal{D}}'(D_{h1}^{1/2})'\, D_{h1}^{1/2}\, A_{\mathcal{D}}\, [A_{\mathcal{D}}'(D_{h1}^{1/2})'\, D_{h1}^{1/2}\, A_{\mathcal{D}}]^{-} \\
&\quad A_{\mathcal{D}}'(D_{h1}^{1/2})'\,(D_{h1}^{-1/2})'\, D_{h2}\, A_{\mathcal{D}}\, [A_{\mathcal{D}}' D_{h2}\, A_{\mathcal{D}}]^{-}\, A_{\mathcal{D}}' D_{h1}\, A_{\mathcal{D}} \\
&= K_1\, A_{\mathcal{D}}'(D_{h1}^{1/2})'\,(D_{h1}^{-1/2})'\, D_{h2}\, A_{\mathcal{D}}\, [A_{\mathcal{D}}' D_{h2}\, A_{\mathcal{D}}]^{-}\, A_{\mathcal{D}}' D_{h1}\, A_{\mathcal{D}} \\
&= K_1\, A_{\mathcal{D}}'(D_{h2}^{1/2})'\, D_{h2}^{1/2}\, A_{\mathcal{D}}\, [A_{\mathcal{D}}'(D_{h2}^{1/2})'\, D_{h2}^{1/2}\, A_{\mathcal{D}}]^{-} \\
&\quad A_{\mathcal{D}}'(D_{h2}^{1/2})'\,(D_{h2}^{-1/2})\, D_{h1}\, A_{\mathcal{D}} \\
&= K_1\, A_{\mathcal{D}}'(D_{h2}^{1/2})'\,(D_{h2}^{-1/2})\, D_{h1}\, A_{\mathcal{D}} \\
&= K_1\, J_h = L.
\end{aligned}
$$

Hence $\psi_0 \in \Psi^*(\delta, L)$ and $\text{rk}(L_h) = \text{rk}(L) = s$ so that the assertion follows by applying Theorem 8.1 on the modified model. □

A special case of Theorem 8.2 appears when

$$
\int \frac{a(t)\, a(t)'}{|a(t)|^2}\, \delta(dt) = k\, E_r,\ L = E_r,
$$

where k is a scalar and E_r the identity matrix. Note that in any case $k = \frac{1}{r}$. Then setting $h(t) = |a(t)|$ we have

$$
\begin{aligned}
&a_h(t)'\, \mathcal{I}_h(\delta)^{-1}\, L_h'\, [L_h\, \mathcal{I}_h(\delta)^{-1}\, L_h']^{-1}\, L_h\, \mathcal{I}_h(\delta)^{-1}\, a_h(t) \\
&\quad = a_h(t)'\, \mathcal{I}_h(\delta)^{-1}\, \mathcal{I}_h(\delta)\, (J_h^{-1})'\, [J_h^{-1}\, \mathcal{I}_h(\delta)\, \mathcal{I}_h(\delta)^{-1}\, \mathcal{I}_h(\delta)\, (J_h^{-1})']^{-1} \\
&\qquad J_h^{-1}\, \mathcal{I}_h(\delta)\, \mathcal{I}_h(\delta)^{-1}\, a_h(t) \\
&\quad = a_h(t)'\, \mathcal{I}_h(\delta)^{-1}\, a_h(t) \\
&\quad = \frac{a(t)'}{|a(t)|}\, r\, E_r\, \frac{a(t)}{|a(t)|} \\
&\quad = r
\end{aligned}
$$

for all $t \in \mathcal{T}$ so that according to Theorem 2.3 δ is D-optimal for β in $\Delta(\beta)$ within the modified model $Y_h = a_h(t)'\beta + Z$. According to Theorem 8.2 an ALE-test statistic with influence function ψ_0 given by

$$
\psi_0(z, t) = J_h^{-1}\, \mathcal{I}_h(\delta)\, \mathcal{I}_h(\delta)^{-1}\, a_h(t)\, \text{sgn}(z)\, \sqrt{\tfrac{\pi}{2}} = J_h^{-1}\, \tfrac{a(t)}{|a(t)|}\, \text{sgn}(z)\, \sqrt{\tfrac{\pi}{2}}
$$

is an ALE-test statistic with minimum asymptotic bias. In particular, we get also the second part of Theorem 2' of Ronchetti and Rousseeuw (1985) (or Proposition 1(ii) in Hampel et al. (1986), p. 318) concerning estimators with minimum self-standardized gross-error-sensitivity. Compare also with the application of Theorem 7.5.

Lemma 8.1, Theorem 8.1 and Theorem 8.2 do not use the fact that we regard only influence functions ψ which are elements of $\Psi^*(\delta, L)$. But in the following this will be important. For that we set $A_{\mathcal{D}}$, D, $D_1(\rho)$, $D_2(\rho)$ and $M(\rho)$ as in Section 7.2.

Lemma 8.2 *If $supp(\delta) = \{\tau_1, ..., \tau_I\}$, $a(\tau_1), ..., a(\tau_I)$ are linearly independent and $\psi \in \Psi^*(\delta, L)$, then*

$$\begin{aligned} &\psi(z,\tau_i)' \left(\int \psi\, \psi'\, d(P \otimes \delta) \right)^{-1} \psi(z,\tau_i) \\ &= \rho(z,\tau_i)^2\, u_i'\, D_1(\rho)^{-1}\, D^{-1}\, K'\, [K\, D_1(\rho)^{-1}\, D_2(\rho)\, D_1(\rho)^{-1}\, D^{-1}\, K']^{-1} \\ &\quad K\, D_1(\rho)^{-1}\, D^{-1}\, u_i \end{aligned}$$

for all $z \in I\!R$ and $i = 1, ..., I$, where $L = K\, A_{\mathcal{D}}$, $\rho : I\!R \times \mathcal{T} \to I\!R$, $\psi(z,t) = L\, M(\rho)^-\, a(t)\, \rho(z,t)$ for all $(z,t) \in I\!R \times supp(\delta)$ and u_i is the ith unit vector in $I\!R^I$.

Proof. Lemma 7.1 provides $\psi(z,t) = L\, M(\rho)^-\, a(t)\, \rho(z,t)$ for all $(z,t) \in I\!R \times \text{supp}(\delta)$ and

$$\int \psi\, \psi'\, d(P \otimes \delta) = K\, D_1(\rho)^{-1}\, D_2(\rho)\, D_1(\rho)^{-1}\, K'.$$

Lemma 2.1 provides

$$\begin{aligned} &L\, M(\rho)^-\, a(\tau_i) = L\, M(\rho)^-\, A_{\mathcal{D}}\, u_i \\ &= K\, A_{\mathcal{D}}(A_{\mathcal{D}}'\, D_1(\rho)\, D\, A_{\mathcal{D}})^-\, A_{\mathcal{D}}\, u_i = K\, D_1(\rho)^{-1}\, D^{-1}\, u_i \end{aligned}$$

so that the assertion follows. □

For estimation with minimum asymptotic bias we have seen in Section 7.1 that for special design situations the influence function of an AL-estimator with minimum asymptotic bias is not unique. The same holds for testing with minimum asymptotic bias. For designs with s linearly independent regressors two different influence functions of ALE-test statistics with minimum asymptotic bias are ψ_0 and ψ_{01} given by

$$\psi_{01}(z,t) = L\,\mathcal{I}(\delta)^-\, a(t)\, \text{sgn}(z) \sqrt{\frac{\pi}{2}} \qquad \text{for } t \in \mathcal{T} \tag{8.2}$$

and

$$\psi_0(z,t) = \begin{cases} L\,\mathcal{I}(\delta)^-\, a(t)\, \text{sgn}(z)\, \sqrt{\frac{\pi}{2}}, \\ \qquad \text{for } t \in \text{supp}(\delta) \text{ with } \frac{1}{\delta(\{t\})} = b_0^T(\delta, L), \\ L\,\mathcal{I}(\delta)^-\, a(t)\, \text{sgn}(z)\, \frac{\min\{|z|, \sqrt{b_0^T(\delta,L)}\, y_0(t)\}}{2\Phi(\sqrt{b_0^T(\delta,L)}\, y_0(t)) - 1}, \\ \qquad \text{for } t \in \text{supp}(\delta) \text{ with } \frac{1}{\delta(\{t\})} < b_0^T(\delta, L), \\ 0, \qquad \text{for all other } t \in \mathcal{T}. \end{cases} \tag{8.3}$$

This shows the following theorem. Thereby we have

$$y_0(t)^2 = \delta(\{t\})\, g\left(\sqrt{b_0^T(\delta, L)}\, y_0(t)\right) > 0$$

with

$$g(y) := \int \min\{|z|, y\}^2\, P(dz)$$

and

$$b_0^T(\delta, L) = \max\{\frac{1}{\delta(\{t\})};\ t \in \mathrm{supp}(\delta)\}.$$

For the existence and the calculation of $y_0(t)$ see Appendix A.2. The following theorem generalizes Theorem 2 in Müller (1992a), which was given for $L = E_r$, and is analogous to Theorem 7.2. While in Theorem 7.2 we only assume that $\mathrm{supp}(\delta) = \{\tau_1, ..., \tau_I\}$ and $a(\tau_1), ..., a(\tau_I)$ are linearly independent, we here additionally assume $I = s$, where s is the rank of L.

Theorem 8.3 *Let φ be identifiable at $supp(\delta) = \{\tau_1, ..., \tau_s\}$ and $a(\tau_1), ..., a(\tau_s)$ be linearly independent. Then:*

a) The minimum asymptotic bias for testing φ at δ satisfies

$$b_0^T(\delta, L) = \max\{\frac{1}{\delta(\{t\})};\ t \in supp(\delta)\}.$$

b) ψ_{01} given by (8.2) is the influence function of an ALE-test statistic for testing φ at δ with minimum asymptotic bias.

c) ψ_0 given by (8.3) is the influence function of an ALE-test statistic for testing φ at δ with minimum asymptotic bias.

Proof. Because φ is identifiable at $\mathcal{D} = \mathrm{supp}(\delta)$ and $L \in \mathbb{R}^{s \times r}$ is of rank s we have $L = K\, A_{\mathcal{D}}$ for some $K \in \mathbb{R}^{s \times s}$ and K is nonsingular. Hence, according to Lemma 8.2 every $\psi \in \Psi^*(\delta, L)$ satisfies

$$\begin{aligned} & \psi(z,\tau_i)' \left(\int \psi\, \psi'\, d(P \otimes \delta)\right)^{-1} \psi(z, \tau_i) \qquad (8.4) \\ = \;& \rho(z,\tau_i)^2\, u_i'\, D_1(\rho)^{-1}\, D^{-1}\, K'\, [K\, D_1(\rho)^{-1}\, D_2(\rho)\, D_1(\rho)^{-1}\, D^{-1}\, K']^{-1} \\ & K\, D_1(\rho)^{-1}\, D^{-1}\, u_i \\ = \;& \rho(z,\tau_i)^2\, u_i'\, D_2(\rho)^{-1}\, D^{-1}\, u_i \\ = \;& \frac{\rho(z,\tau_i)^2}{\int \rho(y,\tau_i)^2\, P(dy)}\, \frac{1}{\delta(\{\tau_i\})} \end{aligned}$$

for $i = 1, ..., s$. For the one-dimensional model $\overline{Y} = \overline{\beta} + Z$ with $\overline{\beta} \in \mathbb{R}$, $\overline{a}(t) = 1$ and $L = 1$ we have $\frac{\rho(\cdot,\tau_i)}{\int \rho(z,\tau_i)\, z\, P(dz)} \in \overline{\Psi}(1)$, where $\overline{\Psi}(1)$ is defined for the one-dimensional model as in the proof of Theorem 7.10. Then, Lemma 8.1 applied on the one-dimensional model with $s = 1$ provides

$$\max_{z \in \mathbb{R}} \frac{\rho(z, \tau_i)^2}{\int \rho(y, \tau_i)^2 \, P(dy)} \geq 1$$

for all $i = 1, ..., s$. Hence, we have

$$\max_{z \in \mathbb{R}, i=1,...,s} \psi(z, \tau_i)' \left(\int \psi \, \psi' \, d(P \otimes \delta)\right)^{-1} \psi(z, \tau_i) \geq \max\{\tfrac{1}{\delta(\{\tau_i\})}; \, i = 1, ..., s\}$$

for all $\psi \in \Psi^*(\delta, L)$ so that

$$b_0^T(\delta, L) \geq \max\left\{\frac{1}{\delta(\{\tau_i\})}; \, i = 1, ..., s\right\}.$$

It is easy to see that ψ_0 and ψ_{01} are elements of $\Psi^*(\delta, L)$ (for ψ_{01}, the influence function of the L_1 estimator, see also Section 7.1). Moreover, we have that ψ_{01} has the form $\psi_{01}(z, t) = L\, M(\rho_{01})^- \, a(t) \, \rho_{01}(z, t)$ with $\rho_{01}(z, t) = \text{sgn}(z)$ so that (8.4) provides

$$\max_{(z,t) \in \mathbb{R} \times \text{supp}(\delta)} \psi_{01}(z, t)' \left(\int \psi_{01} \, \psi_{01}' \, d(P \otimes \delta)\right)^{-1} \psi_{01}(z, t) = \max\{\tfrac{1}{\delta(\{\tau_i\})}; \, i = 1, ..., s\}.$$

This implies assertion a) and b).

For ψ_0 we have $\psi_0(z, t) = L\, M(\rho_0)^- \, a(t) \, \rho_0(z, t)$ with

$$\rho_0(z, \tau_i) = \begin{cases} \text{sgn}(z) & \text{for } \frac{1}{\delta(\{\tau_i\})} = b_0^T(\delta, L), \\ \text{sgn}(z) \min\{|z|, \sqrt{b_0^T(\delta, L)} \, y_0(\tau_i)\} & \text{for } \frac{1}{\delta(\{\tau_i\})} < b_0^T(\delta, L), \end{cases}$$

so that

$$\max_{z \in \mathbb{R}} \frac{\rho_0(z, \tau_i)^2}{\int \rho_0(y, \tau_i)^2 \, P(dy)} \, \frac{1}{\delta(\{\tau_i\})} = b_0^T(\delta, L)$$

for all $i = 1, ..., s$. Note that

$$\begin{aligned} & \max_{z \in \mathbb{R}} \frac{\rho_0(z, \tau_i)^2}{\int \rho_0(y, \tau_i)^2 \, P(dy)} \, \frac{1}{\delta(\{\tau_i\})} \\ = \; & \max_{z \in \mathbb{R}} \frac{\rho_0(z, \tau_i)^2}{g(\sqrt{b_0^T(\delta, L)} \, y_0(\tau_i))} \, \frac{1}{\delta(\{\tau_i\})} \\ = \; & \frac{b_0^T(\delta, L) \, y_0(\tau_i)^2}{y_0(\tau_i)^2} \end{aligned}$$

for $\frac{1}{\delta(\{\tau_i\})} < b_0^T(\delta, L)$. Hence, the assertion c) is also proved. □

Example 8.1 (One-way lay-out, continuation of Example 2.2)
Example 2.2 provides that the D-optimal design δ_D for $\varphi(\beta) = (\beta_2-\beta_1, \beta_3-\beta_1, \beta_4-\beta_1)'$ in a one-way lay-out model with four levels is

$$\delta_D = \frac{1}{4}(e_1 + e_2 + e_3 + e_4).$$

Theorem 8.1 and Corollary 8.1 provide the minimum asymptotic bias of an ALE-test for testing $H_0 : \varphi(\beta) = (\beta_2 - \beta_1, \beta_3 - \beta_1, \beta_4 - \beta_1)' = 0$, or equivalently $H_0 : \beta_1 = \beta_2 = \beta_3 = \beta_4$, against $H_1 : \varphi(\beta) = (\beta_2 - \beta_1, \beta_3 - \beta_1, \beta_4 - \beta_1)' \neq 0$ in a one-way lay-out model with shrinking contamination. According to this theorem and this corollary the minimum asymptotic bias at the D-optimal design δ_D satisfies

$$b_0^T(\delta_D, L) = b_0^T(\Delta(\varphi), L) = 3$$

and the influence function of an ALE-test statistic for testing $H_0 : \varphi(\beta) = 0$ at δ_D with minimum asymptotic bias has the form

$$\psi_{01}(z,t) = \begin{cases} (-1,-1,-1)' \operatorname{sgn}(z)\, 4\sqrt{\frac{\pi}{2}} & \text{for } t = 1, \\ (1_2(t), 1_3(t), 1_4(t))' \operatorname{sgn}(z)\, 4\sqrt{\frac{\pi}{2}} & \text{for } t \neq 1. \end{cases} \tag{8.5}$$

□

Example 8.2 (Quadratic regression)
In a quadratic regression model the observation at t_{nN} is given by

$$Y_{nN} = \beta_0 + \beta_1 t_{nN} + \beta_2 t_{nN}^2 + Z_{nN} = a(t_{nN})'\beta + Z_{nN}$$

for $n = 1, ..., N$, where $\beta = (\beta_0, \beta_1, \beta_2)' \in I\!R^3$ and $a(t) = (1, t, t^2)' \in I\!R^3$. Assume that $\mathcal{T} = [0,1]$ and that the interesting aspect is $\varphi(\beta) = (\beta_0, \beta_1 + \beta_2)' \in I\!R^2$. In particular, testing the hypothesis $H_0 : \varphi(\beta) = (l_1, 0)'$ means that the hypothesis is tested that the quadratic function is symmetric on $[0,1]$ and is equal to l_1 at the end points $t = 0$ and $t = 1$. If we restrict ourselves to designs with a support included in $\mathcal{D} = \{0,1\}$, then according to Theorem 2.8 the D-optimal design for φ in $\Delta_{\mathcal{D}}$ is

$$\delta_D = \frac{1}{2}(e_0 + e_1)$$

and according to Theorem 2.7 the A-optimal design for φ in $\Delta_{\mathcal{D}}$ is

$$\delta_A = \frac{1}{\sqrt{2}+1}(\sqrt{2}e_0 + e_1).$$

Note that as for linear regression on $\mathcal{T} = [0,1]$ (see Example 2.1) the A-optimal design puts more observations at $t = 0$ than the D-optimal design. As for linear estimation this is due to the fact that at $t = 0$ the component β_0 can be estimated very precisely so that also the estimation of $\beta_1 + \beta_2$, which is confounded with β_0 at $t = 1$, profits by a precise estimation of β_0. Also as for linear regression a design prefering $t = 0$ has no advantage for testing because any hypothesis of the form $H_0 : \varphi(\beta) = (l_1, l_2)'$ is equivalent with $H_0 : \tilde{\varphi}(\beta) = (\beta_0, \beta_0 + \beta_1 + \beta_2)' = (l_1, l_1 + l_2)'$ and even the A-optimal design for $\tilde{\varphi}$ has the form $\frac{1}{2}(e_0 + e_1)$.

Theorem 8.1 and Corollary 8.1 provide the minimum asymptotic bias of an ALE-test for testing φ at δ_D in the quadratic regression model with shrinking contamination. According to this theorem and this corollary the minimum asymptotic bias at the D-optimal design δ_D satisfies

$$b_0^T(\delta_D, L) = b_0^T(\Delta(\varphi), L) = 2$$

and the influence function of an ALE-test statistic for testing φ at δ_D with minimum asymptotic bias has the form

$$\psi_{01}(z,t) = \begin{cases} (1,-1)' \operatorname{sgn}(z)\, 2\, \sqrt{\frac{\pi}{2}} & \text{for } t = 0, \\ (0,1)' \operatorname{sgn}(z)\, 2\, \sqrt{\frac{\pi}{2}} & \text{for } t = 1. \end{cases} \tag{8.6}$$

According to Theorem 8.3 the minimum asymptotic bias at the A-optimal design δ_A is

$$b_0^T(\delta_A, L) = \sqrt{2} + 1$$

and there exists at least two ALE-test statistics with minimum asymptotic bias at δ_A and different influence functions. The different influence functions are ψ_{01} given by

$$\psi_{01}(z,t) = \begin{cases} (1,-1)' \operatorname{sgn}(z)\, \frac{\sqrt{2}+1}{\sqrt{2}}\, \sqrt{\frac{\pi}{2}} & \text{for } t = 0, \\ (0,1)' \operatorname{sgn}(z)\, (\sqrt{2}+1)\, \sqrt{\frac{\pi}{2}} & \text{for } t = 1, \end{cases}$$

and ψ_0 given by

$$\psi_0(z,t) = \begin{cases} (1,-1)' \operatorname{sgn}(z)\, \frac{\sqrt{2}+1}{\sqrt{2}}\, \frac{\min\{|z|, y\sqrt{\sqrt{2}+1}\}}{2\Phi(y\sqrt{\sqrt{2}+1})-1} & \text{for } t = 0, \\ (0,1)' \operatorname{sgn}(z)\, (\sqrt{2}+1)\, \sqrt{\frac{\pi}{2}} & \text{for } t = 1, \end{cases}$$

where $y^2 = \frac{\sqrt{2}}{\sqrt{2}+1}\, g\left(y\sqrt{\sqrt{2}+1}\right)$, i.e. $y \approx 0.366$. □

8.2 Optimal Tests and Designs for a Bias Bound

For a robust test the asymptotic bias of the first error should be bounded. For an ALE-test statistic with influence function ψ this is according to Corollary 6.1 the case if and only if $\psi' C(\psi)^{-1} \psi$ is bounded. In Section 8.1

we derived ALE-test statistics and designs with minimum asymptotic bias of the first error of the test. But the power of the test should be also not too bad.

Because ALE-test statistics have asymptotically a chi-squared distribution (see Theorem 6.1), the power of an ALE-test depends on the noncentrality parameter which appears for contiguous alternatives. According to Theorem 6.1 contiguous alternatives of the form $\beta_N = \beta + N^{-1/2}\overline{\beta}$ with $\varphi(\beta_N) = l + N^{-1/2}\gamma$ provide a noncentrality parameter of

$$[\gamma + \sigma\, b(\psi, (Q_{N,\delta})_{N\in\mathbb{N}})]'\, [\sigma^2\, C(\psi)]^{-1}\, [\gamma + \sigma\, b(\psi, (Q_{N,\delta})_{N\in\mathbb{N}})] \qquad (8.7)$$

for shrinking contamination. Thereby the power of the test increases when the noncentrality parameter increases. Hence the noncentrality parameter given by (8.7) should be large for all $\gamma \in \mathbb{R}^s$ and all contamination distributions $Q_{N,\delta}$. In particular, it should be large for the ideal distribution, i.e. for $Q_{N,\delta} = P_\delta$. Then, because of $b(\psi, (P_\delta)_{N\in\mathbb{N}}) = 0$ the noncentrality parameter

$$\gamma'\, [\sigma^2\, C(\psi)]^{-1}\, \gamma$$

should be large for all $\gamma \in \mathbb{R}$. This is the case if and only if the asymptotic covariance matrix $C(\psi)$ of the underlying AL-estimator is small with respect to the ordering of matrices, where $C_1 \leq C_2$ if and only if $C_2 - C_1$ is positive-semidefinite. With respect to this ordering of matrices $C(\psi)$ can be minimized at a given design δ if there is no bound for the asymptotic bias. For, $C(\psi)$ is minimized by ψ_∞ given by $\psi_\infty(z,t) = L\mathcal{I}(\delta)^-\, a(t)\, z$, which is the influence function of the Gauss-Markov estimator for φ (see (5.15) in Section 5.2). Because the F-test is based on the Gauss-Markov estimator this means that the F-test is asymptotically most powerful within all ALE-tests. But the asymptotic bias of the first error of the F-test is unbounded because

$$\max_{(z,t)\in\mathbb{R}\times\text{supp}(\delta)}\, \psi_\infty(z,t)\, C(\psi_\infty)^{-1}\, \psi_\infty(z,t) = \infty.$$

Hence, to derive optimal robust tests we should try to minimize $C(\psi)$ with respect to ψ under the side condition that the asymptotic bias is bounded by some bias bound b. ALE-test statistics, which solve this problem with respect to the ordering of matrices, i.e. which minimize $C(\psi)$ in the positive-semidefinite sense, are called *asymptotically U-optimal ALE-test statistics for the bias bound b* (compare also with the definition of asymptotically U-optimal estimators in Section 7.2 and of U-optimal designs in Section 2.2). Because only for very special design situations asymptotically U-optimal ALE-test statistics for a bias bound b can be found we define also *admissible ALE-test statistics for a bias bound b*. We define also *D-optimal ALE-test statistics for a bias bound b* which minimize the determinant of the asymptotic covariance matrix under the side condition of bounded asymptotic bias. Thereby we use the determinant criterion because it is invariant with respect to one-to-one linear transformations of the aspect φ which is very important for testing (see also Section 2.2). We also will derive *D-optimal designs for testing with bias bound* b which minimize within a given class Δ of designs the determinant of the asymptotic covariance matrix of D-optimal ALE-test statistics for a bias bound b.

Definition 8.3 (Optimal ALE-test statistics and designs for a bias bound b)
a) An ALE-test statistic $\widehat{\tau}_N$ for testing $H_0 : \varphi(\beta) = l$ with influence function $\psi_{b,\delta}$ is asymptotically U-optimal for testing φ at δ with bias bound b if

$$\psi_{b,\delta} \in \arg\min\{\textstyle\int \psi\psi' \, d(P\otimes\delta);\ \psi \in \Psi^*(\delta, L) \ \textit{with} \ \|\psi' \, C(\psi)^{-1} \, \psi\|_\delta \leq b\}.$$

b) An ALE-test statistic $\widehat{\tau}_N$ for testing $H_0 : \varphi(\beta) = l$ with influence function $\psi_{b,\delta}$ is asymptotically admissible for testing φ at δ with bias bound b if

$$\int \psi\psi' \, d(P\otimes\delta) = \int \psi_{b,\delta}\psi'_{b,\delta} \, d(P\otimes\delta)$$

for all $\psi \in \Psi^*(\delta, L)$ with $\|\psi' \, C(\psi)^{-1} \, \psi\|_\delta \leq b$ and $\int \psi\psi' \, d(P \otimes \delta) \leq \int \psi_{b,\delta}\psi'_{b,\delta} \, d(P\otimes\delta)$.

c) An ALE-test statistic $\widehat{\tau}_N$ for testing $H_0 : \varphi(\beta) = l$ with influence function $\psi_{b,\delta}$ is asymptotically D-optimal for testing φ at δ with bias bound b if

$$\psi_{b,\delta} \in \arg\min\left\{\det\left(\int \psi\psi' \, d(P\otimes\delta)\right);\ \psi \in \Psi^*(\delta, L) \right.$$
$$\left. \textit{with} \ \|\psi' \, C(\psi)^{-1} \, \psi\|_\delta \leq b\right\}.$$

d) A design δ_b is asymptotically D-optimal in Δ for testing φ with bias bound b if

$$\delta_b \in \arg\min\left\{\min\left\{\det\left(\int \psi\psi' \, d(P\otimes\delta)\right);\ \psi \in \Psi^*(\delta, L)\right.\right.$$
$$\left.\left. \textit{with} \ \|\psi' \, C(\psi)^{-1} \, \psi\|_\delta \leq b\right\};\delta \in \Delta\right\}.$$

Note that $\delta_b = \arg\min\{\det(\int \psi_{b,\delta}\psi'_{b,\delta} \, d(P\otimes\delta));\ \delta \in \Delta\}$ if $\psi_{b,\delta}$ is the influence function of an asymptotically D-optimal ALE-test statistic for testing φ at δ with bias bound b.

By generalizing a result of Krasker and Welsch (1982) concerning admissible estimators for estimating β with a self-standardized gross-error-sensitivity bounded by b we can characterize the influence function of admissible ALE-tests for a bias bound b. This shows the following theorem which uses the following matrices:

$$
\begin{aligned}
M_b(B) &:= \int a(t)a(t)' \left[2\Phi\left(\tfrac{\sqrt{b}}{|Ba(t)|}\right) - 1\right] \delta(dt),\\
Q_b(B) &:= \int a(t)a(t)' \, g\left(\tfrac{\sqrt{b}}{|Ba(t)|}\right) \delta(dt),\\
M_{b,0}(B) &:= \int a(t)a(t)' \tfrac{\sqrt{b}}{|Ba(t)|}\sqrt{\tfrac{2}{\pi}}\, \delta(dt),\\
Q_{b,0}(B) &:= \int a(t)a(t)' \left(\tfrac{\sqrt{b}}{|Ba(t)|}\right)^2 \delta(dt).
\end{aligned}
$$

Theorem 8.4 *Let φ be identifiable at δ, and*

$$\psi_{b,\delta}(z,t) := L\, M_b(B_b)^- \, a(t) \min\left\{|z|, \tfrac{\sqrt{b}}{|B_b a(t)|}\right\} \operatorname{sgn}(z),$$

where

$$B_b = [L\, M_b(B_b)^- Q_b(B_b) M_b(B_b)^- L']^{-1/2} L\, M_b(B_b)^-,$$

or

$$\psi_{b,\delta}(z,t) := L\, M_{b,0}(B_b)^- \, a(t) \tfrac{\sqrt{b}}{|B_b a(t)|} \operatorname{sgn}(z),$$

where

$$B_b = [L\, M_{b,0}(B_b)^- Q_{b,0}(B_b) M_{b,0}(B_b)^- L']^{-1/2} L\, M_{b,0}(B_b)^-.$$

Then the ALE-test statistic with influence function $\psi_{b,\delta}$ is asymptotically admissible for testing φ at δ with bias bound b.

Proof. Because $L = K\, M_b(B_b)$ and $L = K_0\, M_{b,0}(B_b)$ for some $K,\ K_0 \in \mathbb{R}^{s\times r}$, which follows similarly as the assertion of Lemma 1.3, it is easy to see that $\psi_{b,\delta}$ is an element of $\Psi^*(\delta, L)$. To show that $\psi_{b,\delta}$ is admissible the proof of Krasker and Welsch (1982) is generalized:

At first assume

$$\psi_{b,\delta}(z,t) := L\, M_b(B_b)^- \, a(t) \min\left\{|z|, \tfrac{\sqrt{b}}{|B_b a(t)|}\right\} \operatorname{sgn}(z).$$

Setting

$$
\begin{aligned}
V &:= L\, M_b(B_b)^- \, Q_b(B_b)\, M_b(B_b)^- \, L',\\
D &:= B_b\, M_b(B_b)
\end{aligned}
$$

and

$$\psi_B(z,t) := B_b a(t) \min\left\{|z|, \tfrac{\sqrt{b}}{|B_b a(t)|}\right\} \operatorname{sgn}(z)$$

one gets

$$\begin{aligned}
\psi_B &= V^{-1/2}\psi_{b,\delta}, \\
\int & \psi_B(z,t)a(t)'z\, P(dz)\delta(dt) = D, \\
\int & \psi_B\psi_B' d\,(P\otimes\delta) = B_b Q_b(B_b) B_b' \\
&= V^{-1/2} L\, M_b(B_b)^- Q_b(B_b) M_b(B_b)^- L' V^{-1/2} = E_{s\times s}
\end{aligned}$$

and (see Theorem 7.7)

$$\psi_B = \arg\min\{\mathrm{tr}\int \psi\psi'\, d(P\otimes\delta);\ \int \psi(z,t)a(t)'z\, P(dz)\delta(dt) = D, \quad \|\psi'\psi\|_\delta \le b\}. \tag{8.8}$$

Assume that there exist ψ_0 with

$$\begin{aligned}
\textstyle\int \psi_0\psi_0'\, d(P\otimes\delta) < \int \psi_{b,\delta}\psi_{b,\delta}'\, d(P\otimes\delta), \\
\textstyle\int \psi_0(z,t)a(t)'z\, P(dz)\delta(dt) = L
\end{aligned} \tag{8.9}$$

and

$$\|\psi_0 \left(\textstyle\int \psi_0\psi_0'\, d(P\otimes\delta)\right)^{-1} \psi_0\|_\delta \le b. \tag{8.10}$$

Then

$$\textstyle\int V^{-1/2}\psi_0(z,t)a(t)'z\, P(dz)\delta(dt) = V^{-1/2}L = B_b M_b(B_b) = D$$

and

$$\begin{aligned}
& V^{-1/2}\int \psi_0\psi_0'\, d(P\otimes\delta)\, V^{-1/2} < V^{-1/2}\int \psi_{b,\delta}\psi_{b,\delta}'\, d(P\otimes\delta)\, V^{-1/2} \\
&= \int \psi_B\psi_B'\, d(P\otimes\delta) = E_{s\times s}
\end{aligned}$$

which implies with (8.10)

$$\begin{aligned}
& \|\psi_0' V^{-1/2} V^{-1/2}\psi_0\|_\delta \\
&\le \left\| \psi_0' V^{-1/2}\left(V^{-1/2}\int \psi_0\psi_0'\, d(P\otimes\delta)\, V^{-1/2}\right)^{-1} V^{-1/2}\psi_0 \right\|_\delta \\
&= \left\| \psi_0'\left(\int \psi_0\psi_0'\, d(P\otimes\delta)\right)^{-1}\psi_0 \right\|_\delta \le b.
\end{aligned}$$

Hence property (8.8) provides for $V^{-1/2}\psi_0$

$$\begin{aligned}
& \mathrm{tr}\int V^{-1/2}\psi_0\psi_0' V^{-1/2}\, d(P\otimes\delta) \ge \mathrm{tr}\int \psi_B\psi_B'\, d(P\otimes\delta) \\
&= \mathrm{tr}\, V^{-1/2}\int \psi_{b,\delta}\psi_{b,\delta}'\, d(P\otimes\delta)\, V^{-1/2}
\end{aligned}$$

which is a contradiction to (8.9).

For

$$\psi_{b,\delta}(z,t) := L\, M_{b,0}(B_b)^{-}\, a(t)\, \frac{\sqrt{b}}{|B_b a(t)|}\, \operatorname{sgn}(z)$$

the assertion follows as above. Setting

$$\psi_B(z,t) := B_b\, a(t)\, \frac{\sqrt{b}}{|B\, a(t)|}\, \operatorname{sgn}(z).$$

only $M_b(B_b)$ and $Q_b(B_b)$ have to be replaced by $M_{b,0}(B_b)$ and $Q_{b,0}(B_b)$. □

For $s = 1$ the admissible solution for a bias bound b is also U-optimal for the bias bound b. In particular, for the one-dimensional location case we get the following corollary.

Corollary 8.2 *If $a(t) = 1$ for all $t \in \mathcal{T}$ and $L = 1$, then an ALE-test statistic with influence function ψ_b given by*

$$\psi_b(z) = \begin{cases} sgn(z)\,\sqrt{\frac{\pi}{2}} & \text{for } b = 1, \\ sgn(z)\,\frac{\min\{|z|,\sqrt{b}\,y\}}{2\Phi(\sqrt{b}\,y)-1} & \text{for } b > 1, \end{cases}$$

where $y^2 = g(\sqrt{b}\,y) > 0$, is asymptotically U-optimal for testing β at δ with bias bound b. In particular, setting

$$\rho_b(z) = \begin{cases} sgn(z) & \text{for } b = 1, \\ sgn(z)\,\min\{|z|,\sqrt{b}\,y\} & \text{for } b > 1, \end{cases}$$

we have

$$\frac{\int \rho_b(z)^2\, P(dz)}{(\int \rho_b(z)\, z\, P(dz))^2} \le \frac{\int \rho(z)^2\, P(dz)}{(\int \rho(z)\, z\, P(dz))^2}$$

for all $\rho : \mathbb{R} \to \mathbb{R}$ with $\max_{z\in\mathbb{R}} \frac{\rho(z)^2}{\int \rho(y)^2\, P(dy)} \le b$, $\int \rho(z)\, P(dz) = 0$ and $\int \rho(z)^2\, P(dz) < \infty$.

Proof. For $b > 1$ set $B_b = \frac{1}{y}$. Then we have

$$Q_b(B_b) = g\left(\frac{\sqrt{b}}{B_b}\right) = g(\sqrt{b}\,y) = y^2$$

so that

$$B_b = \sqrt{\frac{1}{Q_b(B_b)}} = [M_b(B_b)^{-1}\, Q_b(B_b)\, M_b(B_b)^{-1}]^{-1/2}\, M_b(B_b)^{-1}$$

and

$$\psi_b(z) = M_b(B_b)^{-1}\, \operatorname{sgn}(z)\, \min\left\{|z|, \frac{\sqrt{b}}{|B_b|}\right\}.$$

For $b = 1$ we can choose B_b arbitrarily because

$$Q_{b,0}(B_b) = \frac{b}{B_b^2} = \frac{1}{B_b^2}.$$

Again we have

$$B_b = \sqrt{\frac{1}{Q_{b,0}(B_b)}} = [M_{b,0}(B_b)^{-1}\, Q_{b,0}(B_b)\, M_{b,0}(B_b)^{-1}]^{-1/2}\, M_{b,0}(B_b)^{-1}$$

and

$$\psi_b(z) = M_{b,0}(B_b)^{-1}\,\mathrm{sgn}(z)\,\frac{\sqrt{b}}{|B_b|}.$$

Hence, the assertion follows from Theorem 8.4. □

If the regressors are linearly independent on the support of the design, then Corollary 8.2 provides an explicit characterization of the influence functions of asymptotically U-optimal ALE-test statistics for testing with a bias bound. This result is analogous to Theorem 7.10 which deals with asymptotically U-optimal estimation with bias bound. But additionally to estimation we need that the support of the design has only s elements if the rank of $L \in \mathbb{R}^{s\times r}$ is s. Then the influence function of the asymptotically U-optimal ALE-test statistic is given by

$$\psi_{b,\delta}(z,t) = \begin{cases} L\,\mathcal{I}(\delta)^-\, a(t)\,\mathrm{sgn}(z)\,\sqrt{\frac{\pi}{2}}, & \\ \qquad \text{for } t \in \mathrm{supp}(\delta) \text{ with } \frac{1}{\delta(\{t\})} = b, & \\ L\,\mathcal{I}(\delta)^-\, a(t)\,\mathrm{sgn}(z)\,\frac{\min\{|z|,\sqrt{b}\,y(t)\}}{2\Phi(\sqrt{b}\,y(t))-1}, & \\ \qquad \text{for } t \in \mathrm{supp}(\delta) \text{ with } \frac{1}{\delta(\{t\})} < b, & \\ 0, \qquad \text{for all other } t \in \mathcal{T}, & \end{cases} \tag{8.11}$$

where

$$y(t)^2 = \delta(\{t\})\, g\left(\sqrt{b}\, y(t)\right) > 0.$$

Thereby $\psi_{b,\delta}$ exists if and only if

$$b \geq \max\{\frac{1}{\delta(\{t\})};\ t \in \mathrm{supp}(\delta)\}$$

because $y(t)$ exists if and only if $b > \frac{1}{\delta(\{t\})}$ (see Appendix A.2).

Theorem 8.5 *If φ is identifiable at supp(δ) $= \{\tau_1, ..., \tau_s\}$, $a(\tau_1), ..., a(\tau_s)$ are linearly independent and $b \geq \max\{\frac{1}{\delta(\{t\})};\ t \in supp(\delta)\}$, then $\psi_{b,\delta}$ given by (8.11) is the influence function of an asymptotically U-optimal ALE-test statistic for testing φ at δ with bias bound b.*

Proof. Set

$$\rho_{b,\delta}(z,t) = \begin{cases} \operatorname{sgn}(z)\sqrt{\frac{\pi}{2}}, & \text{for } t \in \operatorname{supp}(\delta) \text{ with } \frac{1}{\delta(\{t\})} = b, \\ \operatorname{sgn}(z)\frac{\min\{|z|,\sqrt{b}\,y(t)\}}{2\Phi(\sqrt{b}\,y(t))-1}, & \text{for } t \in \operatorname{supp}(\delta) \text{ with } \frac{1}{\delta(\{t\})} < b, \\ 0, & \text{for all other } t \in \mathcal{T}, \end{cases} .$$

Then we have $\psi_{b,\delta}(\cdot,\cdot) = L\,M(\rho_{b,\delta})^-\,a(\cdot)\,\rho_{b,\delta}(\cdot,\cdot)$. Because φ is identifiable at $\mathcal{D} = \operatorname{supp}(\delta)$ there exist a nonsingular $K \in I\!R^{s\times s}$ with $L = K\,A_{\mathcal{D}}$. Then Lemma 7.1 provides for any $\psi(\cdot,\cdot) = L\,M(\rho)^-\,a(\cdot)\,\rho(\cdot,\cdot) \in \Psi^*(\delta, L)$

$$\int \psi\,\psi'\,d(P\otimes\delta) = K\,D_1(\rho)^{-1}\,D_2(\rho)\,D_1(\rho)^{-1}\,D^{-1}\,K',$$

and Lemma 8.2 provides

$$\begin{aligned} &\psi(z,\tau_i)'\left(\int \psi\,\psi'\,d(P\otimes\delta)\right)^{-1}\psi(z,\tau_i) \\ &= \rho(z,\tau_i)^2\,u_i'\,D_1(\rho)^{-1}\,D^{-1}\,K'\,[K\,D_1(\rho)^{-1}\,D_2(\rho)\,D_1(\rho)^{-1}\,D^{-1}\,K']^{-1} \\ &\quad K\,D_1(\rho)^{-1}\,D^{-1}\,u_i \\ &= \rho(z,\tau_i)^2\,u_i'\,D_2(\rho)^{-1}\,D^{-1}\,u_i \end{aligned}$$

for $i = 1, ..., s$. Hence, for every $\psi \in \Psi^*(\delta, L)$ with $||\psi'(\int \psi\psi' d(P\otimes\delta))^{-1}\psi||_\delta \leq b$ we obtain

$$\max_{z\in I\!R}\frac{\rho(z,\tau_i)^2}{\int \rho(y,\tau_i)^2\,P(dy)} \leq b\,\delta(\{\tau_i\}) \tag{8.12}$$

for $i = 1, ..., s$. Because $y(t)^2 = \delta(\{t\})\,g(\sqrt{b}\,y(t))$ if and only if

$$\frac{y(t)^2}{\delta(\{t\})} = g\left(\sqrt{b\,\delta(\{t\})}\,\frac{y(t)}{\sqrt{\delta(\{t\})}}\right)$$

Corollary 8.2 provides for $i = 1, ..., s$ that the i'th diagonal element of $D_1(\rho)^{-2}\,D_2(\rho)\,D^{-1}$ is minimized by $\rho_{b,\delta}(\cdot,\tau_i)$ under the side condition (8.12). Hence,

$$D_1(\rho_{b,\delta})^{-2}\,D_2(\rho_{b,\delta})\,D^{-1} \leq D_1(\rho)^{-2}\,D_2(\rho)\,D^{-1}$$

and therefore

$$K\,D_1(\rho_{b,\delta})^{-2}\,D_2(\rho_{b,\delta})\,D^{-1}\,K' \leq K\,D_1(\rho)^{-2}\,D_2(\rho)\,D^{-1}\,K',$$

for all $\psi \in \Psi^*(\delta, L)$ with $\|\psi'(\int \psi\psi' d(P\otimes\delta))^{-1}\psi\|_\delta \leq b$. □

Theorem 8.4 can also be used to show that at a D-optimal design δ_D the influence function of an admissible ALE-test statistic for testing with a bias bound has a simple form. These influence functions are given by

$$\psi_{b,\delta_D}(z,t) = \begin{cases} L\mathcal{I}(\delta_D)^- a(t)\,\text{sgn}(z)\sqrt{\frac{\pi}{2}}, & \text{for } b = s, \\ L\mathcal{I}(\delta_D)^- a(t)\,\text{sgn}(z)\,\frac{\min\{|z|,\sqrt{b}\,y_b\}}{2\Phi(\sqrt{b}\,y_b)-1}, & \text{for } b > s, \end{cases} \tag{8.13}$$

where

$$y_b^2 = \frac{1}{s}\, g\left(\sqrt{b}\, y_b\right) > 0.$$

This shows the following theorem. This theorem is analogous to Theorem 7.11 which shows that for estimation with bias bound at A-optimal designs the influence functions of A-optimal AL-estimators have a simple form. Thereby instead of the trace of the asymptotic covariance matrix we have a simple formula for the asymptotic covariance matrix of the underlying AL-estimator. For that define $\overline{W} : (0,\infty)^2 \to I\!R$ by $\overline{W}(c,y) = \frac{1}{c}\, g(y) - y^2$ and $\overline{w}(c)$ implicitely by $\overline{W}(c,\overline{w}(c)) = 0$. Note that y_b of (8.13) satisfies $y_b = \overline{w}\left(\frac{s}{b}\right)\frac{1}{\sqrt{b}}$, and $y(t)$ of (8.11) satisfies $y(t) = \overline{w}\left(\frac{1}{b\,\delta(\{t\})}\right)\frac{1}{\sqrt{b}}$. Set also

$$\overline{v}(c) = \begin{cases} \frac{(2\Phi(\overline{w}(c))-1)^2}{g(\overline{w}(c))} & \text{for } c < 1 \\ \frac{2}{\pi} & \text{for } c = 1 \end{cases}.$$

Theorem 8.6 *If φ is identifiable at $\mathcal{D}$, δ_D is D-optimal in $\Delta_\mathcal{D}$ and $b \geq s$, then an ALE-test statistic with influence function ψ_{b,δ_D} given by (8.13) is asymptotically admissible for testing φ with bias bound b at δ_D and its asymptotic covariance matrix satisfies*

$$\int \psi_{b,\delta_D}\,\psi'_{b,\delta_D}\, d(P\otimes\delta_D) = \frac{1}{\overline{v}\left(\frac{s}{b}\right)}\, L\mathcal{I}(\delta_D)^-\, L'.$$

Proof. Set

$$B_b = \begin{cases} (L\mathcal{I}(\delta_D)^-\,L')^{-1/2}\, L\mathcal{I}(\delta_D)^-\,\frac{1}{\sqrt{s}\,y_b} & \text{for } b > s, \\ (L\mathcal{I}(\delta_D)^-\,L')^{-1/2}\, L\mathcal{I}(\delta_D)^- & \text{for } b = s. \end{cases}$$

Then the equivalence theorem for D-optimality (Theorem 2.5) provides

$$\begin{aligned} |B_b\, a(t)| &= \sqrt{a(t)'\,\mathcal{I}(\delta_D)^-\,L'\,[L\mathcal{I}(\delta_D)^-\,L']^{-1}\,L\mathcal{I}(\delta_D)^-\,a(t)}\;\frac{1}{\sqrt{s}\,y_b} \\ &= \frac{1}{y_b} \end{aligned}$$

for $b > s$ and similarly $|B_b\, a(t)| = \sqrt{s}$ for $b = s$ for all $t \in \mathcal{D}$ so that for $b > s$

$$\begin{aligned} M_b(B_b) &= (2\Phi(\sqrt{b}\, y_b) - 1)\, \mathcal{I}(\delta_D), \\ Q_b(B_b) &= g(\sqrt{b}\, y_b)\, \mathcal{I}(\delta_D) = s\, y_b^2\, \mathcal{I}(\delta_D) \end{aligned}$$

and for $b = s$

$$\begin{aligned} M_{b,0}(B_b) &= \sqrt{\frac{b}{s}}\sqrt{\frac{2}{\pi}}\,\mathcal{I}(\delta_D) = \sqrt{\frac{2}{\pi}}\,\mathcal{I}(\delta_D), \\ Q_{b,0}(B_b) &= \frac{b}{s}\,\mathcal{I}(\delta_D) = \mathcal{I}(\delta_D). \end{aligned}$$

Hence for $b > s$

$$B_b = [L\, M_b(B_b)^- Q_b(B_b) M_b(B_b)^- L']^{-1/2}\, L\, M_b(B_b)^-$$

and

$$\psi_{b,\delta_D}(z,t) := L\, M_b(B_b)^-\, a(t)\, \min\left\{|z|, \tfrac{\sqrt{b}}{|B_b a(t)|}\right\} \operatorname{sgn}(z),$$

and for $b = s$

$$B_b = [L\, M_{b,0}(B_b)^- Q_{b,0}(B_b) M_{b,0}(B_b)^- L']^{-1/2}\, L\, M_{b,0}(B_b)^-$$

and

$$\psi_{b,\delta_D}(z,t) := L\, M_b(B_b)^-\, a(t)\, \tfrac{\sqrt{b}}{|B_b a(t)|} \operatorname{sgn}(z),$$

so that the admissibility of ψ_{b,δ_D} follows from Theorem 8.4. Moreover, we obtain for $b > s$ because of $\sqrt{b}\, y_b = \overline{w}\left(\frac{s}{b}\right)$

$$\begin{aligned} &\int \psi_{b,\delta}\, \psi_{b,\delta}\, d(P \otimes \delta_D) \\ &= L\, M_b(B_b)^-\, Q_b(B_b)\, M_b(B_b)^-\, L' \\ &= \frac{1}{\overline{v}\left(\frac{s}{b}\right)}\, L\, \mathcal{I}(\delta_D)^-\, L' \end{aligned}$$

and for $b = s$

$$\begin{aligned} &\int \psi_{b,\delta}\, \psi_{b,\delta}\, d(P \otimes \delta_D) \\ &= L\, M_{b,0}(B_b)^-\, Q_{b,0}(B_b)\, M_{b,0}(B_b)^-\, L' \\ &= \frac{1}{\overline{v}(1)}\, L\, \mathcal{I}(\delta_D)^-\, L'. \square \end{aligned}$$

Note that for $\mathcal{D} = \{\tau_1, ..., \tau_I\}$, where $I = s$ and $a(\tau_1), ..., a(\tau_I)$ are linearly independent, a D-optimal design is given by $\delta = \frac{1}{s} \sum_{i=1}^{s} e_{\tau_i}$ (see Theorem

2.8). Hence, in this case the assertion of Theorem 8.6 follows also from Theorem 8.5, where Theorem 8.5 provides besides admissibility also optimality. If $I > s$ but $a(\tau_1), ..., a(\tau_I)$ are linearly independent, then we still have D-optimality of an ALE-test statistic with influence function $\psi_{b,\delta}$ for testing with bias bound b. Moreover, the D-optimal design is also D-optimal for testing with bias bound b in $\Delta_{\mathcal{D}}$ and this holds not only within $\Delta_{\mathcal{D}}$ with linearly independent regressors but also for any D-optimal design within designs of $\Delta(\varphi)$ with finite support. Therefore we have a theorem which is analogous to Theorem 7.12 for robust estimation. To show this theorem we need the following lemma. Therefor set

$$v_\rho(t) \;:=\; \frac{(\int \rho(z,t)\, z\, P(dz))^2}{\int \rho(z,t)^2 P(dz)}$$

and

$$\tilde{\delta}(\{t\}) \;:=\; \frac{1}{\sum_{\tau \in \mathcal{D}} v_\rho(\tau)\, \delta(\{\tau\})}\, v_\rho(t)\, \delta(\{t\})$$

and recall from Lemma 7.1 that for every $\psi \in \Psi^*(\delta, L)$ a $\rho : I\!R \times \mathcal{T} \to I\!R$ exists such that $\psi(z,t) = L\, M(\rho)^-\, a(t)\, \rho(z,t)$.

Lemma 8.3 *If φ is identifiable at $\mathcal{D} = \{\tau_1, ..., \tau_I\}$ and $a(\tau_1), ..., a(\tau_I)$ are linearly independent, then*

$$\int \psi\, \psi'\, d(P \otimes \delta) \geq \frac{1}{\overline{v}\left(\frac{s}{b}\right)}\; L\, \mathcal{I}(\tilde{\delta})^-\, L'$$

for all $\psi \in \Psi^*(\delta, L)$ with $\|\psi'(\int \psi\psi' d(P\otimes\delta))^{-1}\psi\|_\delta \leq b$ and all $\delta \in \Delta_{\mathcal{D}}$.

Proof. Regard any $\psi \in \Psi^*(\delta, L)$ with $\|\psi'(\int \psi\psi' d(P\otimes\delta))^{-1}\psi\|_\delta \leq b$ and any $\delta \in \Delta_{\mathcal{D}}$. Setting

$$\tilde{D} := \text{diag}(\tilde{\delta}(\{\tau_1\}), ..., \tilde{\delta}(\{\tau_I\})),$$

Lemma 7.1 and Lemma 2.1 provide with $L = K\, A_{\mathcal{D}}$

$$\int \psi\, \psi'\, d(P\otimes\delta) = K\, D_1(\rho)^{-1}\, D_2(\rho)\, D_1(\rho)^{-1}\, D^{-1}\, K' \tag{8.14}$$

$$= K \frac{1}{\sum_{t \in \mathcal{D}} v_\rho(t)\, \delta(\{t\})} \cdot \left(\frac{1}{\sum_{t \in \mathcal{D}} v_\rho(t)\, \delta(\{t\})}\, D_1(\rho)\, D_2(\rho)^{-1}\, D_1(\rho)\, D\right)^{-1} K'$$

$$= \frac{1}{\sum_{t \in \mathcal{D}} v_\rho(t)\, \delta(\{t\})}\, K\, \tilde{D}\, K'$$

$$= \frac{1}{\sum_{t \in \mathcal{D}} v_\rho(t)\, \delta(\{t\})}\, L\, \mathcal{I}(\tilde{\delta})^-\, L'.$$

Moreover, Lemma 8.2 and Lemma 2.1 provide for all $z \in \mathbb{R}$ and all $i = 1, ..., I$

$$\begin{aligned}
b \geq \psi(z,\tau_i)' & \left(\int \psi\,\psi'\, d(P \otimes \delta)\right)^{-1} \psi(z,\tau_i) \\
= \; & \rho(z,\tau_i)^2\, u_i'\, D_1(\rho)^{-1}\, D^{-1}\, K' \,[K\, D_1(\rho)^{-1}\, D_2(\rho)\, D_1(\rho)^{-1}\, D^{-1}\, K']^{-1} \\
& K\, D_1(\rho)^{-1}\, D^{-1}\, u_i \\
= \; & \rho(z,\tau_i)^2\, u_i'\, D_1(\rho)^{-1}\, D^{-1}\, K' \left(\sum\nolimits_{t\in\mathcal{D}} v_\rho(t)\, \delta(\{t\})\right) [L\,\mathcal{I}(\tilde{\delta})^-\, L']^{-1} \\
& K\, D_1(\rho)^{-1}\, D^{-1}\, u_i \\
= \; & \frac{\rho(z,\tau_i)^2}{\int \rho(y,\tau_i)^2 P(dy)}\, \frac{\int \rho(y,\tau_i)^2 P(dy)}{(\int \rho(y,\tau_i)\, y\, P(dy))^2}\, \frac{\sum_{t\in\mathcal{D}} v_\rho(t)\, \delta(\{t\})}{\delta(\{\tau_i\})^2} \\
& \cdot u_i'\, K' \,[L\,\mathcal{I}(\tilde{\delta})^-\, L']^{-1}\, K\, u_i \\
= \; & \frac{\rho(z,\tau_i)^2}{\int \rho(y,\tau_i)^2 P(dy)}\, \frac{1}{\tilde{\delta}(\{\tau_i\})\, \delta(\{\tau_i\})}\, u_i'\, K' \,[L\,\mathcal{I}(\tilde{\delta})^-\, L']^{-1}\, K\, u_i \\
= \; & \frac{\rho(z,\tau_i)^2}{\int \rho(y,\tau_i)^2 P(dy)}\, \frac{\tilde{\delta}(\{\tau_i\})}{\delta(\{\tau_i\})}\, u_i'\, \tilde{D}^{-1}\, K' \,[L\,\mathcal{I}(\tilde{\delta})^-\, L']^{-1}\, K\, \tilde{D}^{-1}\, u_i \\
= \; & \frac{\rho(z,\tau_i)^2}{\int \rho(y,\tau_i)^2 P(dy)}\, \frac{\tilde{\delta}(\{\tau_i\})}{\delta(\{\tau_i\})}\, a(\tau_i)'\, \mathcal{I}(\tilde{\delta})^-\, L' [L\,\mathcal{I}(\tilde{\delta})^-\, L']^{-1}\, L\,\mathcal{I}(\tilde{\delta})^-\, a(\tau_i).
\end{aligned}$$

Setting

$$b_i := \frac{b\, \delta(\{\tau_i\})}{\tilde{\delta}(\{\tau_i\})\, a(\tau_i)'\, \mathcal{I}(\tilde{\delta})^-\, L' \,[L\,\mathcal{I}(\tilde{\delta})^-\, L']^{-1}\, L\,\mathcal{I}(\tilde{\delta})^-\, a(\tau_i)}$$

we have for $i = 1, ..., I$

$$\max_{z\in\mathbb{R}} \frac{\rho(z,\tau_i)^2}{\int \rho(y,\tau_i)^2 P(dy)} \leq b_i$$

so that Lemma 8.1 applied to the one dimensional case provides $b_i \geq 1$. According to Corollary 8.2 we obtain for $i = 1, ..., I$

$$v_\rho(\tau_i) \leq v_{\rho_0}(\tau_i)$$

for

$$\rho_0(z,\tau_i) := \begin{cases} \operatorname{sgn}(z) & \text{for } b_i = 1, \\ \operatorname{sgn}(z)\, \min\{|z|, \sqrt{b_i}\, y_i\} & \text{for } b_i > 1, \end{cases}$$

where $y_i^2 = g(\sqrt{b_i}\, y_i) > 0$. Note that $\sqrt{b_i}\, y_i = \overline{w}\left(\frac{1}{b_i}\right)$ so that

$$v_{\rho_0}(\tau_i) = \overline{v}\left(\frac{1}{b_i}\right).$$

Because $\overline{v}$ is concave (see Lemma A.12 in Appendix A.2) we have

$$\begin{aligned}
\sum_{i=1}^{I} v_\rho(\tau_i)\,\delta(\{\tau_i\}) &\leq \sum_{i=1}^{I} \overline{v}\left(\frac{1}{b_i}\right)\,\delta(\{\tau_i\}) \\
&\leq \overline{v}\left(\sum_{i=1}^{I} \frac{1}{b_i}\,\delta(\{\tau_i\})\right) \\
&= \overline{v}\left(\sum_{i=1}^{I} \frac{\tilde{\delta}(\{\tau_i\})\,a(\tau_i)'\mathcal{I}(\tilde{\delta})^-\,L'\,[L\mathcal{I}(\tilde{\delta})^-\,L']^{-1}\,L\mathcal{I}(\tilde{\delta})^-\,a(\tau_i)}{b\,\delta(\{\tau_i\})}\right. \\
&\qquad \left.\cdot\,\delta(\{\tau_i\})\right) \\
&= \overline{v}\left(\frac{1}{b}\,\mathrm{tr}\left(\sum_{i=1}^{I}[L\mathcal{I}(\tilde{\delta})^-\,L']^{-1}\,L\mathcal{I}(\tilde{\delta})^-\,a(\tau_i)\,a(\tau_i)'\mathcal{I}(\tilde{\delta})^-\,L'\,\tilde{\delta}(\{\tau_i\})\right)\right) \\
&= \overline{v}\left(\frac{1}{b}\,\mathrm{tr}\left([L\mathcal{I}(\tilde{\delta})^-\,L']^{-1}\,[L\mathcal{I}(\tilde{\delta})^-\,L']\right)\right) \\
&= \overline{v}\left(\frac{1}{b}\,\mathrm{tr}(E_s)\right) = \overline{v}\left(\frac{s}{b}\right).
\end{aligned}$$

Then (8.14) implies

$$\begin{aligned}
&\int \psi\,\psi'\,d(P\otimes\delta) \\
&\quad = \frac{1}{\sum_{t\in\mathcal{D}} v_\rho(t)\,\delta(\{t\})}\,L\mathcal{I}(\tilde{\delta})^-\,L' \\
&\quad \geq \frac{1}{\overline{v}\left(\frac{s}{b}\right)}\,L\mathcal{I}(\tilde{\delta})^-\,L'.\square
\end{aligned}$$

Theorem 8.7 *If $\Delta = \{\delta \in \Delta(\varphi);\ supp(\delta)$ is finite $\}$, $b \geq s$ and δ_D is D-optimal for φ in Δ, then δ_D is asymptotically D-optimal in Δ for testing φ with bias bound b and an ALE-test statistic with influence function ψ_{b,δ_D} given by (8.13) is asymptotically D-optimal for testing φ at δ_D with bias bound b.*

Proof. Regard any $\delta \in \Delta$ with $\mathrm{supp}(\delta) = \{\tau_1, ..., \tau_I\}$ and any $\psi \in \Psi^*(\delta, L)$ with $\|\psi'(\int \psi\psi' d(P\otimes\delta))^{-1}\psi\|_\delta \leq b$. If $a(\tau_1), ..., a(\tau_I)$ are not linearly independent, then we can extend the regressors by $\tilde{a}(t)$ so that $\overline{a}(\tau_1), ..., \overline{a}(\tau_I)$ are linearly independent, where $\overline{a}(t) = (a(t)', \tilde{a}(t)')'$. Then we have $\psi \in \overline{\Psi}^*(\delta, \overline{L})$ for some $\tilde{L}$, where $\overline{L} = (L, \tilde{L})$ and $\overline{\Psi}^*(\delta, \overline{L})$ is defined for the extended model given by $\overline{Y} = \overline{a}(t)'\overline{\beta} + Z$ (see also the proof of Theorem 7.12). Denoting $\overline{\mathcal{I}}(\delta) = \int \overline{a}(t)\,\overline{a}(t)'\,\delta(dt)$ Lemma 8.3 provides

$$\begin{aligned}
&\det\left(\int \psi\,\psi'\,d(P\otimes\delta)\right) \qquad (8.15)\\
&\quad \geq \left(\frac{1}{\overline{v}\left(\frac{s}{b}\right)}\right)^s \det(\overline{L}\,\overline{\mathcal{I}}(\tilde{\delta})^-\,\overline{L}')
\end{aligned}$$

$$\geq \left(\frac{1}{\overline{v}\left(\frac{s}{b}\right)}\right)^{s} \det(L\mathcal{I}(\tilde{\delta})^{-}\, L')$$

$$\geq \left(\frac{1}{\overline{v}\left(\frac{s}{b}\right)}\right)^{s} \det(L\mathcal{I}(\delta_D)^{-}\, L').$$

For the property $\overline{L\mathcal{I}}(\tilde{\delta})^{-}\overline{L}' \geq L\mathcal{I}(\tilde{\delta})^{-}\, L'$ see Lemma A.1. According to Theorem 8.6 the lower bound in (8.15) is attained by an ALE-test statistic with influence function ψ_{b,δ_D} at δ_D. □

Theorem 8.7 in particular shows that the asymptotically D-optimal design for a bias bound does not depend on the bias bound b. But the asymptotically D-optimal estimators at the asymptotically D-optimal designs depend on the bias bound b. Hence we need a rule for choosing the bias bound b. For estimation we chose the bias bound b by minimizing the asymptotic mean squared error. Here for testing one possibility of choosing b is to choose b such that the relative power of the test at contiguous alternatives compared with the asymptotic bias of the first error is large. Hence, we should maximize

$$\frac{\gamma'\,(\int \psi\,\psi'\, d(P\otimes\delta))^{-1}\,\gamma}{\|\psi'\,(\int \psi\,\psi'\, d(P\otimes\delta))^{-1}\,\psi\|_{\delta}}. \tag{8.16}$$

Because it already was difficult to maximize the absolute power, which means that $\int \psi\,\psi'\, d(P\otimes\delta)$ should be minimized in the positive-semidefinite sense, we minimized the determinant of the covariance matrix $\int \psi\,\psi'\, d(P\otimes \delta)$. Also for the relative power it is more easy to maximize

$$R(\psi,\delta) := \frac{1}{(\det(\int \psi\,\psi'\, d(P\otimes\delta)))^{1/s}\ \|\psi'\,(\int \psi\,\psi'\, d(P\otimes\delta))^{-1}\,\psi\|_{\delta}}$$

instead of (8.16). Thereby we take the sth root of the determinant of the covariance matrix to ensure that an improvement of the covariance matrix by a factor c provides also an improvement of the relative power value by the factor c. Note also that the sth root of the determinant is often used as a measure for the entropy and that it is the geometric mean of the diagonal elements if the covariance matrix is a diagonal matrix.

For estimation with bias bound b we could show that the influence function $\psi_{b,\delta}$ of an asymptotically A-optimal AL-estimator is unique. But for testing with bias bound b it is not clear whether the influence function of an asymptotically D-optimal ALE-test statistic is unique. In particular, it is not clear whether there exists an influence function ψ of an asymptotically D-optimal ALE-test statistic with $\|\psi' C(\psi)^{-1}\psi\|_{\delta} < b$. Hence for testing we cannot derive ALE-test statistics which maximize $R(\psi,\delta)$ within $\Psi^{*}(\delta, L)$ as it was possible for deriving MSE minimizing AL-estimators. We only can derive ALE-test statistics which maximize $R(\psi_{b,\delta_D},\delta_D)$ with respect to b, where δ_D is a D-optimal design. These ALE-test statistics are called *relative power maximizing ALE-test statistics*, or briefly *RP maximizing ALE-test statistics*.

Definition 8.4 (RP maximizing ALE-test statistic)
An asymptotically D-optimal ALE-test statistic for testing φ at a D-optimal design δ_D with bias bound b_* is relative power maximizing (briefly RP maximizing) if its influence function ψ_{b_*,δ_D} is given by (8.13) with

$$b_* \in \arg\min \left\{ b \cdot \left(\det \left(\int \psi_{b,\delta_D} \psi'_{b,\delta_D} \, d(P \otimes \delta_D) \right) \right)^{1/s} ; \; b \geq s \right\}.$$

The following theorem shows that the bias bound b_* providing a RP maximizing ALE-test statistic is always equal to the minimum asymptotic bias $b_0^T(\delta_D, L)$ which is equal to s according to Theorem 8.1. This result is in opposite to the optimal bias bound for estimation, which defined by minimizing the mean squared error is always greater than the minimum asymptotic bias $b_0^E(\delta, L)$ for estimation (see Section 7.2).

Theorem 8.8 *An ALE-test statistic for testing $H_0 : \varphi(\beta) = l$ at a D-optimal design δ_D with influence function ψ_{b,δ_D} given by (8.13) with $b = s$ is RP maximizing as well as an ALE-test statistic with minimum asymptotic bias.*

Proof. According to Theorem 8.6 we have

$$\begin{aligned} & b \cdot \left(\det \left(\int \psi_{b,\delta_D} \psi'_{b,\delta_D} \, d(P \otimes \delta_D) \right) \right)^{1/s} \qquad (8.17) \\ = \; & b \, \frac{1}{\overline{v}\left(\frac{s}{b}\right)} \left(\det(L \mathcal{I}(\delta_D)^- L') \right)^{1/s} \\ = \; & \frac{b}{s} \, \frac{1}{\overline{v}\left(\frac{s}{b}\right)} \, s \left(\det(L \mathcal{I}(\delta_D)^- L') \right)^{1/s} \end{aligned}$$

for $b \geq s$. According to Lemma A.12 of Appendix A.2 the function $t_1 : [1, \infty) \to I\!R$ given by

$$t_1(b) := b \, \frac{1}{\overline{v}\left(\frac{1}{b}\right)}$$

is a convex function with $\lim_{b \downarrow 1} t_1'(b) = 0$, where t' here denotes the first derivative of t_1. Hence t_1 attains its minimum at $b = 1$ so that (8.17) is minimized by $b = s$. □

If we would not take the sth root of the determinant, then we would try to minimize

$$b \cdot \det \left(\int \psi_{b,\delta_D} \psi'_{b,\delta_D} \, d(P \otimes \delta_D) \right)$$

$$= \frac{b}{s} \left(\frac{1}{\overline{v}\left(\frac{s}{b}\right)} \right)^s s \det(L\mathcal{I}(\delta_D)^- L')$$

for $b \geq s$. According to Lemma A.12 of Appendix A.2 the function $t_s : [1, \infty) \to I\!R$ given by

$$t_s(b) := b \left(\frac{1}{\overline{v}\left(\frac{1}{b}\right)} \right)^s$$

is also a convex function. But now for $s > 1$ its minimum is attained at a value greater than 1 so that the optimal bias bound would be greater than the minimum bias bound $b_0^T(\delta_D, L) = s$.

Example 8.3 (One-way lay-out, continuation of Example 8.1) For testing in a one-way lay-out model with 4 levels $H_0 : \varphi(\beta) = (\beta_2 - \beta_1, \beta_3 - \beta_1, \beta_4 - \beta_1)' = 0$ against $H_1 : \varphi(\beta) = (\beta_2 - \beta_1, \beta_3 - \beta_1, \beta_4 - \beta_1)' \neq 0$, which is equivalent with testing $H_0 : \beta_1 = \beta_2 = \beta_3 = \beta_4$, at the D-optimal design $\delta_D = \frac{1}{4}(e_1 + e_2 + e_3 + e_4)$ for φ the minimum asymptotic bias is $b_0^T(\delta_D, L) = b_0^T(\Delta(\varphi), L) = s = 3$ (see Example 8.1). According to Theorem 8.7 for every $b \geq b_0^T(\delta_D, L) = 3$ the D-optimal design δ_D is asymptotically D-optimal for testing φ with bias bound b and the influence function of an asymptotically D-optimal ALE-test statistic for testing φ at δ_D with bias bound b is given by (8.13) according to Theorem 8.7. For $b = b_0^T(\delta_D, L) = s = 3$ it coincides with ψ_{01} given by (8.5) so that it has the form

$$\psi_{b,\delta_D}(z,t) = \psi_{01}(z,t)$$

$$= \begin{cases} (-1,-1,-1)' \operatorname{sgn}(z)\, 4\, \sqrt{\frac{\pi}{2}}, & \text{for } t = 1, \\ (1_{\{2\}}(t), 1_{\{3\}}(t), 1_{\{4\}}(t))' \operatorname{sgn}(z)\, 4\, \sqrt{\frac{\pi}{2}}, & \text{for } t \neq 1. \end{cases}$$

For $b > b_0^T(\delta_D, L) = 3$ it has the form

$$\psi_{b,\delta_D}(z,t)$$

$$= \begin{cases} (-1,-1,-1)' \operatorname{sgn}(z)\, 4\, \frac{\min\{|z|, \sqrt{b}\, v_b\}}{2\Phi(\sqrt{b}\, v_b) - 1}, & \text{for } t = 1, \\ (1_{\{2\}}(t), 1_{\{3\}}(t), 1_{\{4\}}(t))' \operatorname{sgn}(z)\, 4\, \frac{\min\{|z|, \sqrt{b}\, v_b\}}{2\Phi(\sqrt{b}\, v_b) - 1}, & \text{for } t \neq 1, \end{cases}$$

with $v_b^2 = \frac{1}{3}\, g(\sqrt{b}\, v_b) > 0$, i.e. $v_b = \overline{w}\left(\frac{3}{b}\right) \frac{1}{\sqrt{b}}$. The RP maximizing ALE-test statistic is an asymptotically D-optimal ALE-test statistic for the bias bound $b_* = 3$, where the relative asymptotic power value $R(\psi_{3,\delta_D}, \delta_D)$ is given by $R(\psi_{3,\delta_D}, \delta_D) = \frac{1}{6\pi\, 2^{1/s}} \approx 0.042$. That the D-optimal design provides a more robust analysis also Büning (1994) found in his simulation study although he used other tests. □

Example 8.4 (Quadratic regression, continuation of Example 8.2) For testing in a quadratic regression model $H_0 : \varphi(\beta) = (\beta_0, \beta_1 + \beta_2)' = (l_1, 0)'$ at the D-optimal design $\delta_D = \frac{1}{2}(e_0 + e_1)$ for φ the minimum asymptotic bias is $b_0^T(\delta_D, L) = b_0^T(\Delta(\varphi), L) = s = 2$ (see Example 8.2). According to Theorem 8.7 for every $b \geq b_0^T(\delta_D, L) = 2$ the D-optimal design δ_D is asymptotically D-optimal for testing φ with bias bound b and the influence function of an asymptotically D-optimal ALE-test statistic for testing φ at δ_D with bias bound b is given by (8.13) according to Theorem 8.7. According to Theorem 8.5 an ALE-test statistic with this influence function is also asymptotically U-optimal for testing φ at δ_D with bias bound b. For $b = b_0^T(\delta_D, L) = s = 2$ this influence function coincides with ψ_{01} given by (8.6) so that it has the form

$$\psi_{b,\delta_D}(z,t) = \psi_{01}(z,t) = \begin{cases} (1,-1)' \operatorname{sgn}(z)\, 2\sqrt{\frac{\pi}{2}} & \text{for } t = 0, \\ (0,1)' \operatorname{sgn}(z)\, 2\sqrt{\frac{\pi}{2}} & \text{for } t = 1. \end{cases}$$

For $b > b_0^T(\delta_D, L) = 2$ it has the form

$$\psi_{b,\delta_D}(z,t) = \begin{cases} (1,-1)' \operatorname{sgn}(z)\, 2\, \frac{\min\{|z|, \sqrt{b}\, v_b\}}{2\Phi(\sqrt{b}\, v_b) - 1}, & \text{for } t = 0, \\ (0,1)' \operatorname{sgn}(z)\, 2\, \frac{\min\{|z|, \sqrt{b}\, v_b\}}{2\Phi(\sqrt{b}\, v_b) - 1}, & \text{for } t = 1, \end{cases}$$

with $v_b^2 = \frac{1}{2}\, g(\sqrt{b}\, v_b) > 0$, i.e. $v_b = \overline{w}\left(\frac{2}{b}\right) \frac{1}{\sqrt{b}}$. For example, for $b = 3$ we have $v_b = 0.377$. The RP maximizing ALE-test statistic is an asymptotically D-optimal ALE-test statistic for the bias bound $b_* = 2$, where the relative asymptotic power value $R(\psi_{2,\delta_D}, \delta_D)$ is given by $R(\psi_{2,\delta_D}, \delta_D) = \frac{1}{2\pi} \approx 0.159$. For $b = 3$ the relative asymptotic power value is $R(\psi_{3,\delta_D}, \delta_D) \approx 0.139$.

For the A-optimal design $\delta_A = \frac{1}{\sqrt{2}+1}(\sqrt{2} e_0 + e_1)$ for φ Example 8.2 provided a minimum asymptotic bias of $b_0^T(\delta_A, L) = (\sqrt{2} + 1)$. For $b \geq (\sqrt{2} + 1)$ the influence function of an asymptotically U-optimal ALE-test statistic for testing φ at δ_A with the bias bound b is given by Theorem 8.5. For $b = (\sqrt{2} + 1)$ it coincides with ψ_0 given in Example 8.2 so that it has the form

$$\begin{aligned} \psi_{b,\delta_A}(z,t) &= \psi_0(z,t) \\ &= \begin{cases} (1,-1)' \operatorname{sgn}(z)\, \frac{\sqrt{2}+1}{\sqrt{2}}\, \frac{\min\{|z|, y\sqrt{\sqrt{2}+1}\}}{2\Phi(y\sqrt{\sqrt{2}+1}) - 1} & \text{for } t = 0, \\ (0,1)' \operatorname{sgn}(z)\, (\sqrt{2}+1)\sqrt{\frac{\pi}{2}} & \text{for } t = 1, \end{cases} \end{aligned}$$

where $y^2 = \frac{\sqrt{2}}{\sqrt{2}+1}\, g\left(y\sqrt{\sqrt{2}+1}\right)$, i.e. $y \approx 0.366$. For $b > b_0^T(\delta_A, L) =$

$(\sqrt{2}+1)$ it has the form

$$\psi_{b,\delta_A}(z,t) = \begin{cases} (1,-1)' \operatorname{sgn}(z) \frac{\sqrt{2}+1}{\sqrt{2}} \frac{\min\{|z|, \sqrt{b}\, v_b\}}{2\Phi(\sqrt{b}\, v_b)-1}, & \text{for } t = 0, \\ (0,1)' \operatorname{sgn}(z) (\sqrt{2}+1) \frac{\min\{|z|, \sqrt{b}\, w_b\}}{2\Phi(\sqrt{b}\, w_b)-1}, & \text{for } t = 1, \end{cases}$$

with $v_b^2 = \frac{\sqrt{2}}{\sqrt{2}+1}\, g(\sqrt{b}\, v_b) > 0$, i.e. $v_b = \overline{w}\left(\frac{\sqrt{2}+1}{\sqrt{2}\, b}\right) \frac{1}{\sqrt{b}}$ and $w_b^2 = \frac{1}{\sqrt{2}+1}\, g(\sqrt{b}\, w_b) > 0$, i.e. $w_b = \overline{w}\left(\frac{\sqrt{2}+1}{b}\right) \frac{1}{\sqrt{b}}$. For example, for $b = 3$ we have $v_b \approx 0.503$ and $w_b \approx 0.215$, and the relative asymptotic power value is $R(\psi_{3,\delta_A}, \delta_A) \approx 0.134$. □

9

High Breakdown Point and High Efficiency

While it is known that for robustness concepts based on the influence function (or on shrinking contamination neighbourhoods) high robustness and high efficiency can be combined by constrained problems (see in particular Hampel et al. (1986) and Chapter 7) there exists a recent discussion whether there is a conflict between efficiency and high breakdown point. Morgenthaler (1991), Stefanski (1991) and Coakley et al. (1994) showed that estimators with positive breakdown point have very low efficiencies compared with the least squares estimator. Therefore Davies (1993, 1994) proposed for desirable properties of estimators mainly robustness properties and no efficiency property. But as Rousseeuw (1994) argued this depends on the assumption of outliers in the independent variables (x-variables, experimental conditions) and the assumption of equal variances at all independent variables and in particular at leverage points. Hence, the conflict appears only in artificial situations. If the independent variables are random with possible outliers then a better model will be a multivariate model and in such model there will be no conflict between efficiency and positive breakdown (see He (1994)). For fixed independent variables as appear in planned experiments there is also no conflict as Morgenthaler (1994) noticed. But there is a conflict between high breakdown point designs and high efficient designs, which is shown in this chapter. At first basing on results of Section 4.3 in Section 9.1 trimmed weighted L_p estimators and corresponding designs are derived which maximize the breakdown point. Then Section 9.2 concerns the combination of high breakdown point and high efficiency. Thereby the breakdown point for finite samples as defined in Section 4.2 is used.

9.1 Breakdown Point Maximizing Estimators and Designs

An estimator which maximizes the breakdown point within all regression equivariant estimators for φ is called a *breakdown point maximizing estimator* and its breakdown point is called the *maximum breakdown point.*

And a design at which the breakdown point of a breakdown point maximizing estimator is maximal within some set $\Delta \subset \Delta_N(\varphi)$ is called a *breakdown point maximizing design* (see also Müller (1995b, 1996a,b)).

Definition 9.1 (Breakdown point maximizing estimator and design)
a) The maximum breakdown point $\epsilon_\varphi^*(d_N)$ for φ at d_N is defined as

$$\epsilon_\varphi^*(d_N) := \max\{\epsilon^*(\widehat{\varphi}_N, d_N);\ \widehat{\varphi}_N \text{ is a regression equivariant estimator for } \varphi \text{ at } d_N\}.$$

b) An estimator $\widehat{\varphi}_N^*$ for φ at d_N is a breakdown point maximizing estimator at d_N if

$$\epsilon^*(\widehat{\varphi}_N^*, d_N) = \epsilon_\varphi^*(d_N).$$

c) A design $d_N^* \in \Delta$ is breakdown point maximizing for φ in Δ if

$$d_N^* = \arg\max\{\epsilon_\varphi^*(d_N);\ d_N \in \Delta\}.$$

In Section 4.2 it was shown that at a design d_N the upper bound for the breakdown point is $\frac{1}{N}\left\lfloor\frac{N-\mathcal{N}_\varphi(d_N)+1}{2}\right\rfloor$. And in Section 4.3 it was shown that the h-trimmed L_p estimator for φ with $\lfloor\frac{N+\mathcal{N}_\varphi(d_N)+1}{2}\rfloor \leq h \leq \lfloor\frac{N+\mathcal{N}_\varphi(d_N)+2}{2}\rfloor$ attains this upper bound. Hence this estimator is breakdown point maximizing and the maximum breakdown point at the design d_N is

$$\epsilon_\varphi^*(d_N) = \frac{1}{N}\left\lfloor\frac{N-\mathcal{N}_\varphi(d_N)+1}{2}\right\rfloor.$$

Thereby the maximum breakdown point $\epsilon_\varphi^*(d_N)$ depends on the design d_N only via $\mathcal{N}_\varphi(d_N)$ such that the breakdown point maximizing design is a design which minimizes $\mathcal{N}_\varphi(d_N)$. Therefore the following theorem is obvious.

Theorem 9.1 *A h-trimmed L_p estimator $\widehat{\varphi}_{h,p}$ for φ with $\lfloor\frac{N+\mathcal{N}_\varphi(d_N)+1}{2}\rfloor \leq h \leq \lfloor\frac{N+\mathcal{N}_\varphi(d_N)+2}{2}\rfloor$ is breakdown point maximizing for φ at d_N, the maximum breakdown point for φ at d_N satisfies*

$$\epsilon_\varphi^*(d_N) = \frac{1}{N}\left\lfloor\frac{N-\mathcal{N}_\varphi(d_N)+1}{2}\right\rfloor$$

and a design $d_N^* \in \Delta$ is breakdown point maximizing for φ in Δ if and only if it minimizes $\mathcal{N}_\varphi(d_N)$ within all $d_N \in \Delta$, i.e.

$$d_N^* = \arg\min\{\mathcal{N}_\varphi(d_N);\ d_N \in \Delta\}.$$

According to Theorem 4.7 a breakdown maximizing design d_N^* maximizes not only the breakdown point of a breakdown point maximizing estimator but also the breakdown point of every h-trimmed L_p estimator with $\mathcal{N}_\varphi(d_N^*) < h \leq \frac{N+\mathcal{N}_\varphi(d_N^*)+1}{2}$. Hence a breakdown point maximizing design d_N^* maximizes the breakdown point of several estimators.

Because the maximum breakdown point depends only on $\mathcal{N}_\varphi(d_N)$ and $\frac{1}{N}\mathcal{N}_\varphi(d_N) = \mathcal{M}_\varphi(\delta_N)$, where $\mathcal{M}_\varphi(\delta)$ is the maximum mass of a design measure δ on a nonidentifying set (see Section 4.2), we can construct a breakdown point maximizing design by deriving a design measure δ which minimizes $\mathcal{M}_\varphi(\delta)$. Then we can approximate this design measure δ by a concrete design with N design points. Because $\mathcal{M}_\varphi(\delta)$ is a convex function $\mathcal{M}_\varphi(\delta)$ can be minimized similarly as for classical design criteria (see Chapter 2, Fedorov (1972), Bandemer et al. (1977, 1980), Silvey (1980), Pukelsheim (1993)). In particular we can facilitate the minimization problem by reduction by invariance which successfully was applied for classical design problems for example by Kiefer (1959), Kiefer and Wolfowitz (1959), Atwood (1969), Pukelsheim (1987), (1993) and Schwabe (1996). This reduction by invariance can be applied for minimizing $\mathcal{M}_\varphi(\delta)$ if we have a finite group $\mathcal{G}$ of mappings $g : \mathcal{T} \to \mathcal{T}$ with $\#\mathcal{G}$ elements and $\mathcal{M}_\varphi(\delta)$ is invariant with respect to the group $\mathcal{G}$, i.e. $\mathcal{M}_\varphi(\delta^g) = \mathcal{M}_\varphi(\delta)$ for all $g \in \mathcal{G}$ and $\delta \in \Delta(\varphi)$. Then every design δ can be improved by the symmetrized design $\overline{\delta} = \frac{1}{\#\mathcal{G}} \sum_{g\in\mathcal{G}} \delta^g$. This means in particular that for minimizing $\mathcal{M}_\varphi(\delta)$ we have to regard only those design measures δ which are invariant with respect to $\mathcal{G}$, i.e. which satisfy $\delta^g = \delta$ for all $g \in \mathcal{G}$. Hence the class of $\mathcal{G}$-invariant design measures is complete. This shows the following theorem which bases on the fact that $\mathcal{M}_\varphi(\cdot)$ is a convex function on Δ_0, the set of all design measures on $\mathcal{T}$.

Lemma 9.1 *$\mathcal{M}_\varphi(\cdot)$ is a convex function on Δ_0.*

Proof. Let be $\alpha \in [0, 1]$ and $\delta_1, \delta_2 \in \Delta_0$ arbitrary. Then we have

$$
\begin{aligned}
&\mathcal{M}_\varphi(\alpha\,\delta_1 + (1-\alpha)\,\delta_2)\\
&= \sup\{(\alpha\,\delta_1 + (1-\alpha)\,\delta_2)(\mathcal{D});\ \varphi \text{ is not identifiable at } \mathcal{D}\}\\
&\leq \alpha \sup\{\delta_1(\mathcal{D});\ \varphi \text{ is not identifiable at } \mathcal{D}\}\\
&\quad +(1-\alpha) \sup\{\delta_2(\mathcal{D});\ \varphi \text{ is not identifiable at } \mathcal{D}\}\\
&= \alpha\,\mathcal{M}_\varphi(\delta_1) + (1-\alpha)\,\mathcal{M}_\varphi(\delta_2)
\end{aligned}
$$

so that $\mathcal{M}_\varphi(\cdot)$ is convex. □

Theorem 9.2 *Let be $\mathcal{G}$ a finite group of mappings $g : \mathcal{T} \to \mathcal{T}$ so that $\mathcal{M}_\varphi(\delta^g) = \mathcal{M}_\varphi(\delta)$ for all $g \in \mathcal{G}$ and all $\delta \in \Delta(\varphi)$. Then for every design $\delta \in \Delta(\varphi)$ there exists a design $\overline{\delta} \in \Delta(\varphi)$ with*

$$\mathcal{M}_\varphi(\overline{\delta}) \leq \mathcal{M}_\varphi(\delta)$$

and $\overline{\delta}^g = \overline{\delta}$ for all $g \in \mathcal{G}$.

Proof. The proof is standard. Because $\mathcal{M}_\varphi(\cdot)$ is convex and $\mathcal{G}$-invariant for every design δ the symmetrization $\overline{\delta} = \frac{1}{\#\mathcal{G}} \sum_{g\in\mathcal{G}} \delta^g$ satisfies

$$\mathcal{M}_\varphi(\overline{\delta}) \leq \frac{1}{\#\mathcal{G}} \sum\nolimits_{g\in\mathcal{G}} \mathcal{M}_\varphi(\delta^g) = \mathcal{M}_\varphi(\delta).$$

Moreover obviously the symmetrized design satisfies $\overline{\delta}^g = \overline{\delta}$ for all $g \in \mathcal{G}$.
□

Because breakdown point maximizing designs are symmetric, as Theorem 9.2 shows, they are often different from classical optimal designs as A-optimal designs or designs with minimum support. This is shown by the following examples. In particular the examples show how different the maximum breakdown points of classical optimal designs and breakdown point maximizing designs can be.

Example 9.1 (One-way lay-out, continuation of Example 2.2)
For estimating $\varphi(\beta) = (\beta_2 - \beta_1, \beta_3 - \beta_1, ..., \beta_I - \beta_1)$ in the one-way lay-out model with $\mathcal{T} = \{1, ..., I\}$, $a(t) = a(i) = (1_1(i), ..., 1_I(i))'$ and $\beta = (\beta_1, ..., \beta_I)'$, where the first level is the control level, the A-optimal design within $\Delta_N(\varphi)$ is a design which approximates $N\frac{1}{\sqrt{I-1}+I-1}(\sqrt{I-1}e_1 + \sum_{i=2}^I e_i)$ (see Example 2.2). For example, for $I = 5$ and $N = 6\,K$ the A-optimal design d_N has $2\,K$ observations at the level 1 and K observations at each level i with $i = 2, ..., 5$. For this design any set $\mathcal{D} \subset \mathcal{T}$ with $\mathcal{D} \neq \mathcal{T}$ is a nonidentifying set and any subset with $1 \in \mathcal{D} \subset \mathcal{T}$ and $\#\mathcal{D} = 4$ contains the maximal number of design points of the design d_N, namely $2K+3K$ design points. Hence we have $\mathcal{N}_\varphi(d_N) = 5\,K$ so that the maximum breakdown point at this design is

$$\begin{aligned} \epsilon^*_\varphi(d_N) &= \frac{1}{N}\left\lfloor \frac{N-5K+1}{2} \right\rfloor \\ &= \left\lfloor \frac{6K-5K+1}{12\,K} \right\rfloor \stackrel{K\to\infty}{\longrightarrow} \frac{1}{12}. \end{aligned}$$

If $N = K\,I$, then a design d_N^* with K observations at each level $i = 1, ..., I$ is invariant with respect to the group of permutations. Hence, according to Theorem 9.2 $\mathcal{N}_\varphi(d_N^*) = K(I-1)$ is the minimum possible value for $\mathcal{N}_\varphi(d_N)$

so that d_N^* is breakdown point maximizing within $\Delta_N(\varphi)$ with maximum breakdown point of

$$\begin{aligned}\epsilon_\varphi^*(d_N^*) &= \frac{1}{N}\left\lfloor\frac{N-(KI-K)+1}{2}\right\rfloor \\ &= \left\lfloor\frac{IK-IK+K+1}{2IK}\right\rfloor \overset{K\to\infty}{\longrightarrow} \frac{1}{2I}.\end{aligned}$$

Note that the breakdown point maximizing design is also D-optimal within $\Delta_N(\varphi)$. □

In the following example the maximum breakdown point is derived for designs at which the whole parameter β is not identifiable.

Example 9.2 (Two-way lay-out)
Assume that in a two-way lay-out model

$$y(i,j) = \alpha_i + \beta_j + z(i,j)$$

with $t=(i,j) \in \mathcal{T}=\{1,2,3\} \times \{1,2\}$, $a(t)=(1_1(i), 1_2(i), 1_3(i), 1_1(j), 1_2(j))'$ and $\beta = (\alpha_1, \alpha_2, \alpha_3, \beta_1, \beta_2)' \in I\!R^5$ the aspect $\varphi(\beta) = (\alpha_2 - \alpha_1, \alpha_3 - \alpha_1)' \in I\!R^2$ shall be estimated, so that the level 1 of the first factor is the control level and the second factor is a nuisance parameter. Then a design d_N with equal numbers of observations, say K observations, at the experimental conditions $(1,1), (2,1), (3,2), (1,2)$ is a design with minimum support. Hence we have $\mathcal{N}_\varphi(d_N) = 3K$ so that the maximum breakdown point at this design is equal to

$$\begin{aligned}\epsilon_\varphi^*(d_N) &= \frac{1}{N}\left\lfloor\frac{N-3K+1}{2}\right\rfloor \\ &= \frac{1}{4K}\left\lfloor\frac{4K-3K+1}{2}\right\rfloor \overset{K\to\infty}{\longrightarrow} \frac{1}{8}.\end{aligned}$$

The design d_N is not invariant with respect to all permutations of the levels of the first factor. Hence according to Theorem 9.2 the breakdown point maximizing design d_N^* in $\Delta_N(\varphi)$ is the design with K observations at each possible experimental condition $(i,j) \in \mathcal{T}$, i.e. with K observations at $(1,1), (2,1), (3,1), (1,2), (2,2), (3,2)$. Because any set $\mathcal{D} \subset \mathcal{T}$ in which one level of the first factor does not appear, as for example $\mathcal{D} = \{(1,1), (2,1), (1,2), (2,2)\}$, is nonidentifying for φ we have $\mathcal{N}_\varphi(d_N^*) = 4K$ such that the maximum breakdown point at d_N^* is

$$\begin{aligned}\epsilon_\varphi^*(d_N) &= \frac{1}{N}\left\lfloor\frac{N-4K+1}{2}\right\rfloor \\ &= \frac{1}{6K}\left\lfloor\frac{6K-4K+1}{2}\right\rfloor \overset{K\to\infty}{\longrightarrow} \frac{1}{6}.\end{aligned}$$

Note if it is possible to fix the second factor at one level, say level 1, then the breakdown point maximizing design is given by Example 9.1 and has a maximum breakdown point of $\frac{1}{3K}\lfloor\frac{3K-2K+1}{2}\rfloor \to \frac{1}{6}$ for $K \to \infty$. Hence, this maximum breakdown point is equal to the maximum breakdown point in the case where the second factor cannot be fixed to one level for all observations. This example shows that increasing the dimension of the parameter β does not always reduce the breakdown point. □

The following obvious theorem which generalizes Lemma 5.1 in Müller (1995b) shows that designs with $\mathcal{N}_\varphi(d_N) = s-1$ are always breakdown point maximizing. In particular for estimating β the designs with experimental conditions in general position, i.e. $\mathcal{N}_\beta(d_N) = r-1$, are breakdown point maximizing. Hence the often used assumption that the experimental conditions are in general position (see Rousseeuw and Leroy (1987)) ensured that the breakdown points are as high as possible.

Theorem 9.3 *If Δ contains a design d_N with $\mathcal{N}_\varphi(d_N) = s-1$, then d_N is a breakdown point maximizing design in Δ.*

The condition $\mathcal{N}_\varphi(d_N) = s-1$ in particular means that all N experimental conditions are different. But in designed experiments often the number of different experimental conditions is restricted by some bound J, say, which is less than N. Hence the set of possible designs is restricted to

$$\Delta_J := \{d_N \in \Delta_N(\varphi);\ \#\{t_{1N}, ..., t_{NN}\} \leq J\},$$

where $\#S$ denotes the number of elements of the set S. Then in this set Δ_J a breakdown point maximizing design should be found.

Example 9.3 (Linear regression, continuation of Example 2.1)
In the linear regression model

$$y_{nN} = \beta_0 + \beta_1 t_{nN} + z_{nN},$$

where $a(t) = (1,t)'$ and $\beta = (\beta_0, \beta_1)'$ a classical optimal design for estimating β within the experimental region $\mathcal{T} = [-1,1]$ is $d_N = (t_{1N}, ..., t_{NN})'$ with $t_{nN} = -1$ for $n \leq \frac{N}{2}$ and $t_{nN} = 1$ for $n > \frac{N}{2}$. In particular this design is A- and D-optimal in $\Delta = \Delta_N(\beta)$ (at least for even or large N) and hence also optimal for robust estimation and testing in models with shrinking contamination as was shown in Section 7.2 and Section 8.2. For this design we get $\mathcal{N}_\beta(d_N) = \lfloor\frac{N+1}{2}\rfloor$ and hence

$$\begin{aligned}\epsilon^*_\beta(d_N) &= \frac{1}{N}\left\lfloor\frac{N - \lfloor\frac{N+1}{2}\rfloor + 1}{2}\right\rfloor \\ &= \begin{cases} \frac{1}{N}\lfloor\frac{N+2}{4}\rfloor \overset{N\to\infty}{\longrightarrow} \frac{1}{4} & \text{if } N \text{ is even,} \\ \frac{1}{N}\lfloor\frac{N+3}{4}\rfloor \overset{N\to\infty}{\longrightarrow} \frac{1}{4} & \text{if } N \text{ is odd.} \end{cases}\end{aligned}$$

For L_1 estimators and $N = 4K$ the design d_N besides other symmetric designs provides a breakdown point of also $\frac{1}{4}$ (see Ellis and Morgenthaler (1992)) so that according to Theorem 4.10 this design is breakdown point maximizing for L_1 estimators. But this design is not breakdown point maximizing in the sense of Definition 9.1, i.e. breakdown point maximizing for all regression equivariant estimators.

For a design d_N^* with N different design points $t_{1N}, ..., t_{nN} \in [-1, 1]$ we obtain $\mathcal{N}_\beta(d_N^*) = 1$ so that according to Theorem 9.3 this design is breakdown point maximizing in $\Delta = \Delta_N(\beta)$ with $\epsilon_\beta^*(d_N^*) = \frac{1}{N}\lfloor\frac{N}{2}\rfloor$. If only at most J different experimental conditions are possible any design d_N^J with at most $\lfloor\frac{N+J-1}{J}\rfloor$ replications at every of J different design points $\tau_1, ..., \tau_J$ is breakdown point maximizing in Δ_J with

$$\epsilon_\beta^*(d_N^J) = \frac{1}{N}\left\lfloor\frac{N - \lfloor\frac{N+J-1}{J}\rfloor + 1}{2}\right\rfloor \overset{N\to\infty}{\longrightarrow} \frac{1}{2} - \frac{1}{2J}.$$

Then the only problem is to choose the J different experimental conditions. This problem also appear if we can choose N different experimental condition, i.e. for the special case $J = N$.

If the minimum distance between two different design points is limited by some value, m_d say, and $J\, m_d \leq 2$, then we can use $-1, -1 + m_d, -1 + 2m_d, ..., -1 + \frac{J-2}{2}m_d, 0, 1 - \frac{J-2}{2}m_d, ..., 1 - 2m_d, 1 - m_d, 1$ for even J and $-1, -1+m_d, -1+2m_d, ..., -1+\frac{J-1}{2}m_d, 1-\frac{J-1}{2}m_d, ..., 1-2m_d, 1-m_d, 1$ for odd J as the $J+1$ different design points. This provides that the breakdown point maximizing design is as efficient as possible in the classical sense (see Section 2.2).

Another possibility is to choose the J different experimental conditions so that the bias $b_M(\widehat{\beta}_{h,p}, d_N, y_N)$ of the breakdown point maximizing estimator is minimized for all $M < \epsilon_\beta^*(d_N^J)$. To facilitate the task we can try to minimize a bound for the bias $b_M(\widehat{\beta}_{h,p}, d_N, y_N)$. According to Lemma 4.10 the quantity

$$\frac{R(y_N)\, S(d_N)\, (W + 1)}{\rho_\beta(d_N^J)^2}$$

is a bound for the bias $b_M(\widehat{\beta}_{h,p}, d_N^J, y_N)$ so that $\rho_\beta(d_N^J)$ should be maximized within all designs with J different design points.

For $J = 4$ this can be solved by the following calculations: Because of symmetry the four design points should be $1, x, -x$ and -1 so that only x has to be determined. Let be $v_{1,1}$ the projection of $(1, x)'$ on $\{h(1, 1)';\ h \in \mathbb{R}\}$ and $v_{1,-x}$ the projection of $(1, x)'$ on $\{h(1, -x)';\ h \in \mathbb{R}\}$. Then x should be determined so that $|(1, x)' - v_{1,1}| = |(1, x)' - v_{1,-x}|$. Because of $|(1, x)' - v_{1,1}|^2 = |(1, x)'|^2 - |v_{1,1}|^2$ and $|(1, x)' - v_{1,-x}|^2 = |(1, x)'|^2 - |v_{1,-x}|^2$ this equivalent to determine the value x so that $|v_{1,1}|^2 = |v_{1,-x}|^2$. Calculating

$$|v_{1,1}| = \left| \begin{pmatrix} 1 \\ 1 \end{pmatrix} \left((1,1)' \begin{pmatrix} 1 \\ 1 \end{pmatrix} \right)^{-1} (1,1) \begin{pmatrix} 1 \\ x \end{pmatrix} \right|$$
$$= \left| \begin{pmatrix} 1 \\ 1 \end{pmatrix} \frac{1}{2}(1+x) \right| = \frac{1+x}{\sqrt{2}}$$

and

$$|v_{1,-x}| = \left| \begin{pmatrix} 1 \\ -x \end{pmatrix} \left((1,-x)' \begin{pmatrix} 1 \\ -x \end{pmatrix} \right)^{-1} (1,-x) \begin{pmatrix} 1 \\ x \end{pmatrix} \right|$$
$$= \left| \begin{pmatrix} 1 \\ -x \end{pmatrix} \frac{1-x^2}{1+x^2} \right| = \frac{1-x^2}{\sqrt{1+x^2}} = \frac{(1-x)(1+x)}{\sqrt{1+x^2}}$$

the value x should satisfy

$$1 + x^2 = 2(1 - 2x + x^2).$$

This is equivalent with $0 = x^2 - 4x + 1$ so that $x = 2 - \sqrt{3} \approx 0.268$ is the only solution within $(0,1)$. Hence $\rho_\beta(d_N^J)$ is maximal if d_N^J bases on 1, $2 - \sqrt{3}$, $\sqrt{3} - 2$ and -1. □

9.2 Combining High Breakdown Point and High Efficiency

As for linear regression for many other linear models the classical optimal designs, which minimize some function of the covariance matrix of the least squares estimator, base on some very few different experimental conditions which are repeated several times (see Chapter 2 or Fedorov (1972), Silvey (1980), Pukelsheim (1993)). But because of the high number of repetitions all these designs provide a very low maximum breakdown point and the beakdown point maximizing designs base on many different experimental conditions. Hence, we have a conflict between classical optimality and high breakdown point.

This contradiction is in particular grave because classically optimal designs provide also high efficiency of some robust procedures. In particular rank-based estimators as regarded in Hössjer (1994) and Huber M-estimators have an asymptotic covariance matrix which coincides with the covariance matrix of the least squares estimator except for some constant which does not depend on the design (see Hössjer (1994), Bickel (1975), Yohai and Maronna (1979) and Section 5.2). Only by restricting to L_1 estimators the classical optimal design coincides with the breakdown point maximizing design (see Example 9.3). But L_1 estimator are neither breakdown point maximizing within all regression equivariant estimators nor most efficient within all robust estimators.

Most efficient robust estimators were derived in Section 7.2 by minimizing the trace of the asymptotic covariance matrix under the side condition that the asymptotic bias in shrinking neighbourhoods is bounded by some constant b. Thereby it was shown that the optimal designs for these optimal robust estimators are the classical A-optimal designs. Moreover, in Section 7.1 it was shown that at the A-optimal designs also the maximum asymptotic bias in shrinking neighbourhoods is minimized. Hence, classical optimal designs are also optimal for efficient robust estimation and also provide some special robustness properties.

Therefore we should look for designs which combine efficiency and high breakdown point. We also should look for estimators which combine high efficiency and high breakdown point.

Estimators which combine high breakdown point and high efficiency can be easily constructed by one-step M-estimators

$$\widehat{\beta}_N(y_N, d_N) = \widehat{\beta}_N^0(y_N, d_N) + \frac{1}{N}\sum\nolimits_{n=1}^{N} \psi(y_{nN} - a(t_{nN})'\widehat{\beta}_N^0(y_N, d_N), t_{nN}),$$

where the initial estimator $\widehat{\beta}_N^0$ is a high breakdown point estimator and the score function ψ provides high asymptotic efficiency. If the initial estimator is asymptotically normally distributed with convergence rate of $N^{-1/2}$ as many h-trimmed L_p-estimators (see Hössjer (1994)), then the finite sample breakdown point of $\widehat{\beta}_N$ coincides with that of the initial estimator $\widehat{\beta}_N^0$ and the asymptotic covariance matrix is only determined by the score function ψ. See Section 5.2, Rousseeuw (1984), Jurečková and Portnoy (1987), Jurečková and Sen (1990) and Simpson et al. (1992). For other estimators combining high breakdown point and high efficiency see also Yohai (1987), Yohai and Zamar (1988), He (1991), Naranjo and Hettmansperger (1994), Behnen (1994) and Hennig (1995).

To find a design which combines high breakdown point and high efficiency is more difficult. There are situations where even designs exist which are classically optimal as well as breakdown point maximizing. This shows the following example which already was given in Müller (1996a).

Example 9.4 (Multiple regression on balls)

In a r-dimensional multiple regression problem without constant term the observations are given by

$$y_{nN} = a(t_{nN})'\beta + z_{nN} = \sum\nolimits_{i=1}^{r} \beta^i t_{nN}^i + z_{nN}$$

for $n = 1, ..., N$, i.e. $\beta = (\beta^1, ..., \beta^r)' \in I\!R^r$ and $a(t) = t = (t^1, ..., t^r)'$. Here we assume that the experimental region is a ball, i.e.

$$t_{nN} \in \mathcal{T} = \{t \in I\!R^r;\ \sum\nolimits_{i=1}^{r} (t^i)^2 \leq R^2\}.$$

Without loss of generality we can assume that the radius of the ball is $R = 1$. Then the equivalence theorems (see Chapter 2) provide that a design

$d_N \in \mathcal{T}^N$ satisfying $A'_{d_N} A_{d_N} = \frac{N}{r} E_r$, where E_r denotes the $r \times r$-identity matrix, is A-, D- and E-optimal in $\Delta_N(\beta) \subset \mathcal{T}^N$, i.e. classically optimal. In particular if the sample size N satisfies $N = K\, r$, where K is an integer, then the design given by $d_N = (E_r|...|E_r)'$, where the identity matrix E_r is repeated K times, is classically optimal because $A'_{d_N} A_{d_N} = \frac{N}{r} E_r$. This design has minimum support but for $r > 1$ it provides a low maximum breakdown point of

$$\epsilon^*(d_N) = \frac{1}{N} \left\lfloor \frac{N - (r-1)K + 1}{2} \right\rfloor \overset{N \to \infty}{\longrightarrow} \frac{1}{2r}$$

because $\mathcal{N}_\beta(d_N) = (r-1)K = N\frac{r-1}{r}$. But instead of repeating the identity matrix K times we can use K different orthonormal matrices $O_1, ..., O_K \in \mathbb{R}^{r \times r}$ with $O_k\, O_k' = E_r$ for $k = 1, ..., K$ so that the r unit vectors are rotated to different positions by keeping the angles between them. It is always possible to find K positions so that the design d_N^* with $d_N^* = (O_1|...|O_K)' = A_{d_N^*}$ satisfies $\mathcal{N}_\beta(d_N^*) = r - 1$ so that according to Lemma 9.3 d_N^* is breakdown point maximizing in $\Delta(\beta)$. In particular for $r = 2$ it is always possible to find $N = 2K$ different experimental conditions t_n with $|t_n| = 1$ so that every t_n is othogonal to one another experimental condition t_m and no t_n, t_m with $n \neq m$ are lying on a line through the origin. Hence, we have the following result.

If $d_N^* = (O_1|...|O_K)'$ with $O_k\, O_k' = E_r$ for $k = 1, ..., K$ and $\mathcal{N}_\beta(d_N^*) = r - 1$, then d_N^* is classically optimal in $\Delta_N(\beta)$ as well as breakdown point maximizing in $\Delta_N(\beta)$ with maximum breakdown point of

$$\epsilon^*(d_N^*) = \frac{1}{N} \left\lfloor \frac{N - r + 2}{2} \right\rfloor \overset{N \to \infty}{\longrightarrow} \frac{1}{2}.$$

A design $d_N^* = (O_1|...|O_K)'$ with $O_k\, O_k' = E_r$ for $k = 1, ..., K$ and $\mathcal{N}_\beta(d_N^*) = r - 1$ always exists. □

But the situations, where a classical optimal design exists which is also breakdown point maximizing, are rare. More typical is the situation of linear regression, where classically optimal designs are very different from the breakdown point maximizing designs. In all these situations constrained problems will be a good compromise to combine high breakdown point and high asymptotic efficiency.

One constrained problem is to minimize a functional ϕ of the covariance matrix $(A'_{d_N} A_{d_N})^{-1}$ under the side condition that the breakdown point is not too bad, i.e.

$$\begin{aligned} d_N^{**} &= \arg\min\{\phi((A'_{d_N} A_{d_N})^{-1});\ d_N \in \Delta_N(\varphi) \text{ with } \epsilon^*(d_N) \geq B\} \\ &= \arg\min\{\phi((A'_{d_N} A_{d_N})^{-1});\ d_N \in \Delta_N(\varphi) \text{ with } \mathcal{N}_\varphi(d_N) \leq \overline{B}\}. \end{aligned}$$

For the particular case that $\mathcal{N}_\varphi(d_N)$ is the maximal number of repetitions as for linear regression this problem was already treated by Wynn (1982),

Wierich (1985) and Fedorov (1989) by constructing optimal designs with bounded number of repetitions.

Another constrained problem is to maximize the breakdown point under the side condition that the functional ϕ of the covariance matrix $(A'_{d_N}A_{d_N})^{-1}$ is not too large, i.e.

$$\begin{aligned} d_N^{**} &= \arg\max\{\epsilon^*(d_N);\ d_N \in \Delta_N(\varphi) \text{ with } \phi((A'_{d_N}A_{d_N})^{-1}) \leq B\} \\ &= \arg\min\{\mathcal{N}_\varphi(d_N);\ d_N \in \Delta_N(\varphi) \text{ with } \phi((A'_{d_N}A_{d_N})^{-1}) \leq B\}. \end{aligned}$$

In particular we can set

$$B = \min\{\phi((A'_{d_N}A_{d_N})^{-1});\ d_N \in \Delta_N(\varphi)\}$$

so that we restrict ourselves to ϕ-optimal designs. This of course makes only sense if we have several ϕ-optimal designs. For example for linear regression this makes no sense because the A-, D- and E-optimal design is unique. But there are other problems for which we have several classically optimal designs as Example 9.4 and the following example show.

Example 9.5 (Weighing designs for four objects)
Here we regard the problem of weighing four objects with a chemical balance. The weights $\beta^1, ..., \beta^4$ of the four objects are unknown and we can use N weighing realizations for determining the weights of the objects. For each weighing realization for each object we have to decide on which side of the balance the object should be put. Hence for the nth weighing realization we have to choose

$$t_{nN} \in \mathcal{T} = \{-1, 0, 1\}^4,$$

where $t^i_{nN} = 0$ means that in the nth realization the ith object does not participate while $t^i_{nN} = -1$ and $t^i_{nN} = 1$ means that in the nth realization the ith object is put on the lefthand side or righthand side of the balance, respectively. Then the result of the nth weighing realization will be

$$y_{nN} = a(t_{nN})'\beta + z_{nN} = \beta^1 t^1_{nN} + ... + \beta^4 t^4_{nN} + z_{nN},$$

where $\beta = (\beta^1, \beta^2, \beta^3, \beta^4)' \in I\!R^4$, $a(t) = t = (t^1, t^2, t^3, t^4)' \in \mathcal{T}$ and z_{nN} is an error which may be also an outlier if some extraordinary shake happens.

If a design d_N satisfies $A'_{d_N}A_{d_N} = N\,E_4$, where E_4 denotes the 4×4 identity matrix, then as for multiple regression on balls d_N is A-, D- and E-optimal in $\Delta_N(\beta) \subset \mathcal{T}^N$, i.e. classically optimal (see Chapter 2). In particular if the sample size is $N = 4K$, where K is an integer, then a classically optimal design is $d_N = (X_0|...|X_0)'$, where the matrix $X_0 = (1_4|X)' \in I\!R^{4\times 4}$ is repeated K-times, $1_4 = (1, 1, 1, 1)'$ and

$$X = \begin{pmatrix} 1 & 1 & 1 \\ 1 & -1 & -1 \\ -1 & 1 & -1 \\ -1 & -1 & 1 \end{pmatrix}.$$

The design d_N has minimum support but it provides a maximum breakdown point of

$$\epsilon^*(d_N) = \frac{1}{N}\left\lfloor\frac{N-3K+1}{2}\right\rfloor \overset{N\to\infty}{\longrightarrow} \frac{1}{8}$$

because $\mathcal{N}_\beta(d_N) = 3K = \frac{3N}{4}$. In the case of $N = 8\,K$, where K is an integer, the breakdown point is improved by a design $d_N^* = (X_1|...|X_1)'$, where the matrix $X_1 = ((1_4|X)'|(1_4|-X)') \in I\!R^{4\times 8}$ is repeated K times. In $\Delta_N(\beta)$ this design is classical optimal and breakdown point maximizing within all classical optimal designs. The maximum breakdown point of this design is

$$\epsilon^*(d_N^*) = \frac{1}{N}\left\lfloor\frac{N-4K+1}{2}\right\rfloor = \frac{1}{4}.$$

See Müller (1996a,b). □

For weighing problems with more than four objects the following theorem shows how designs can be constructed which are breakdown point maximizing within the classical optimal designs. The theorem also can be use to construct general 2^r factorial designs which are breakdown point maximizing within the classical optimal designs.

Theorem 9.4 *Let be $\mathcal{T} = \{-1,1\}^r$, $a(t) = t$, $\varphi(\beta) = \beta \in I\!R^r$ and*

$$d_2^* = \begin{pmatrix} 1 & 1 \\ 1 & -1 \end{pmatrix}.$$

Then for $r \geq 2$ and $N = 2^{r-1}$ the design recursively given by

$$d_r^* = \begin{pmatrix} d_{r-1}^* & | & 1_{2^{r-2}} \\ d_{r-1}^* & | & -1_{2^{r-2}} \end{pmatrix}'$$

is breakdown point maximizing in $\Delta_N(\beta)$ and the maximum breakdown point is $\epsilon^*(d_N^*) = \frac{1}{4}$ for $r \geq 3$.

Proof. The assertion can be extended by the assertion $\mathcal{N}_\beta(d_r^*) = 2^{r-2}$ which at once provides for $r \geq 3$ the proposed maximum breakdown point according to Theorem 9.1. According to Example 9.5 the extended assertion holds for $r = 3$ and obviously for $r = 2$. Hence we can proof the assertion per induction. Assume that the assertion holds for $r-1$, i.e. $\mathcal{N}_\beta(d_{r-1}^*) = 2^{r-3}$ and, to ensure identifiability, that we have $\text{rk}(A_{d_{r-1}^*}) = \text{rk}(d_{r-1}^*) = r-1$. Then we have

$$\text{rk}((d_{r-1}^*|1_{2^{r-2}})) = r-1 \qquad (9.1)$$

so that $\text{rk}(d_r^*) = r$ and $\mathcal{N}_\beta(d_r^*) \geq 2^{r-2}$.

To show $\mathcal{N}_\beta(d_r^*) \leq 2^{r-2}$ take any $t_1, .., t_{(2^{r-3}+1)} \in \{1\} \times \{-1,1\}^{r-2} \times \{1\}$. Then because of $\mathcal{N}_\beta(d_{r-1}^*) = 2^{r-3}$ we have $\text{rk}(t_1, ..., t_{(2^{r-3}+1)}) = r-1$. Without loss of generality set $(d_{r-1}^* | 1_{2^{r-2}}) = (t_1, ..., t_{(2^{r-3}+1)}, t_{(2^{r-3}+2)}, ..., t_{(2^{r-2})})'$. Then $t_1, ..., t_{(2^{r-2})}$ and $t_1, ..., t_{(2^{r-3}+1)}$ generate the same $(r-1)$-dimensional subspace of $\mathbb{R}^r$. Moreover, every $v_1 \in \{1\} \times \{-1,1\}^{r-2} \times \{-1\}$ is not lying in this subspace because $v_1 = \lambda_1 t_1 + ... + \lambda_{(2^{r-2})} t_{(2^{r-2})}$ would imply $1 = \lambda_1 + ... + \lambda_{(2^{r-2})} = -1$. Hence for any $v_2, ..., v_{(2^{r-3})} \in \{1\} \times \{-1,1\}^{r-1}$ we have $\text{rk}(t_1, ..., t_{(2^{r-3}+1)}, v_1, v_2, ..., v_{(2^{r-3})}) = r$ so that $2 \cdot 2^{r-3} + 1 = 2^{r-2} + 1$ experimental conditions are not lying in a $(r-1)$-dimensional subspace. The same holds if $2^{r-3}+1$ experimental conditions are lying in $\{1\} \times \{-1,1\}^{r-2} \times \{-1\}$. Hence, we have $\mathcal{N}_\varphi(d_r^*) = 2^{r-2}$.

Now let be $\overline{d}_r \in \Delta_N(\beta)$ another design with $\overline{d}_r \neq d_r$. Because for every $t \in \{-1\} \times \{-1,1\}^{r-1}$ we have $-t \in \{1\} \times \{-1,1\}^{r-1}$ we can assume without loss of generality that there exist $t_1, ..., t_{(2^{r-2})} \in \{1\} \times \{-1,1\}^{r-2} \times \{1\}$ and $v_1, ..., v_{(2^{r-2}-1)} \in \{1\} \times \{-1,1\}^{r-1}$ such that $\overline{d}_r = (t_1, t_1, t_2, ..., t_{(2^{r-2})}, v_1, v_2, ..., v_{(2^{r-2}-1)})'$. Because of (9.1) we have $\text{rk}(t_1, ..., t_{(2^{r-2})}) \leq r-1$ which implies $\mathcal{N}_\beta(\overline{d}_r) \geq 2^{r-2}$. □

Note that the design regarded in Theorem 9.4 can be constructed by product designs (see for example Schwabe (1996a)). Often the product designs are criticized that they have too much support points. But for the breakdown point this is an advantage so that the breakdown point provides a justification of product designs.

Outlook

Using the robustness criterion which is based on shrinking neighbourhoods (or the influence function) we obtained in Chapter 7 and 8 very satisfying results concerning designs. Namely, it turned out that the D-optimal designs play an important role for robust testing while the A-optimal designs play an important role for robust estimation. This is in particular nice because the D-optimal designs are invariant with respect to one-to-one linear transformations of the linear aspect in the hypothesis and this property is very important for testing. Moreover, D-optimal designs for the whole parameter vector are equileverage designs so that they have some other robustness properties, in particular concerning outlier robustness of designs if least squares estimators are used (see Huber (1981), Section 7.2, Staudte and Sheather (1990), Section 7.3). In opposite to D-optimal designs A-optimal designs are not invariant to one-to-one linear transformations of the linear aspect. But in estimation problems this sensitivity can be also an advantage. Less satisfying is the fact that in opposite to estimation optimal tests could be only derived at D-optimal designs. But there is not much hope to achieve more. Moreover, also in opposite to estimation, the optimal bias bound for tests is the minimum bias bound so that the corresponding test statistics are not Fréchet differentiable. This may depend on the relative power criterion which was used for choosing the optimal bias bound so that other criteria would provide other results.

The optimality and robustness of the derived estimators, tests and designs hold only asymptotically. Almost nothing is known about the behaviour of optimal robust estimators and tests for finite samples. Only Rieder (1989) gave some results for the one-parameter regression through the origin. Therefore Monte-Carlo simulations are important. For robust estimation of linear aspects in linear models Müller (1993a) presented some simulation results which confirmed the asymptotic results for finite samples. But for robust testing and for robust estimation of nonlinear aspects and in nonlinear models such simulations are still missing.

The results concerning optimal estimators, tests and designs were obtained by using for the shrinking neighbourhoods the conditional contamination neighbourhoods. But there are also other neighbourhoods as those in Bickel (1984), Rieder (1985, 1987, 1994), Davies (1993) and Behnen (1994) so that in further studies those neighbourhoods can be investigated. Thereby it would be also interesting to investigate the role of Hadamard differentiability in this context. The Hadamard differentiability is in particular interesting because Ren (1994) involved with its help a connection between some M-estimators and R-estimators which already was discovered by Jurečková (1977) and used in Heiler (1992). The results of Heiler (1992) show that probably this connection can be extended to the optimal M-estimators as the Hampel-Krasker estimator and the Krasker-Welsch estimator so that R-estimators with the same efficiency can be derived. Then a high asymptotic breakdown point can be combined with asymptotic optimality as a high breakdown point for finite samples can be combined with asymptotic optimality by one-step M-estimators.

But to make such considerations also more investigations should be done concerning the asymptotic breakdown point in planned experiments. Because of the first results in Section 4.1 about the asymptotic breakdown point, it seems that similar results for it are possible as for the finite sample breakdown point.

Designs which maximizes the finite sample breakdown point and designs which combine a high finite sample breakdown point with a high efficiency were investigated in Chapter 9. Because these designs differ very much from the classical optimal designs special considerations are necessary to construct such designs. Here this was done only for special problems so that for other problems further investigations are necessary.

Moreover, Mizera and Müller (1996) discovered that the maximum finite sample breakdown point is not only attained by trimmed weighted L_p estimators (which are special R-estimators) but also by some M-estimators. For example the Cauchy M-estimator is such an M-estimator. This is in particular surprising because former results of Maronna et al. (1979) have shown that the breakdown point of M-estimators with convex score function is as bad as possible, namely $\frac{1}{N}$. But they did not regard planned experiments and assumed random experimental conditions so that their breakdown point was the breakdown which here was called the breakdown point for contaminated experimental conditions. If we are regarding planned experiments we now can pose the question whether other estimators can attain the maximum finite sample breakdown point. Other possible estimators are for example other R-estimators, the L-, S-, GS-, GM-, MM-estimators and the regression quantiles. See for example Huber (1981), Rousseeuw and Yohai (1984), Hettmansperger (1984), Hampel et al. (1986), Yohai (1987), Rousseeuw and Leroy (1987), Heiler and Willers (1988), Jurečková and Welsh (1990), Davies (1990), Simpson et al. (1992), Heiler (1992), Gutenbrunner and Jurečková (1992), Jurečková (1992), Hössjer et al. (1994), Croux et al. (1994, 1996).

Besides the breakdown point more investigations should be made concerning bias curves for planned experiments. In particular as for the bias bound in Pronzato and Walter (1991) it is not clear that the bound for the bias given in Section 4.3 is sharp.

Starting from a notion of test resistance introducd by Ylvisaker (1977) recently several breakdown point concepts were involved for tests (see for example He et al. (1990), He (1991), Coakley and Hettmansperger (1992, 1994), Markatou and He (1994)). But until now it is an open problem which role in these breakdown point concepts the designs play.

Moreover, in this book only tests of Wald-type were investigated with respect to the bias in shrinking contamination neighbourhoods. But many other robust tests were already developed as likelihood-ratio-type tests (or drop-in-dispersion tests, τ-tests), score-tests, and tests based on ranks (see for example Ronchetti (1982), Hampel et al. (1986), Markatou et al. (1991), Akritas (1991b), Büning (1991), Silvapulle (1992a,b), Gutenbrunner et al. (1993), García Pérez (1993), Heritier and Ronchetti (1994) and Markatou and He (1994)). For these tests it will be also interesting to see how the design influences the robustness and efficiency properties.

For dealing with nonlinear problems in this book as a first step the problem of estimating a nonlinear aspect in a linear model was regarded. In Section 4.4 breakdown points of estimators of nonlinear aspects were regarded and perhaps these first considerations can be used for further investigations. Thereby a more model oriented approach as the asymptotic breakdown and its relation to the finite sample breakdown would be very helpful. For such investigations the first results of Stromberg and Ruppert (1992) and Sakata and White (1995) concerning breakdown points in nonlinear models should be taken into account.

As for the breakdown point also for other robustness concepts only few results exist concerning robust estimation in nonlinear models (see Stefanski et al. (1986), Wang and Carroll (1993, 1995), Carroll and Pederson (1993), Liese and Vajda (1994)). Here the general robustness concept based on shrinking neighbourhoods were regarded. For that some results concerning optimal robust estimators and designs for nonlinear problems could be found. For robust estimation of nonlinear aspects in linear models maximin efficient designs could be derived in Section 7.3. But besides this result only locally optimal robust estimators and locally optimal robust designs could be derived in Section 7.3 and 7.4 for estimating nonlinear aspects in linear models and for estimation in nonlinear models, respectively. Because locally optimal designs are also only possible for the ideal model there is no hope to achieve more for designs for robust estimation. But it is not very satisfying that for robust estimation only locally optimal estimators are available. Maybe this can be improved by special one-step estimators at least at designs where the locally optimal estimators have a simple form. Perhaps also results about robust tests for nonlinear problems and corresponding designs are possible. First results concerning robust tests were given in Heritier and Ronchetti (1994) and Markatou and Manos (1996).

Besides in nonlinear models also the role of designs can be investigated for robust estimation in variance component problems by extending and applying results of Bednarski and Zontek (1994, 1996), Zmyślony and Zontek (1994) and Huggins (1993). Furthermore, repeated measurement models can be a field of further work. For that more detailed investigations of robust scale and covariance estimators in experiments with different experimental conditions will be necessary.

An interesting question is also whether optimal designs for outlier robust analysis can be used for outlier detection. There exists a rich literature on detecting outliers and influential points (see for example Box and Draper (1975), Cook and Weisberg (1982), Atkinson (1982, 1985, 1988b), Ghosh and Kipngeno (1985), Rousseeuw and Leroy (1987), Chatterjee and Hadi (1988), Wu and Luo (1993), McKean et al. (1993, 1994)). But the most approaches deal with the influential points of designs, if the least squares estimator is used, or with the detection properties of special estimators without considering different design constellations. Since often it turns out that estimators with high breakdown points have good outlier detection properties it could be that breakdown point maximizing designs improve the ability of outlier detection.

In this book only robustness against deviations from the assumed distribution of the errors and in particular outlier robustness were regarded. But other deviations from the ideal model are possible and were already regarded in robustness studies.

For example, deviations from the assumed response function are possible. A rich literature exists concerning model robust designs, i.e. designs which provide a good behaviour of statistical procedures if there are some deviations from the response function. If there are some arbitrary but small deviations from the assumed form of the response function then see for example Huber (1975), Marcus and Sacks (1976), Sacks and Ylvisaker (1978, 1984), Pesotchinsky (1982, 1984), Li and Notz (1982), Li (1984), Spruill (1985), Notz (1989), Jost (1991) and Wiens (1992). For deviations, which can be described by an extended model, design considerations are made for example in Studden (1982), Khuri and Cornell (1987), Hedayat and Majumdar (1988), Draper and Sanders (1988), Mathew and Bhaumik (1989), DeFeo and Myers (1992), Benda (1992, 1996), Schwabe (1995, 1996b) and Dette (1993, 1994). For an overview see also Atkinson (1988a). Thereby mainly the designs are derived for the Gauss-Markov estimators. Only in few considerations as in Draper and Sanders (1988) and Benda (1992, 1996) special estimators are used.

A rich literature exists also for choosing designs which are robust against the unavailability of data, i.e. for choosing designs if some observations may be missing. See for example Ghosh et al. (1983), Baksalary and Tabis (1987), Bhaumik and Whittinghill III (1991), Sathe and Satam (1992) and Jensen and Ramirez (1993), Dey (1993) and Notz et al. (1994). Less is known about the behaviour of statistical methods and designs if there are some deviations from the equality of the error variances or from the independence of the errors. For robustness against heteroscedasticity see for example Shao and Wu (1987), Wong and Cook (1993) and for robustness against dependent errors see for example Bickel and Herzberg (1979), Constantine (1989), Beran (1992), Martin et al. (1993), Künsch et al. (1993), Müller (1993b) and Brunner and Denker (1994). It is also possible that there are some known or unknown deviations from the planned experimental conditions. For robustness against such deviations see for example Vuchkov and Boyadjieva (1983) and Carroll et al. (1993). Besides these robustness approaches there exist also robustness approaches against deviations from a prior distribution in Bayesian analysis (see Das Gupta and Studden (1991) and Polasek and Pötzelberger (1994)) and robustness approaches against different design criteria (see Herzberg (1982)).

For all such robustness approaches there exists the question whether outlier robustness is connected with robustness against other deviations. Because breakdown point maximizing designs are those designs where the number of experimental conditions in sets is minimized at which the linear aspect is not identifiable (see Section 9.1) there will be a close connection to the robustness of designs against the unavailability of data. In particular, breakdown point maximizing designs can work with a maximum number of missing data. For special design situations a similar result was also obtained by Kuwada (1993) and Notz et al. (1994). Hence, the theory about high breakdown point designs may profit from results concerning robustness against the unavailability of data. Because breakdown point maximizing designs also minimize the number of repeated experimental conditions there will be also a relation to model robust designs. But until now there is no formal reason for this relation.

If outlier robustness and robustness against other deviations need different methods then there is also the question how to combine the methods. A first step of combining outlier robustness with other robustness properties was done by Wiens (1994, 1996) by combining outlier robustness with model robustness. Thereby he investigated the behaviour of M-estimators in the presence of deviations from the response function. Because often the deviations from the response function are given by shrinking neighbourhoods and shrinking neighbourhoods are also regarded in outlier robustness more rigorous approaches should be possible.

Appendix

A.1 Asymptotic Linearity of Fréchet Differentiable Functionals

Proof of Theorem 5.1.

For simplicity we write as in Section 3.1 Q_{N,ξ_N} instead of Q_{N,θ,ξ_N}, i.e. we drop θ. Define

$$N(t,d_N) := \sum_{n=1}^{N} 1_{\{t\}}(t_{nN}) = N\,\delta_N(\{t\}),$$
$$y_{(N,t)} := (y_{n(1)N},...,y_{n(N(t,d_N))N})'$$

if $\{t_{n(1)N},...,t_{n(N(t,d_N))N}\} = \{t_{nN};\ t_{nN} = t \text{ and } n = 1,...,N\}$ and

$$P_{y_{(N,t)}} := \frac{1}{N(t,d_N)} \sum_{n=1}^{N} e_{y_{nN}}\ 1_{\{t\}}(t_{nN})$$

the empirical distribution of $y_{(N,t)}$. Because of $P_{y_N,t} = P_{y_{(N,t)}}$ and $N(t,d_N) \to \infty$ for $N \to \infty$ we can apply the Kolmogorov-Smirnov theorem. Because the Kolmogorov-Smirnov statistic is uniformly bounded in probability for all distributions (see for example Billingsley (1968), p. 103/104, Fisz (1963), p. 394, Noether (1963)) for all $\epsilon_0 > 0$ there exists $K_0 \in I\!R$ and N_0 such that for deterministic designs d_N we have

$$\begin{aligned}
&Q^N(\sqrt{N}\max_{y\in I\!R}\delta_N(\{t\})|G_{N,t}(y) - F_{y_N,t}| > K_0) \qquad \text{(A.1)}\\
&= \left(\bigotimes_{i=1}^{N(t,d_N)} Q_{N,t}\right)(\sqrt{N}\max_{y\in I\!R}\delta_N(\{t\})|G_{N,t}(y) - F_{y_N,t}| > K_0)\\
&\le \left(\bigotimes_{i=1}^{N(t,d_N)} Q_{N,t}\right)(\sqrt{N(t,d_N)}\max_{y\in I\!R}|G_{N,t}(y) - F_{y_{(N,t)}}| > K_0)\\
&\le \frac{\epsilon_0}{2I}
\end{aligned}$$

for all $N \ge N_0$ and $t \in \mathcal{T}$. Additionally because of assumption (5.4) and (5.8) K_0 and N_0 can be chosen such that $\sqrt{N}d_K(Q_{N,\xi_N}, P_{\theta,\delta}) \le I\,K_0$ and with Lemma 3.1a)

$$\begin{aligned}
&\max_{t\in \text{supp}(\delta)} \sqrt{N}|\delta_N(\{t\}) - \xi_N(\{t\})| \\
&\quad < \max_{t\in \text{supp}(\delta)} \sqrt{N}|\delta_N(\{t\}) - \delta(\{t\})| \\
&\qquad + \max_{t\in \text{supp}(\delta)} \sqrt{N}|\delta(\{t\}) - \xi_N(\{t\})| \\
&\quad \le \max_{t\in \text{supp}(\delta)} \sqrt{N}|\delta_N(\{t\}) - \delta(\{t\})| + \sqrt{N}\, d_K(Q_{N,\xi_N}, P_{\theta,\delta}) \\
&\quad \le I\,K_0
\end{aligned}$$

for all $N \ge N_0$. Then Lemma 3.1c) provides

$$\begin{aligned}
&Q^N(\sqrt{N} d_K(P_{\theta,\delta}, P_{y_N,d_N}) > 3\,I\,K_0) \qquad\qquad (A.2)\\
&\quad \le Q^N(\sqrt{N} d_K(Q_{N,\xi_N}, P_{y_N,d_N}) > 2\,I\,K_0) \\
&\quad \le Q^N(\sqrt{N}\, \max_{(y,t)\in \mathbb{R}\times \text{supp}(\delta)} \delta_N(\{t\})|G_{N,t}(y) - F_{y_N,t}| > I\,K_0) \\
&\quad \le \sum_{t\in \text{supp}(\delta)} Q^N(\sqrt{N}\, \max_{y\in \mathbb{R}} \delta_N(\{t\})|G_{N,t}(y) - F_{y_N,t}| > K_0) \\
&\quad \le \epsilon_0
\end{aligned}$$

for all $N \ge N_0$.

For random designs D_N set

$$\begin{aligned}
Q^{N|d_N} &:= \bigotimes_{n=1}^{N} Q_{N,t_{nN}}, \\
\xi^N &:= \bigotimes_{n=1}^{N} \xi_N, \\
\mathcal{D}_0 &:= \{d_N \in \mathcal{T}^N;\ N(t,d_N) < N_0\}
\end{aligned}$$

and

$$\mathcal{D}_1 := \{d_N \in \mathcal{T}^N;\ \max_{t\in\mathcal{T}} \sqrt{N}|\delta_N(\{t\}) - \xi_N(\{t\})| > I\,K_0\}.$$

Then the constants K_0 and N_0 can be chosen such that (A.1) holds for all d_N with $N(t,d_N) \ge N_0$ and that additionally $\xi^N(\mathcal{D}_0) \le \frac{\epsilon_0}{4}$ and $\xi^N(\mathcal{D}_1) \le \frac{\epsilon_0}{4}$. Note that δ_N is given by d_N and d_N is given by ξ^N such that we always have $\text{supp}(\delta_N) \subset \text{supp}(\xi_N) \subset \text{supp}(\delta)$. Then similarly as for deterministic designs we have for random designs

$$\begin{aligned}
&Q^N(\sqrt{N} d_K(P_{\theta,\delta}, P_{y_N,d_N}) > 3\,I\,K_0) \qquad\qquad (A.3)\\
&\quad = \int \left(\bigotimes_{n=1}^{N} Q_{N,t_{nN}}\right)(\sqrt{N} d_K(P_{\theta,\delta}, P_{y_N,d_N}) > 3\,I\,K_0) \\
&\qquad\qquad \left(\bigotimes_{n=1}^{N} \xi_N\right)(d(d_N))
\end{aligned}$$

$$\leq \int Q^{N|d_N}(\sqrt{N} d_K(Q_{N,\xi_N}, P_{y_N,d_N}) > 2\, I\, K_0)\ \xi^N(d(d_N))$$

$$\leq \int Q^{N|d_N}(\sqrt{N}\ \max_{(y,t)\in \mathbb{R}\times \mathcal{T}}\ \delta_N(\{t\})|G_{N,t}(y) - F_{y_N,t}| > I\, K_0)\ \xi^N(d(d_N))$$

$$+ \int Q^{N|d_N}(\max_{t\in\mathcal{T}} \sqrt{N}|\delta_N(\{t\}) - \xi_N(\{t\})| > I\, K_0)\ \xi^N(d(d_N))$$

$$\leq \sum_{t\in \text{supp}(\delta)} \int_{\{d_N;\ N(t,d_N)\geq N_0\}} Q^{N|d_N}(\sqrt{N}\ \max_{y\in\mathbb{R}}\ \delta_N(\{t\})|G_{N,t}(y) - F_{y_N,t}| > K_0)\ \xi^N(d(d_N)) + \xi^N(\mathcal{D}_0) + \xi^N(\mathcal{D}_1)$$

$$\leq \frac{\epsilon_0}{2} + \frac{\epsilon_0}{4} + \frac{\epsilon_0}{4} = \epsilon_0.$$

Now the Fréchet differentiability and condition (5.1) implies

$$\sqrt{N}|\widehat{\zeta}_N(y_N, d_N) - \zeta(\theta) - \frac{1}{N}\sum_{n=1}^N \psi_\theta(y_{nN}, t_{nN})|$$
$$= \sqrt{N}\, d_K(P_{\theta,\delta}, P_{y_N,d_N}) \cdot \frac{|\widehat{\zeta}(P_{y_N,d_N}) - \widehat{\zeta}(P_{\theta,\delta}) - \int \psi_\theta(y,t)\, P_{y_N,d_N}(dy,dt)|}{d_K(P_{\theta,\delta}, P_{y_N,d_N})}$$
$$\stackrel{N\to\infty}{\longrightarrow} 0$$

in probability for $(Q_{N,\xi_N})_{N\in\mathbb{N}}$.

Fréchet differentiability and conditions (5.1) and (5.4) also imply

$$\sqrt{N}|\zeta(\theta_N) - \zeta(\theta) - \int \psi_\theta(y,t) P_{\theta_N,\delta}(dy,dt)|$$
$$= \sqrt{N}\, d_K(P_{\theta,\delta}, P_{\theta_N,\delta})\, \frac{|\widehat{\zeta}(P_{\theta_N,\delta}) - \widehat{\zeta}(P_{\theta,\delta}) - \int \psi_\theta(y,t) P_{\theta_N,\delta}(dy,dt)|}{d_K(P_{\theta,\delta}, P_{\theta_N,\delta})}$$
$$\stackrel{N\to\infty}{\longrightarrow} 0.$$

Now note that condition (5.5) and integration by parts yield for alsolutely continuous $P_{\theta_N,t}$

$$\int \psi_{\overline{\theta}}(y,t)\, P_{\theta_N,t}(dy) = \psi_{\overline{\theta}}(\infty,t) - \int \frac{\partial}{\partial \overline{y}} \psi_{\overline{\theta}}(\overline{y},t)/_{\overline{y}=y}\, F_{\theta_N,t}(y)\ \lambda(dy)$$

and for the discrete measures as $P_{y_N,t}$ we have

$$\int \psi_{\overline{\theta}}(y,t)\, P_{y_N,t}(dy) = \sum_y \psi_{\overline{\theta}}(y,t)\, P_{y_N,t}(\{y\})$$

$$= \psi_{\overline{\theta}}(\infty,t) - \sum\nolimits_y \int 1_{[y,\infty)}(z)\, \frac{\partial}{\partial \overline{y}} \psi_{\overline{\theta}}(\overline{y},t)/_{\overline{y}=z}\, \lambda(dz) P_{y_N,t}(\{y\})$$
$$= \psi_{\overline{\theta}}(\infty,t) - \int \frac{\partial}{\partial \overline{y}} \psi_{\overline{\theta}}(\overline{y},t)/_{\overline{y}=y}\, F_{y_N,t}(y)\, \lambda(dy).$$

Then conditions (5.6) and (5.7) provide for deterministic designs with (A.2) and Lemma 3.1

$$\begin{aligned}
&\sqrt{N}\left|\int (\psi_{\theta_N}(y,t) - \psi_\theta(y,t))\,(P_{\theta_N,\delta} - P_{y_N,d_N})\,(dy,dt)\right| \\
&\leq \sqrt{N}\left|\int (\psi_{\theta_N}(y,t) - \psi_\theta(y,t))\,(P_{\theta_N,\delta} - P_{\theta_N,\delta_N})\,(dy,dt)\right| \\
&\quad + \sqrt{N}\left|\int (\psi_{\theta_N}(y,t) - \psi_\theta(y,t))\,(P_{\theta_N,\delta_N} - P_{y_N,d_N})\,(dy,dt)\right| \\
&= \sqrt{N}\left|\sum\nolimits_{t\in T} (\delta(\{t\}) - \delta_N(\{t\})) \int (\psi_{\theta_N}(y,t) - \psi_\theta(y,t))\, P_{\theta_N,t}\,(dy)\right| \\
&\quad + \sqrt{N}\left|\sum\nolimits_{t\in T} \delta_N(\{t\}) \int (\psi_{\theta_N}(y,t) - \psi_\theta(y,t))\,(P_{\theta_N,t} - P_{y_N,t})(dy)\right| \\
&\leq \sqrt{N}\sum\nolimits_{t\in T} |\delta(\{t\}) - \delta_N(\{t\})| \int |\psi_{\theta_N}(y,t) - \psi_\theta(y,t)|\, P_{\theta_N,t}\,(dy) \\
&\quad + \sqrt{N}\sum\nolimits_{t\in T} \delta_N(\{t\}) \int \left|\frac{\partial}{\partial \overline{y}}\psi_{\theta_N}(\overline{y},t)/_{\overline{y}=y} - \frac{\partial}{\partial \overline{y}}\psi_\theta(\overline{y},t)/_{\overline{y}=y}\right| \\
&\qquad \cdot |F_{\theta_N,t}(y) - F_{y_N,t}(y)|\, \lambda(dy) \\
&\leq I\, K|\theta_N - \theta|\, \sqrt{N}\, d_K(P_{\theta_N,\delta}, P_{y_N,d_N}) \\
&\qquad + K|\theta_N - \theta|\, \frac{2\sqrt{N}\, d_K(P_{\theta_N,\delta}, P_{y_N,d_N})}{\min\{\delta(\{t\});\ t \in \mathrm{supp}(\delta)\}} \\
&\overset{N\to\infty}{\longrightarrow}\ 0
\end{aligned}$$

in probability for $(Q^N)_{N\in\mathbb{N}}$. For random designs the same holds with (A.3). We only additionally have to observe that

$$\begin{aligned}
&\sqrt{N}\sum\nolimits_{t\in T} |\delta(\{t\}) - \delta_N(\{t\})| \\
&\leq \sqrt{N}\sum\nolimits_{t\in T} |\delta(\{t\}) - \xi_N(\{t\})| + \sqrt{N}\sum\nolimits_{t\in T} |\xi_N(\{t\}) - \delta_N(\{t\})|
\end{aligned}$$

is bounded in probability for $(Q^N)_{N\in\mathbb{N}}$. Hence we have

$$\begin{aligned}
&\sqrt{N}\left|\widehat{\zeta}(P_{y_N,d_N}) - \zeta(\theta_N) - \frac{1}{N}\sum\nolimits_{n=1}^N \psi_{\theta_N}(y_{nN}, t_{nN})\right| \\
&\leq \sqrt{N}\left|\widehat{\zeta}(P_{y_N,d_N}) - \zeta(\theta) - \int \psi_\theta(y,t) P_{y_N,d_N}(dy,dt)\right| \\
&\quad + \sqrt{N}\left|\zeta(\theta) - \zeta(\theta_N) - \int (\psi_{\theta_N}(y,t) - \psi_\theta(y,t)) P_{\theta_N,\delta}(dy,dt)\right|
\end{aligned}$$

$$+ \sqrt{N}| \int (\psi_{\theta_N}(y,t) - \psi_\theta(y,t)) P_{\theta_N,\delta}(dy,dt)$$
$$- \int (\psi_{\theta_N}(y,t) - \psi_\theta(y,t)) P_{y_N,d_N}(dy,dt)|$$
$$\stackrel{N\to\infty}{\longrightarrow} \quad 0$$

in probability for $(Q^N)_{N\in I\!N}$. □

A.2 Properties of Special Matrices and Functions

Lemma A.1 *Let be* $A = (E|F)$ *with* $E \in I\!R^{I\times r}$ *and* $F \in I\!R^{I\times p}$. *Then for all* $K \in I\!R^{s\times I}$ *we have*

$$K\,A\,(A'A)^-\,A'\,K' \geq K\,E\,(E'E)^-\,E'\,K'.$$

Proof. Set $C := K\,E$, $B := K\,F$ and $Q := F'F - F'E\,(E'E)^-\,E'F$. Because of

$$(A'A)^- = \begin{pmatrix} E'E & E'F \\ F'E & F'F \end{pmatrix}^-$$
$$= \begin{pmatrix} [(E'E)^- + (E'E)^- E'F\,Q^- F'E\,(E'E)^-] & [-(E'E)^- E'F\,Q^-] \\ [-Q^- F'E\,(E'E)^-] & [Q^-] \end{pmatrix}$$

(see e.g. Ben-Israel and Greville (1974), p. 197) we have

$$\begin{aligned} & K\,A\,(A'A)^-\,A'\,K' \qquad\qquad (A.4)\\ = \;& (C|B)\,(A'A)^- \begin{pmatrix} C' \\ B' \end{pmatrix} \\ = \;& C(E'E)^-C' + C(E'E)^-E'F\,Q^-F'E(E'E)^-C' \\ & - C(E'E)^-E'F\,Q^-B' - B\,Q^-F'E(E'E)^-C' + B\,Q^-B' \\ = \;& C(E'E)^-C' + [C(E'E)^-E'F - B]\,Q^-[F'E(E'E)^-C' - B']. \end{aligned}$$

Because K is arbitrary, equation (A.4) holds in particular for $\overline{K} := K\,E\,(E'E)^-\,E' - K$. With $\overline{B} := \overline{K}\,F$ and

$$\overline{C} := \overline{K}\,E = K\,E - K\,E = 0$$

one obtains for $\overline{K}$

$$\begin{aligned} 0 \leq\; & \overline{K}\,A\,(A'A)^-A'A\,(A'A)^-\,A'\,\overline{K}' \\ =\; & \overline{K}\,A\,(A'A)^-\,A'\,\overline{K}' \\ =\; & \overline{C}(E'E)^-\overline{C}' + [\overline{C}(E'E)^-E'F - \overline{B}]\,Q^-[F'E(E'E)^-\overline{C}' - \overline{B}'] \\ =\; & \overline{B}\,Q^-\,\overline{B}' \\ =\; & [K\,E\,(E'E)^-E' - K]\,F\,Q^-F'[E\,(E'E)^-E'K' - K'] \\ =\; & [C(E'E)^-E'F - B]\,Q^-[F'E(E'E)^-C' - B']. \end{aligned}$$

This implies

$$K\,A\,(A'A)^-\,A'\,K' \geq C\,(E'E)^-C' = K\,E\,(E'E)^-\,E'\,K'.\square$$

Let Φ be the distribution function of the normal distribution $n_{(0,1)}$ with mean 0 and variance 1. In the following considerations we denote with f' and f'' the first and the second derivative of f.

Lemma A.2

a) $$\int |z|\,n_{(0,1)}(dz) = 2\,\Phi'(0) = \sqrt{\frac{2}{\pi}}.$$

b) $$\int |z|\,\min\{|z|, y\}\,n_{(0,1)}(dz) = 2\,\Phi(y) - 1.$$

c) $$\int \min\{|z|, y\}^2\,n_{(0,1)}(dz) = 2\,y^2\,\Phi(-y) + 2\,\Phi(y) - 1 - 2\,y\,\Phi'(y).$$

Proof. For $x \geq y \geq 0$ we have

$$\int_y^x |z|\,n_{(0,1)}(dz) = \Phi'(y) - \Phi'(x)$$

and

$$\int_{-y}^y z^2\,n_{(0,1)}(dz) = 2\,\Phi(y) - 1 - 2\,y\,\Phi'(y)$$

which imply the assertions. $\square$

Set for $y \geq 0$

$$\begin{aligned} l(y) &:= 2\,\Phi(y) - 1 - 2\,y\,\Phi'(y), \\ h(y) &:= y\,\Phi(-y) - \Phi'(y), \\ g(y) &:= \int \min\{|z|, y\}^2 n_{(0,1)}(dz) \end{aligned}$$

and for $y > 0$

$$q(y) := \frac{g(y)}{(2\Phi(y) - 1)^2}.$$

Lemma A.3

a) $l(y) > 0$ *for* $y > 0$.

b) $h(y) < 0$ *and* $h'(y) = \Phi(-y) > 0$ *for* $y \geq 0$.

c) $g'(y) = 4\,y\,\Phi(-y) = 4\,(h(y) + \Phi'(y))$.

d) $q'(y) = \dfrac{4\,h(y)\,l(y)}{[2\Phi(y) - 1]^3} < 0$ *for* $y > 0$.

e) $\lim_{y \downarrow 0} q(y) = \dfrac{\pi}{2}$ *and* $\lim_{y \to \infty} q(y) = 1$.

Proof.
a) $l(0) = 0$ and $l'(y) = 2\,y^2\Phi'(y) > 0$ for $y > 0$ imply $l(y) > 0$ for $y > 0$.
b) $h(0) < 0$, $h'(y) > 0$ and $\lim_{y\to\infty} h(y) = 0$ imply $h(y) < 0$ for $y \geq 0$.
c) Because of $g(y) = 2\,y\,h(y) + 2\,\Phi(y) - 1$ we have $g'(y) = 2\,h(y) + 2\,y\,\Phi(-y) - 2\,\Phi'(y) + 4\,\Phi'(y)$.
d) Assertion c) and its proof imply

$$
\begin{aligned}
q'(y) &= \frac{-4\,\Phi'(y)\,[2\,y\,h(y) + 2\,\Phi(y) - 1] + [2\,\Phi(y) - 1]\,[4\,(h(y) + \Phi'(y))]}{[2\,\Phi(y) - 1]^3} \\
&= \frac{4\,h(y)\,[2\,\Phi(y) - 1 - 2\,y\,\Phi'(y)]}{[2\,\Phi(y) - 1]^3}.
\end{aligned}
$$

e) Assertion c) and the rule of L'Hospital provide

$$
\begin{aligned}
\lim_{y\downarrow 0} q(y) &= \lim_{y\downarrow 0} \frac{g'(y)}{2\,[2\Phi(y) - 1]\,2\,\Phi'(y)} \\
&= \lim_{y\downarrow 0} \frac{y\,\Phi(-y)}{[2\Phi(y) - 1]\,\Phi'(y)} \\
&= \lim_{y\downarrow 0} \frac{\Phi(-y) - y\,\Phi'(y)}{-y\,\Phi'(y)\,[2\Phi(y) - 1] + 2\,\Phi'(y)^2} \\
&= \frac{\Phi(0)}{2\Phi'(0)^2} = \frac{\pi}{2}.
\end{aligned}
$$

Because $\lim_{y\to\infty} g(y) = 1$ the assertion is proved. □

Lemma A.4 *The function $\omega : [0,\infty) \to [0,\infty)$ given by*

$$
\omega(y) = c[2\,\Phi(b\,y) - 1] - y \qquad \text{for } b, c > 0
$$

has at most one positive root and this positive root exists if and only if

$$
c \cdot b > \sqrt{\frac{\pi}{2}}.
$$

Proof. Because of $\omega'(y) = b\,c\,2\,\Phi'(b\,y) - 1$ and $\omega''(y) = -b^3\,c\,2\,y\,\Phi'(b\,y) < 0$ we see that ω is a strictly concave function which converges for $y \to \infty$ to $-\infty$. Moreover $\omega(0) = 0$. Hence, ω has a positive root if and only if $\omega'(0) = b\,c\sqrt{\frac{2}{\pi}} - 1 > 0$. □

Lemma A.4 and its proof show that a root of ω can be easily calculated by Newton's method by starting with $y = c$ since $\omega(c) < 0$.

Define $W : [0,\infty)^3 \to I\!R$ as

$$
\begin{aligned}
W(b,c,y) &:= c\,[2\Phi(b\,y) - 1] - y, \\
M &:= \left\{(b,c) \in (0,\infty)^2;\ b \cdot c > \sqrt{\frac{\pi}{2}}\right\}
\end{aligned}
$$

and $w : M \to (0,\infty)$ implicitly by $W(b,c,w(b,c)) = 0$. Set also

$$v_b(c) := \begin{cases} \frac{\pi}{2} & \text{for } c = \frac{\pi}{2b^2}, \\ \frac{c\, g(b\, w(b,\sqrt{c}))}{w(b,\sqrt{c})^2} & \text{for } c > \frac{\pi}{2b^2}, \end{cases}$$

$$s_b(c) := \frac{1}{c}\, v_b(c^2)$$

and for $b \geq \frac{1}{c}\sqrt{\frac{\pi}{2}}$

$$r_c(b) := v_b(c^2).$$

Lemma A.5 *We have*

$$\lim_{c \downarrow \frac{1}{b}\sqrt{\frac{\pi}{2}}} w(b,c) = 0 = \lim_{b \downarrow \frac{1}{c}\sqrt{\frac{\pi}{2}}} w(b,c).$$

In particular, $v_b : [\frac{\pi}{2b^2}, \infty) \to I\!R$, $s_b : [\frac{1}{b}\sqrt{\frac{\pi}{2}}, \infty) \to I\!R$ *and* $r_c : [\frac{1}{c}\sqrt{\frac{\pi}{2}}, \infty) \to I\!R$ *are continuous.*

Proof. Set for $0 < \epsilon < c$

$$b(\epsilon) := \frac{1}{\epsilon}\, \Phi^{-1}\left(\frac{\epsilon + c}{2\,c}\right).$$

Then we have

$$\begin{aligned} \lim_{\epsilon \downarrow 0} b(\epsilon) &= \lim_{\epsilon \downarrow 0} \frac{\partial}{\partial \epsilon} \Phi^{-1}\left(\frac{\epsilon + c}{2\,c}\right) \\ &= \lim_{\epsilon \downarrow 0} \frac{1}{\Phi'\left(\Phi^{-1}\left(\frac{\epsilon+c}{2\,c}\right)\right)} \frac{1}{2c} \\ &= \frac{1}{\Phi'\left(\Phi^{-1}\left(\frac{1}{2}\right)\right)} \frac{1}{2c} \\ &= \frac{1}{\Phi'(0)\, 2\, c} \\ &= \frac{1}{c}\sqrt{\frac{\pi}{2}} \end{aligned}$$

and $w(b(\epsilon), c) = \epsilon$ so that

$$\lim_{b \downarrow \frac{1}{c}\sqrt{\frac{\pi}{2}}} w(b,c) = \lim_{\epsilon \downarrow 0} w(b(\epsilon), c) = 0$$

and with Lemma A.3 e)

$$\lim_{b \downarrow \frac{1}{c}\sqrt{\frac{\pi}{2}}} r_c(b) = \lim_{b \downarrow \frac{1}{c}\sqrt{\frac{\pi}{2}}} \frac{c^2\, g(b\, w(b,c))}{w(b,c)^2}$$

$$\begin{aligned}
&= \lim_{b \downarrow \frac{1}{c}\sqrt{\frac{\pi}{2}}} \frac{g(b\,w(b,c))}{[2\Phi(b\,w(b,c)) - 1]^2} \\
&= \lim_{\epsilon \downarrow 0} \frac{g(b(\epsilon)\,\epsilon)}{[2\Phi(b(\epsilon)\,\epsilon) - 1]^2} \\
&= \lim_{\epsilon \downarrow 0} q(b(\epsilon)\,\epsilon) = \frac{\pi}{2}.
\end{aligned}$$

Similarly for $c(\epsilon) := \epsilon\,[2\Phi(b\epsilon) - 1]^{-1}$ it holds $w(b, c(\epsilon)) = \epsilon$ and

$$\lim_{c \downarrow \frac{1}{b}\sqrt{\frac{\pi}{2}}} w(b,c) = \lim_{\epsilon \downarrow 0} w(b, c(\epsilon)) = 0$$

so that

$$\begin{aligned}
\lim_{c \downarrow \frac{1}{b}\sqrt{\frac{\pi}{2}}} s_b(c) &= \lim_{c \downarrow \frac{1}{b}\sqrt{\frac{\pi}{2}}} \frac{c\,g(b\,w(b,c))}{w(b,c)^2} \\
&= \lim_{c \downarrow \frac{1}{b}\sqrt{\frac{\pi}{2}}} \frac{g(b\,w(b,c))}{c\,[\Phi(b\,w(b,c)) - 1]^2} \\
&= \lim_{c \downarrow \frac{1}{b}\sqrt{\frac{\pi}{2}}} \frac{q(b\,w(b,c))}{c} \\
&= \lim_{\epsilon \downarrow 0} \frac{q(b\,\epsilon)}{c(\epsilon)} = b\sqrt{\frac{\pi}{2}}
\end{aligned}$$

and

$$\begin{aligned}
\lim_{c \downarrow \frac{1}{b}\sqrt{\frac{\pi}{2}}} v_b(c^2) &= \lim_{c \downarrow \frac{1}{b}\sqrt{\frac{\pi}{2}}} \frac{c^2\,g(b\,w(b,c))}{w(b,c)^2} \\
&= \lim_{c \downarrow \frac{1}{b}\sqrt{\frac{\pi}{2}}} q(b\,w(b,c)) \\
&= \lim_{\epsilon \downarrow 0} q(b\,\epsilon) = \frac{\pi}{2}. \square
\end{aligned}$$

Lemma A.6 *$s_b : [\frac{1}{b}\sqrt{\frac{\pi}{2}}, \infty) \to I\!R$ and $r_c : [\frac{1}{c}\sqrt{\frac{\pi}{2}}, \infty) \to I\!R$ are strictly decreasing and strictly convex functions.*

Proof. Set $w := w(b,c)$. With Lemma A.3 we have

$$\begin{aligned}
&\frac{\partial}{\partial y} W(b,c,y) \Big/_{y=w} \\
&= 2\,b\,c\,\Phi'(b\,y) - 1 \big/_{y=w} \\
&= \frac{2\,b\,w\,c\,\Phi'(b\,w) - w}{w} \\
&= \frac{c\,2\,b\,w\,\Phi'(b\,w) - c\,[2\Phi(b\,w) - 1] + c\,[2\Phi(b\,w) - 1] - w}{w} \\
&= \frac{-c\;l(b\,w)}{w} < 0
\end{aligned}$$

so that the implicit function theorem provides for every $c > \frac{1}{b}\sqrt{\frac{\pi}{2}}$

$$\begin{aligned}\frac{\partial}{\partial c} w(b,c) &= \left. \frac{\frac{\partial}{\partial c} W(b,c,y)}{-\frac{\partial}{\partial y} W(b,c,y)} \right/_{y=w} \\ &= \frac{2\Phi(b\,w) - 1}{1 - 2\,b\,c\,\Phi'(b\,w)} = \frac{w\,[2\Phi(b\,w) - 1]}{c\,l(b\,w)} \\ &> 0\end{aligned}$$

and for every $b > \frac{1}{c}\sqrt{\frac{\pi}{2}}$

$$\begin{aligned}\frac{\partial}{\partial b} w(b,c) &= \left. \frac{\frac{\partial}{\partial b} W(b,c,y)}{-\frac{\partial}{\partial y} W(b,c,y)} \right/_{y=w} \\ &= \frac{2\,c\,w\,\Phi'(b\,w)}{1 - 2\,b\,c\,\Phi'(b\,w)} \\ &> 0.\end{aligned}$$

Note also that $w(b,c) > 0$ for $b\,c > \sqrt{\frac{\pi}{2}}$ according to Lemma A.4.

Set $k(c) := w(b,c)$ for fixed $b > 0$. Then with $s_b(c) = \frac{1}{c}\,q(b\,k(c))$ and Lemma A.3 we have

$$\begin{aligned}s_b'(c) &= \frac{q'(b\,k(c))\,b\,k'(c)}{c} - \frac{q(b\,k(c))}{c^2} \\ &= \frac{q'(b\,k(c))\,b\,[2\,\Phi(b\,k(c)) - 1]\,k(c)}{c^2\,l(b\,k(c))} - \frac{q(b\,k(c))}{c^2} \\ &= \frac{4\,h(b\,k(c))\,b\,k(c)}{[2\,\Phi(b\,k(c)) - 1]^2\,c^2} - \frac{g(b\,k(c))}{k(c)^2} \\ &= \frac{4\,h(b\,k(c))\,b\,k(c) \;-\; 2\,b\,k(c)\,h(b\,k(c)) \;-\; [2\,\Phi(b\,k(c)) - 1]}{k(c)^2} \\ &= \frac{2\,b\,h(b\,k(c))}{k(c)} - \frac{1}{c\,k(c)} \\ &< 0\end{aligned}$$

and

$$\begin{aligned}& s_b''(c) \\ &\quad= \frac{2\,b^2\,h'(b\,k(c))\,k'(c)\,k(c) \;-\; 2\,b\,h(b\,k(c))\,k'(c)}{k(c)^2} \\ &\qquad+ \frac{c\,k'(c) + k(c)}{c^2\,k(c)^2} \\ &\quad> 0.\end{aligned}$$

Set now $k(b) := w(b,c)$ for fixed $c > 0$. Then with $r_c(b) = q(b\,k(b))$ and Lemma A.3 we have

$$r_c'(b) = q'(b\,k(b))\,[k(b) + b\,k'(b)]$$

$$
\begin{aligned}
&= \frac{4\,h(b\,k(b))\;l(b\,k(b))}{[2\,\Phi(b\,k(b))-1]^3}\cdot\left[k(b)+\frac{2\,b\,c\;k(b)\;\Phi'(b\,k(b))}{1-2\,b\,c\;\Phi'(b\,k(b))}\right]\\
&= \frac{4\,h(b\,k(b))\;l(b\,k(b))\;k(b)}{[2\,\Phi(b\,k(b))-1]^3\,[1-2\,b\,c\;\Phi'(b\,k(b))]}\\
&= \frac{4\,h(b\,k(b))\;l(b\,k(b))\;k(b)^2}{[2\,\Phi(b\,k(b))-1]^3\,[k(b)-2\,b\;k(b)\;c\;\Phi'(b\,k(b))]}\\
&= \frac{4\,h(b\,k(b))\;l(b\,k(b))\;k(b)^2}{[2\,\Phi(b\,k(b))-1]^3\;c\;l(b\,k(b))}\\
&= \frac{4\,h(b\,k(b))\;c^2}{k(b)}\\
&< 0
\end{aligned}
$$

and

$$
\begin{aligned}
&r_c''(b)\\
&= \frac{c^2\;4\;h'(b\,k(b))\;[k(b)+b\,k'(b)]}{k(b)}-\frac{c^2\;4\;h(b\,k(b))\;k'(b)}{k(b)^2}\\
&> 0.\ \square
\end{aligned}
$$

Lemma A.7 *For all* $a, b, c \in \mathbb{R}$ *with* $a \cdot b \cdot c \geq \sqrt{\frac{\pi}{2}}$ *we have* $v_{b\cdot a}(c^2) = v_b(a^2 \cdot c^2)$.

Proof.

$$
\begin{aligned}
a\,w(b,1) &= \left[2\Phi\left(\frac{b}{a}\,a\,w(b,1)\right)-1\right]\,a\\
&= w\left(\frac{b}{a},a\right)
\end{aligned}
$$

implies

$$
\begin{aligned}
v_{\frac{b}{a}}(a^2) &= \frac{a^2\;g\left(\frac{b}{a}\,w\left(\frac{b}{a},a\right)\right)}{w\left(\frac{b}{a},a\right)^2}\\
&= \frac{g(b\,w(b,1))}{w(b,1)^2} = v_b(1).
\end{aligned}
$$

Then we have

$$
\begin{aligned}
v_{b\cdot a}(c^2) &= v_{b\cdot\frac{a}{c}\cdot c}(c^2) = v_{b\cdot a\cdot c}(1)\\
&= v_{\frac{b\cdot a\cdot c}{a\cdot c}}(a^2\cdot c^2) = v_b(a^2\cdot c^2).\square
\end{aligned}
$$

Lemma A.8 *The function* $\overline{\omega} : [0,\infty) \to [0,\infty)$ *given by*

$$\overline{\omega}(y) = \frac{1}{c} g(y) - y^2 \qquad \text{for } c > 0$$

has at most one positive root and this positive root exists if and only if

$$c < 1.$$

The root lies in $S := \{y \in [0,\infty);\ 2\Phi(-y) \leq c\}$ *and* $\overline{\omega}$ *is concave on* S.

Proof. Lemma A.3 c) provides

$$\overline{\omega}'(y) = \frac{1}{c} g'(y) - 2y = \frac{4\,y\,\Phi(-y)}{c} - 2y,$$

where $\overline{\omega}'(y) > 0$ if and only if $1 \geq 2\Phi(-y) > c$. Hence, because $\overline{\omega}$ is decreasing to $-\infty$ and $\overline{\omega}(0) = 0$, there exists $y > 0$ with $2\Phi(-y) > c$ and therefore a positive root if and only if $c < 1$. Only for y with $2\Phi(-y) \leq c$ the function $\overline{\omega}$ is decreasing so that the root must ly in S. $\overline{\omega}$ is concave on S because of

$$\overline{\omega}''(y) = \frac{4\,\Phi(-y) - 4y\Phi'(y)}{c} - 2 \leq 0$$

if $2\Phi(-y) \leq c$. □

Lemma A.8 and its proof show that a root of $\overline{\omega}$ can be easily calculated by Newton's method by starting with an interior point of S. With the root of $\overline{\omega}$ also any solution of $y_b^2 = \delta\, g(\sqrt{b}\, y_b)$ for $b > \frac{1}{\delta}$ can be easily calculated because $\sqrt{b}\, y_b$ is a root of $\overline{\omega}$ for $c = \frac{1}{b\,\delta}$.

Define $\overline{W} : (0,\infty)^2 \to I\!R$ as

$$\overline{W}(c,y) \quad := \quad \frac{1}{c} g(y) - y^2$$

and $\overline{w} : (0,1) \to (0,\infty)$ implicitly by $\overline{W}(c, \overline{w}(c)) = 0$. Set also

$$\overline{v}(c) \quad := \quad \begin{cases} \frac{2}{\pi} & \text{for } c = 1, \\ \frac{(2\Phi(\overline{w}(c))-1)^2}{g(\overline{w}(c))} & \text{for } 0 < c < 1, \end{cases}$$

$$u(b) \quad := \quad \overline{v}\left(\frac{1}{b}\right)$$

for $b \geq 1$ and

$$t_s(b) := b \left(\frac{1}{\overline{v}\left(\frac{1}{b}\right)} \right)^s$$

for $b \geq 1$ and $s \in I\!N$.

Lemma A.9 *We have* $\lim_{c\uparrow 1}\overline{w}(c) = 0$. *In particular,* $\overline{v} : (0,1] \to I\!R$ *and* $t_s : [1,\infty) \to I\!R$ *are continuous.*

Proof. Set

$$c(\epsilon) := \frac{g(\epsilon)}{\epsilon^2}.$$

Then we have

$$\begin{aligned}\lim_{\epsilon\downarrow 0} c(\epsilon) = \lim_{\epsilon\downarrow 0}\frac{g'(\epsilon)}{2\epsilon} &= \lim_{\epsilon\downarrow 0}\frac{g''(\epsilon)}{2} \\ &= \lim_{\epsilon\downarrow 0}\frac{4\Phi(-y) - 4y\Phi'(y)}{2} \\ &= \frac{4\Phi(0)}{2} = 1\end{aligned}$$

and $\overline{w}(c(\epsilon)) = \epsilon$ so that

$$\lim_{c\uparrow 1}\overline{w}(c) = \lim_{\epsilon\downarrow 0}\overline{w}(c(\epsilon)) = 0$$

and

$$\begin{aligned}\lim_{c\uparrow 1}\overline{v}(c) = \lim_{\epsilon\downarrow 0}\frac{(2\Phi(\overline{w}(c(\epsilon))) - 1)^2}{g(\overline{w}(c(\epsilon)))} &= \lim_{\epsilon\downarrow 0}\frac{(2\Phi(\epsilon) - 1)^2}{g(\epsilon)} = \frac{2}{\pi}\end{aligned}$$

(for the last equality see Lemma A.3 e)). With $\overline{v}$ also t_s is continuous. □

To show that $\overline{v}$ and t_s are concave/convex we first show some properties of Φ and u.

Lemma A.10

$$\begin{aligned} a) f_1(y) &:= y^3\Phi(-y) - 4\,\Phi'(0)[2\Phi(y) - 1 - 2y\Phi'(y)] > 0 \\ &\quad \textit{for } 0 < y < 0.204 \\ b) f_2(y) &:= 8\Phi'(0)\,\Phi(-y) + 5y\,\Phi(-y) - 4\Phi'(y) > 0 \\ &\quad \textit{for } y > 0 \\ c) f_3(y) &:= 5y\Phi(-y) - 4\Phi'(y) > 0 \textit{ for } y > 1.663. \end{aligned}$$

Proof.
a) Setting

$$f_0(y) := 3\Phi(-y) - y\Phi'(y) - 8\Phi'(0)\Phi'(y)$$

we have

$$\begin{aligned} f_1'(y) &= 3y^2\Phi(-y) - y^3\Phi'(y) - 4\Phi'(0)\,2y^2\Phi'(y) \\ &= y^2\, f_0(y). \end{aligned}$$

Because

$$f_0(0) = 3\Phi(0) - 8\Phi'(0)^2 > 1.5 - 1.28 > 0$$

and

$$f_0'(y) = \Phi'(y)[y^2 + 8\Phi'(0)y - 4],$$

f_0 is for $y > 0$ at first strictly decreasing to some negative value and then strictly increasing to zero. Hence $f_1(y) > 0$ for $0 < y < x$ and $f_1(y) < 0$ for $y > x$ for some value x. Because $f_1(0.204) > 0$ the assertion a) follows.

b) Because

$$\begin{aligned} f_2(0) &= 0, \\ \lim_{y\to\infty} f_2(y) &= 0, \\ f_2'(y) &= -8\Phi'(0)\Phi'(y) + 5\Phi(-y) - y\Phi'(y), \\ f_2'(0) &= -8\Phi'(0)^2 + 5\Phi(0) > -1.28 + 2.5 > 0 \end{aligned}$$

and

$$f_2''(y) = \Phi'(y)[y^2 + 8\Phi'(0)y - 6],$$

the assertion b) follows.

c) Because

$$\begin{aligned} f_3(0) &< 0, \\ \lim_{y\to\infty} f_3(y) &= 0, \\ f_3'(y) &= 5\Phi(-y) - 5y\Phi'(y) + 4y\Phi'(y) = 5\Phi(-y) - y\Phi'(y), \\ f_3'(0) &= 5\Phi(0) > 0 \end{aligned}$$

and

$$f_3''(y) = \Phi'(y)[y^2 - 6],$$

$f_3(y)$ is for $0 < y < x$ negative and for $y > x$ positive for some value x. Because $f_3(1.663) > 0$ the assertion c) follows. □

Lemma A.11 *For all $b > 1$ we have*

$$\begin{aligned} &a) \quad u'(b) > 0, \\ &b) \quad u''(b) < 0, \\ &c) \quad u''(b)\,b + 2\,u'(b) \le 0. \end{aligned}$$

Proof. Define $\tilde{W} : (0,\infty)^2 \rightarrow I\!R$ as

$$\tilde{W}(b,y) \quad := \quad b\,g(y) - y^2$$

and $\tilde{w} : (1,\infty) \rightarrow (0,\infty)$ implicitly by $\tilde{W}(b,\tilde{w}(b)) = 0$. Then $\tilde{w}(b) = \overline{w}\left(\frac{1}{b}\right)$ and

$$u(b) \quad = \quad \begin{cases} \frac{2}{\pi} & \text{for } b = 1, \\ \frac{1}{q(\tilde{w}(b))} & \text{for } b > 1, \end{cases}$$

where q is defined as above. The implicit function theorem and Lemma A.3 provide for every $b > 1$ with $w := \tilde{w}(b)$

$$\begin{aligned} \tilde{w}'(b) &= \left. \frac{\frac{\partial}{\partial b}\tilde{W}(b,y)}{-\frac{\partial}{\partial y}\tilde{W}(b,y)} \right/_{y=w} \\ &= \left. \frac{g(y)}{-(b\,4\,y\,\Phi(-y) - 2\,y)} \right/_{y=w} \\ &= \frac{-g(w)\,w}{2\,(b\,2\,w^2\,\Phi(-w) - w^2)} \\ &= \frac{-g(w)\,w}{2\,(b\,2\,w^2\,\Phi(-w) - b\,g(w))} \\ &= \frac{g(w)\,w}{2\,b\,l(w)} > 0 \end{aligned}$$

so that

$$\begin{aligned} u'(b) &= \frac{-1}{q(w)^2}\,q'(w)\,\tilde{w}'(b) \\ &= \frac{-(2\Phi(w)-1)^4}{g(w)^2}\,\frac{4\,h(w)\,l(w)}{(2\Phi(w)-1)^3}\,\frac{g(w)\,w}{2\,b\,l(w)} \\ &= \frac{-(2\Phi(w)-1)\,2\,h(w)\,w}{g(w)\,b} = \frac{-(2\Phi(w)-1)\,2\,h(w)}{w} > 0 \end{aligned}$$

and

$$\begin{aligned} u''(b) &= \frac{\tilde{w}'(b)}{w^2}\,[(2\Phi(w)-1)\,2\,h(w) - w\,2\,\Phi'(w)\,2\,h(w) \\ &\quad - w\,(2\Phi(w)-1)\,2\,\Phi(-w)] \\ &= \frac{\tilde{w}'(b)\,2}{w^2}\,[h(w)\,l(w) - (2\Phi(w)-1)\,w\,\Phi(-w)] < 0. \end{aligned}$$

Hence a) and b) are proved. a) and b) imply

$$\begin{aligned} &u''(b)\,b + 2\,u'(b) \\ &= \frac{g(w)\,w\,2\,b}{2\,b\,l(w)\,w^2}\,[h(w)\,l(w) - (2\Phi(w)-1)\,w\,\Phi(-w)] \end{aligned}$$

$$
\begin{aligned}
& - \frac{2\,(2\Phi(w)-1)\,2\,h(w)}{w} \\
= \; & \frac{1}{w\,l(w)}\,[h(w)\,l(w)\,g(w) - (2\Phi(w)-1)\,w\,\Phi(-w)\,g(w) \\
& \qquad - 4\,(2\Phi(w)-1)\,h(w)\,l(w)] \\
= \; & \frac{1}{w\,l(w)}\,[h(w)\,l(w)\,g(w) - (2\Phi(w)-1)\,f(w)]
\end{aligned}
$$

with

$$
\begin{aligned}
f(y) \; := \; & y\Phi(-y)\,g(y) + 4h(y)l(y) \\
= \; & 2y^3\Phi(-y)^2 + y\Phi(-y)\,(2\Phi(y)-1) - 2y^2\Phi(-y)\Phi'(y) \\
& +4y\Phi(-y)\,(2\Phi(y)-1) - 4y^2\Phi(-y)\,2\,\Phi'(y) - 4\Phi'(y)(2\Phi(y)-1) \\
& +4\cdot 2\,y\Phi'(y)^2 \\
= \; & 2y^3\Phi(-y)^2 + 5y\Phi(-y)\,(2\Phi(y)-1) - 10y^2\Phi(-y)\Phi'(y) \\
& -4\Phi'(y)(2\Phi(y)-1) + 8y\,\Phi'(y)^2 \\
= \; & 2y^3\Phi(-y)^2 + l(y)[5y\Phi(-y) - 4\Phi'(y)] \\
= \; & 2y^3\Phi(-y)^2 + l(y)\,f_3(y) \\
= \; & 2y^3\Phi(-y)^2 - 8\Phi'(0)\,\Phi(-y)\,l(y) \\
& +8\Phi'(0)\,\Phi(-y)\,l(y) + l(y)[5y\Phi(-y) - 4\Phi'(y)] \\
= \; & 2\Phi(-y)[y^3\Phi(-y) - 4\Phi'(0)l(y)] \\
& +l(y)\,[8\Phi'(0)\Phi(-y) + 5y\Phi(-y) - 4\Phi'(y)] \\
= \; & 2\Phi(-y)\,f_1(y) + l(y)\,f_2(y) \\
> \; & 0
\end{aligned}
$$

for $y > 1.663$ and $0 < y < 0.204$ according to Lemma A.10. Note that f_1, f_2 and f_3 are defined as in Lemma A.10.

For $0.2035 \le y \le 1.664$ the derivative is bounded. It can be shown that $|f'(y)| \le 0.85$ for $y > 0$, but it is easier to see $|f'(y)| < 20$ for $0.2035 \le y \le 1.664$. For we have

$$
\begin{aligned}
|f'(y)| \; = \; & |\Phi(-y)^2\,6\,y^2 + \Phi(-y)\,\Phi'(y)\,[6y^3 - 10y] - \Phi'(y)^2\,6\,y^2 \\
& +\Phi(-y)\,(2\Phi(y)-1)\,5 - \Phi'(y)\,(2\Phi(y)-1)\,y| \\
\le \; & |\Phi(-y)^2\,6\,y^2| + |6\,y^2\,\Phi'(y)\,[y\,\Phi(-y) - \Phi'(y)]| \\
& +|\Phi(-y)\,\Phi'(y)\,10y| + |\Phi(-y)\,(2\Phi(y)-1)\,5| \\
& +|\Phi'(y)\,(2\Phi(y)-1)\,y| \\
\le \; & |\Phi(0)^2\,6\,y^2| + |6\,y^2\,\Phi'(0)\,\Phi'(0)| + |\Phi(0)\,\Phi'(0)\,10\,y| \\
& +\Phi(0)\,2\cdot 5 + |\Phi'(0)\,2\,y| \\
\le \; & 5+3+4+5+2 < 20.
\end{aligned}
$$

Calculating $f(y_k)$ with $y_k = 0.2035 + k\cdot 0.000022$ for $k = 0, \ldots, 66\,387$ shows that $f(y_k) > 0.00044$ for $k = 0, \ldots, 66\,387$. Assume that there exists

$y_* \in [0.2035, 1.664]$ with $f(y_*) \leq 0$. Then there exist $k_0 \in \{0, ..., 66\,386\}$ and $c \in (y_{k_0}, y_{k_0+1})$ such that $y_* \in (y_{k_0}, y_{k_0+1})$ and

$$\begin{aligned} 20 &> |f'(c)| = \left| \frac{f(y_*) - f(y_{k_0})}{y_* - y_{k_0}} \right| \\ &\geq \frac{0.00044 - 0}{(k_0 + 1)\, 0.000022 - k_0\, 0.000022} = 20. \end{aligned}$$

Hence, we have a contradiction so that $f(y) > 0$ for all $y \in [0.2035, 1.664]$. This provides $f(y) > 0$ for all $y > 0$ and therefore $u''(b)\, b + 2\, u'(b) < 0$ for all $b > 1$. □

Lemma A.12 *$\overline{v} : (0, 1] \to \mathbb{R}$ is concave and $t_s : [1, \infty) \to \mathbb{R}$ is convex with $\lim_{b \downarrow 1} t_1'(b) = 0$.*

Proof. Because $\overline{v}(c) = u\left(\frac{1}{c}\right)$ we obtain with Lemma A.11

$$\overline{v}'(c) = u'\left(\frac{1}{c}\right) \frac{-1}{c^2}$$

and

$$\overline{v}''(c) = u''\left(\frac{1}{c}\right) \frac{1}{c^4} + u'\left(\frac{1}{c}\right) \frac{2}{c^3} \leq 0.$$

Because $t_s(b) = b \left(\frac{1}{u(b)}\right)^s$ we have with Lemma A.11

$$t_s'(b) = \left(\frac{1}{u(b)}\right)^s + b\, \frac{(-s)\, u'(b)}{u(b)^{s+1}}$$

and

$$\begin{aligned} t_s''(b) &= \frac{(-s)\, u'(b)}{u(b)^{s+1}} + \frac{(-s)\, u'(b)}{u(b)^{s+1}} \\ &\quad + \frac{b\, s\, (s+1)\, (u'(b))^2}{u(b)^{s+2}} - \frac{b\, s\, u''(b)}{u(b)^{s+1}} \\ &= \frac{s}{u(b)^{s+1}} \left[\frac{b(s+1)}{u(b)} (u'(b))^2 - b\, u''(b) - 2\, u'(b) \right] \geq 0. \end{aligned}$$

In particular, with $w = \tilde{w}(b) := \overline{w}(\frac{1}{b})$ we have

$$\begin{aligned} t_1'(b) &= \frac{1}{u(b)} \left(1 + b\, \frac{(2\Phi(w) - 1)\, 2\, h(w)}{w} \, \frac{g(w)}{(2\Phi(w) - 1)^2} \right) \\ &= \frac{1}{u(b)} \left(1 + \frac{2\, h(w)\, w}{2\Phi(w) - 1} \right) \end{aligned}$$

(see the proof of Lemma A.11). Lemma A.9 and its proof provide $\lim_{b\downarrow 1} \tilde{w}(b) = 0$ and $\lim_{b\downarrow 1} u(b) = \frac{2}{\pi}$. Hence with

$$\begin{aligned} &\lim_{w\downarrow 0} \frac{2\,h(w)\,w}{2\Phi(w) - 1} \\ &\quad = \lim_{w\downarrow 0} \frac{2\,h'(w)\,w + 2\,h(w)}{2\Phi'(w)} = -1 \end{aligned}$$

we have $\lim_{b\downarrow 1} t_1'(b) = 0$. □

References

Akritas, M.G. (1991a). Robust M estimation in the two-sample problem. *J. Amer. Statist. Assoc.* **86**, 201-204.

Akritas, M.G. (1991b). An alternative derivation of aligned rank tests for regression. *J. Statist. Plann. Inference* **27**, 171-186.

Atkinson, A.C. (1982). Developments in the design of experiments. *Int. Statist. Rev.* **50**, 161-177.

Atkinson, A.C. (1985). *Plots, Transformations, and Regression: An Introduction to Graphical Methods of Diagnostic Regression Analysis.* Clarendon Press, Oxford.

Atkinson, A.C. (1988a). Recent developments in the methods of optimum and related experimental designs. *Int. Statist. Rev.* **56**, 99-115.

Atkinson, A.C. (1988b). Transformation unmasked. *Technometrics* **30**, 311-318.

Atkinson, A.C. and Donev, A.N. (1992). *Optimum Experimental Designs.* Clarendon Press, Oxford.

Atwood, C.L. (1969). Optimal and efficient designs of experiments. *Ann. Math. Statist.* **40**, 1570-1602.

Baksalary, J.K. and Tabis, Z. (1987). Conditions for the robustness of block designs against the unavailability of data. *J. Statist. Plann. Inference* **16**, 49-54.

Bandemer, H. (1977). *Theorie und Anwendung der optimalen Versuchsplanung I. Handbuch zur Theorie.* Akademie-Verlag, Berlin.

Bandemer, H. and Näther, W. (1980). *Theorie und Anwendung der optimalen Versuchsplanung II. Handbuch zur Anwendung.* Akademie-Verlag, Berlin.

Basu, A., Markatou, M. and Lindsay, B.G. (1993). Robustness via weighted likelihood estimating equations. *Manuscript.*

Bednarski, T. (1985). On minimum bias and variance estimation for parametric models with shrinking contamination. *Probability and Mathematical Statistics* **6**, 121-129.

Bednarski, T., Clarke, B.R. and Kolkiewicz, W. (1991). Statistical expansions and locally uniform Fréchet differentiability. *J. Austral. Math. Soc. A* **50**, 88-97.

Bednarski, T. and Zontek, S. (1994). A note on robust estimation of parameters in mixed unbalanced models. In *Proceedings of the International Conference on Linear Statistical Inference LINSTAT'93*, eds. T. Caliński, R. Kala, Kluwer Academic Publishers, Dordrecht, 87-95.

Bednarski, T. and Zontek, S. (1996). Robust estimation of parameters in a mixed unbalanced model. *Ann. Statist.* **24**, 1493-1510.

Behnen, K. (1994). A modification of least squares with high efficiency and high breakdown point in linear regression. In *Asymptotic Statistics*, eds. P. Mandl, M. Hušková, Physica-Verlag, Heidelberg, 183-194.

Benda, N. (1992). Optimierung von Versuchsplänen bei unterschätzter Wirkungsfunktion im linearen Experiment. *Ph.D. thesis*, Free University of Berlin.

Benda, N. (1996). Pre-test estimation and design in the linear model. *J. Statist. Plann. Inference.* **52**, 225-240.

Ben-Israel, A. and Greville, T.N.E. (1974). *Generalized Inverses: Theory and Applications.* Wiley, New York.

Beran, J. (1992). Statistical methods for data with long-range dependence (with discussion). *Statist. Sci.* **7**, 404-427.

Bhaumik, D.K. and Whittinghill III, D.C. (1991). Optimality and robustness to the unavailability of blocks in block designs. *J. R. Statist. Soc. B* **53**, 399-407.

Bickel, P.J. (1975). One-step Huber estimates in the linear model. *J. Amer. Statist. Assoc.* **70**, 428-434.

Bickel, P.J. (1981). Quelque aspects de la statistique robuste. In École d'Été de Probabilités de St. Flour. *Springer Lecture Notes in Math.* **876**, 1-72.

Bickel, P.J. (1984). Robust regression based on infinitesimal neighbourhoods. *Ann. Statist.* **12**, 1349-1368.

Bickel, P.J. and Herzberg, A.M. (1979). Robustness of design against autocorrelation in time I: Asymptotic theory, optimality for location and linear regression. *Ann. Statist.* **7**, 77-95.

Billingsley, P. (1968). *Convergence of Probability Measures.* John Wiley, New York.

Box, G.E.P. and Draper, N.R. (1975). Robust designs. *Biometrika* **62**, 347-352.

Brunner, E. and Denker, M. (1994). Rank statistics under dependent observations and applications to factorial designs. *J. Statist. Plann. Inference* **42**, 353-378.

Büning, H. (1991). *Robuste und adaptive Tests.* De Gruyter, Berlin.

Büning, H. (1994a). Robust and adaptive tests for the two-sample location problem. *OR Spektrum* **16**, 33-39.

Büning, H. (1994b). Robuste ANOVA. *Diskussionspapier 09/1994,* Fachbereich Wirtschaftswissenschaft der Freien Universität Berlin.

Bunke, H. and Bunke, O. (1986). *Statistical Inference in Linear Models. Statistical Methods of Model Building, Volume I.* Wiley, New York.

Bunke, H. and Bunke, O. (1989). *Nonlinear Regression, Functional Relations and Robust Methods. Statistical Methods of Model Building, Volume II.* Wiley, New York.

Buonaccorsi, J.P. (1986a). Design considerations for calibration. *Technometrics* **28**, 149-155.

Buonaccorsi, J.P. (1986b). Designs for slope ratio assays. *Biometrics* **42**, 875-882.

Buonaccorsi, J.P. and Iyer, H.K. (1986). Optimal designs for ratios of linear combinations in the general linear model. *J. Statist. Plann. Inference* **13**, 345-356.

Carroll, R.J., Eltinge, J.L. and Ruppert, D. (1993). Robust linear regression in replicated measurement error models. *Statist. Probab. Lett.* **16**, 169-175.

Carroll, R.J. and Pederson, S. (1993). On robustness in the logistic regression model. *J. R. Statist. Soc. B* **55**, 693-706.

Carroll, R.J. and Ruppert, D. (1982). Robust estimation in heteroscedastic linear models. *Ann. Statist.* **10**, 429-441.

Carroll, R.J., Ruppert, D. and Stefanski, L.A. (1986). Optimally bounded score functions for generalized linear models with applications to logistic regression. *Biometrika* **73**, 413-424.

Chatterjee, S. and Hadi, A.S. (1988). *Sensitivity Analysis in Linear Regression.* Wiley, New York.

Christensen, R. (1987). *Plane Answers to Complex Questions - The Theory of Linear Models.* Springer, New York.

Christmann, A. (1993). Strong consistency of the least median of weighted squares estimator for large strata. *Research paper 93/12*, University of Dortmund.

Clarke, B.R. (1983). Uniqueness and Fréchet differentiability of functional solutions to maximum likelihood type equations. *Ann. Statist.* **11**, 1196-1205.

Clarke, B.R. (1986). Nonsmooth analysis and Fréchet differentiability of M-functionals. *Probab. Th. Rel. Fields* **73**, 197-209.

Coakley, C.W. (1991). Breakdown points under simple regression with replication. *Technical Report No. 91-21*, Department of Statistics, Virginia Polytechnic Institute and State University.

Coakley, C.W. and Hettmansperger, T.P. (1992). Breakdown bounds and expected resistance. *Journal of Nonparametric Statistics* **1**, 267-276.

Coakley, C.W. and Hettmansperger, T.P. (1994). The maximum resistance of tests. *Austral. J. Statist.* **36**, 225-233.

Coakley, C.W. and Mili, L. (1993). Exact fit points under simple regression with replication. *Statist. Probab. Lett.* **17**, 265-271.

Coakley, C.W., Mili, L. and Cheniae, M.G. (1994). Effect of leverage on the finite sample efficiencies of high breakdown estimators. *Statist. Probab. Lett.* **19**, 399-408.

Constantine, G.M. (1989). Robust designs for serially correlated observations. *Biometrika* **76**, 245-251.

Cook, R.D. and Weisberg, S. (1982). *Residuals and Influence in Regression.* Chapman & Hall, New York.

Croux, C., Rousseeuw, P.J. and Hössjer, O. (1994). Generalized S-estimators. *J. Amer. Statist. Assoc.* **89**, 1271-1281.

Croux, C., Rousseeuw, P.J. and Van Bael, A. (1996). Positive-breakdown regression by minimizing nested scale estimators. *J. Statist. Plann. Inference* **53**, 197-235.

Das Gupta, A. and Studden, W.J. (1991). Robust Bayesian experimental designs in normal linear models. *Ann. Statist.* **19**, 1244-1256.

Davies, P.L. (1990). The asymptotics of S-estimators in the linear regression model. *Ann. Statist.* **18**, 1651-1675.

Davies, P.L. (1993). Aspects of robust linear regression. *Ann. Statist.* **21**, 1843-1899.

Davies, P.L. (1994). Desirable properties, breakdown and efficiency in the linear regression model. *Statist. Probab. Lett.* **19**, 361-370.

DeFeo, P. and Myers, R.H. (1992). A new look at experimental design robustness. *Biometrika 79*, 375-380.

Dette, H. (1993). Bayesian D-optimal and model robust designs in linear regression models. *Statistics* 25, 27-46.

Dette, H. (1994). Robust designs for multivariate polynomial regression on the d-cube. *J. Statist. Plann. Inference* **38**, 105-124.

Dey, A. (1993). Robustness of block designs against missing data. *Statistica Sinica* **3**, 219-231.

Donoho, D.L. and Huber, P.J. (1983). The notion of breakdown point. In *A Festschrift for Erich L. Lehmann* (P.J. Bickel, K.A. Doksum and J.L. Hodges, Jr.,eds.) 157-184. Wadsworth, Belmont, Calif.

Draper, N.R. and Sanders, E.R. (1988). Designs for minimum bias estimation. *Technometrics* **30**, 319-325.

Eicker, F. (1963). Asymptotic normality and consistency of the least-squares estimators for families of linear regressions. *Ann. Math. Statist.* **34**, 447-456.

Elfving, G. (1952). Optimum allocation in linear regression theory. *Ann. Math. Statist.* **23**, 255-262.

Ellis, S.P. and Morgenthaler, S. (1992). Leverage and breakdown in L_1 regression. *J. Amer. Statist. Assoc.* **87**, 143-148.

Fedorov, V.V. (1971). Design of experiments for linear optimality criteria. *Theory Probab. Appl.* **16**, 189-195.

Fedorov, V.V. (1972). *Theory of Optimal Experiments.* Academic Press, New York.

Fedorov, V.V. (1989). Optimal design with bounded density: Optimization algorithms of the exchange type. *J. Statist. Plann. Inference* **22**, 1-13.

Fisz, M. (1963). *Probability Theory and Mathematical Statistics.* Wiley, New York.

Ford, I., Kitsos, C.P. and Titterington, D.M. (1989). Recent advances in nonlinear experimental design. *Technometrics* **31**, 49-60.

Gaffke, N. (1987). Further characterizations of design optimality and admissibility for partial parameter estimation in linear regression. *Ann. Statist.* **15**, 942-957.

Gallant, A.R. (1987). *Nonlinear Statistical Models.* Wiley, New York.

García Pérez, A. (1993). On robustness for hypotheses testing. *Int. Statist. Review* **61**, 369-385.

Ghosh, S. and Kipngeno, A.K. (1985). On the robustness of the optimum balanced 2^m factorial designs of resolution V (given by Srivastava and Chopra) in the presence of outliers. *J. Statist. Plann. Inference* **11**, 119-129.

Ghosh, S., Rao, S.B. and Singhi, N.M. (1983). On a robustness property of PBIBD. *J. Statist. Plann. Inference* **8**, 355-363.

Gutenbrunner, C. and Jurečková, J. (1992). Regression rank scores and regression quantiles. *Ann. Statist.* **20**, 305-330.

Gutenbrunner, C., Jurečková, J., Koenker, R. and Portnoy, S. (1993). Tests of linear hypothesis based on regression rank scores. *J. of Nonparametric Statistics* **2**, 307-331.

Hájek, J. and Šidák, Z. (1967). *Theory of Rank Tests.* Academic Press, New York.

Hampel, F.R. (1968). Contributions to the theory of robust estimation. *Ph. D. thesis.* University of California, Berkeley.

Hampel, F.R. (1971). A general qualitative definition of robustness. *Ann. Math. Statist.* **42**, 1887-1896.

Hampel, F.R. (1974). The influence curve and its role in robust estimation. *J. Amer. Statist. Assoc.* **69**, 383-393.

Hampel, F.R. (1978). Optimally bounding the gross-error-sensitivity and the influence of position in factor space. *Proceedings of the ASA Statistical Computing Section*, ASA, Washington, D.C., 59-64.

Hampel, F.R., Ronchetti, E.M., Rousseeuw, P.J. and Stahel, W.A. (1986). *Robust Statistics - The Approach Based on Influence Functions.* John Wiley, New York.

He, X. (1991). A local breakdown property of robust tests in linear regression. *J. Multivariate Anal.* **38**, 294-305.

He, X. (1994). Breakdown versus efficiency - Your perspective matters. *Statist. Probab. Lett.* **19**, 357-360.

He, X., Jurečková, J., Koenker, R. and Portnoy, S. (1990). Tail behavior of regression estimators and their breakdown points. *Econometrica* **58**, 1195-1214.

He, X., Simpson, D.G. and Portnoy, S.L. (1990). Breakdown robustness of tests. *J. Amer. Statist. Assoc.* **85**, 446-452.

Hedayat, A.S. and Majumdar, D. (1988). Model robust optimal designs for comparing test treatments with a control. *J. Statist. Plann. Inference* **18**, 25-33.

Heiler, S. (1992). Bounded-influence and high breakdown point regression with linear combinations of order statistics and rank statistics. In: *L_1-Statistical Analysis and Related Methods, ed. Y. Dodge*, North-Holland, Amsterdam, 201-215.

Heiler, S. and Willers, R. (1988). Asymptotic normality of R-estimates in the linear model. *Statistics* **19**, 173-184.

Hennig, C. (1995). Efficient high-breakdown-point estimators in robust regression: Which function to choose? *Stat. Decis.* **13**, 221-241.

Heritier, S. and Ronchetti, E. (1994). Robust bounded-influence tests in general parametric models. *J. Amer. Statist. Assoc.* **89**, 897-904.

Herzberg, A.M. (1982). The robust design of experiments. A review. *Serdica* **8**, 223-228.

Hettmansperger, T.P. (1984). *Statistical Inference Based on Ranks.* Wiley, New York.

Hössjer, O. (1994). Rank-based estimates in the linear model with high breakdown point. *J. Amer. Statist. Assoc.* **89**, 149-158.

Hössjer, O., Croux, C. and Rousseeuw, P.J. (1994). Asymptotics of generalized S-estimators. *J. Multivariate Analysis* **51**, 148-177.

Huber, P.J. (1964). Robust estimation of a location parameter. *Ann. Math. Statist.* **35**, 73-101.

Huber, P.J. (1973). Robust regression: Asymptotics, conjectures, and Monte Carlo. *Ann. Statist.* **1**, 799-821.

Huber, P.J. (1975). Robustness and designs. In: J.N. Srivastava (ed.), *A Survey of Statistical Design and Linear Models*, North Holland, Amsterdam, 287-303.

Huber, P.J. (1981). *Robust Statistics.* John Wiley, New York.

Huber, P.J. (1983). Minimax aspects of bounded-influence regression (with discussion). *J. Amer. Statist. Assoc.* **78**, 66-80.

Huggins, R.M. (1993). On the robust analysis of variance components models for pedigree data. *Austral. J. Statist.* **35**, 43-57.

Jennrich, R.I. (1969). Asymptotic properties of non-linear least squares estimators. *Ann. Math. Statist.* **40**, 633-643.

Jensen, D.R. and Ramirez, D.E. (1993). Efficiency comparisons in linear inference. *J. Statist. Plann. Inference* **37**, 51-68.

Jost, S.D. (1991). Relative efficiencies of model robust estimators in two-dimensional situations. *J. Statist. Plann. Inference* **28**, 369-381.

Jurečková, J. (1977). Asymptotic relations of M-estimates and R-estimates in linear regression model. *Ann. Statist.* **5**, 464-472.

Jurečková, J. (1992). Estimation in a linear model based on regression rank scores. *Nonparametric Statistics* **1**, 197-203.

Jurečková, J. and Portnoy, S. (1987). Asymptotics for one-step M-estimators in regression with application to combining efficiency and high breakdown point. *Commun. Statist.-Theory Meth.* **16**, 2187-2199.

Jurečková, J. and Sen, P.K. (1990). Effect of the initial estimator on the asymptotic behavior of one-step M-estimator. *Ann. Inst. Statist. Math.* **42**, 345-357.

Jurečková, J. and Sen, P.K. (1996). *Robust Statistical Procedures. Asymptotics and Interrelations.* Wiley, New York.

Jurečková, J. and Welsh, A.H. (1990). Asymptotic relations between L- and M-estimators in the linear model. *Ann. Inst. Statist. Math.* **42**, 671-698.

Kariya, T. and Sinha, B.K. (1988). *Robustness of Statistical Tests.* Academic Press, New York.

Khuri, A.I. and Cornell, J.A. (1987). *Response Surfaces. Designs and Analysis.* Marcel Dekker, New York.

Kiefer, J. (1959). Optimum experimental designs. *J. R. Statist. Soc. B* **21**, 272-304.

Kiefer, J. (1974). General equivalence theory for optimum designs (approximate theory). *Ann. Statist.* **2**, 849-879.

Kiefer, J. and Wolfowitz, J. (1959). Optimum designs in regression problems. *Ann. Math. Statist.* **30**, 271-294.

Kiefer, J. and Wolfowitz, J. (1960). The equivalence of two extremum problems. *Can. J. Math.* **12**, 363-366.

Kitsos, C.P. (1986). Design and inference for nonlinear problems. *Ph.D. thesis*, University of Glasgow, U.K.

Kitsos, C.P. (1992). Quasi-sequential procedures for the calibration problem. In *COMPSTAT 1992, Vol. 2*, Y. Dodge and J. Whittaker (eds.), 227-231, Physica-Verlag.

Kitsos, C.P. and Müller, Ch.H. (1995a). Robust linear calibration. *Statistics* **27**, 93-106.

Kitsos, C.P. and Müller, Ch.H. (1995b). Robust estimation of non-linear aspects. *MODA 4 - Advances in Model-Oriented Data Analysis*, eds. C.P. Kitsos, W.G. Müller, Physica-Verlag, Heidelberg, 223-233.

Kitsos, C.P., Titterington, D.M. and Torsney, B. (1988). An optimal design problem in rhythmometry. *Biometrics* **44**, 657-671.

Krafft, O. (1978). *Lineare statistische Modelle und optimale Versuchspläne.* Vandenhoeck & Ruprecht, Göttingen.

Krasker, W.S. (1980). Estimation in linear regression models with disparate data points. *Econometrica* **48**, 1333-1346.

Krasker, W.S. and Welsch, R.E. (1982). Efficient bounded-influence regression estimation. *J. Amer. Statist. Assoc.* **77**, 595-604.

Künsch, H., Beran, J. and Hampel, F. (1993). Contrasts under long-range correlations. *Ann. Statist.* **21**, 943-964.

Kurotschka, V. (1972). Optimale Versuchsplanung bei Modellen der Varianzanalyse. *Habilitationsschrift*, Universität Göttingen.

Kurotschka, V. (1978). Optimal design of complex experiments with qualitative factors of influence. *Commun. Statist., Theory Methods A* **7**, 1363-1378.

Kurotschka, V. and Müller, Ch.H. (1992). Optimum robust estimation of linear aspects in conditionally contaminated linear models. *Ann. Statist.* **20**, 331-350.

Kuwada, M. (1993). Robustness of balanced fractional 2^m factorial designs derived from simple arrays. *Discrete Mathematics* **116**, 183-208.

Läuter, H. (1989). Note on the strong consistency of the least squares estimator in nonlinear regression. *Statistics* **20**, 199-210.

Lehmann, E.L. (1959). *Testing Statistical Hypotheses.* Wiley, New York.

Li, K.C. (1984). Robust regression design when the design space consists of finitely many points. *Ann. Statist.* **12**, 269-282.

Li, K.C. and Notz, W. (1982). Robust designs for nearly linear regression. *J. Statist. Plann. Inference* **6**, 135-151.

Liese, F. and Vajda, I. (1994). Consistency of M-estimates in general regression models. *J. Multivariate Analysis* **50**, 93-114.

Lindsay, B.G. (1993). Efficiency versus robustness: The case for minimum Hellinger distance and related methods. *Manuscript*, Pennsylvania State University.

Maercker, G. (1992). Hadamard-Differenzierbarkeit statistischer Funktionale. *Diplomarbeit*, Freiburg i. Br.

Malinvaud, E. (1970). *Statistical Methods of Econometrics*. North-Holland, Amsterdam.

Mandal, N.K. and Heiligers, B. (1992). Minimax designs for estimating the optimum point in a quadratic response surface. *J. Statist. Plann. Inference* **31**, 235-244.

Marcus, M.B. and Sacks, J. (1976). Robust designs for regression problems. In: Gupta, S.S. and D.S. Moore (eds.), *Statistical Theory and Related Topics II*, Academic Press, New York, 245-268.

Markatou, M. and He, X. (1994). Bounded influence and high breakdown point testing procedures in linear models. *J. Amer. Statist. Assoc.* **89**, 543-549.

Markatou, M. and Hettmansperger, T.P. (1990). Robust bounded- influence tests in linear models. *J. Amer. Statist. Assoc.* **85**, 187-190.

Markatou, M. and Manos, G. (1996). Robust tests in nonlinear regression models. *J. Statist. Plann. Inference* **55**, 205-217.

Markatou, M., Stahel, W.A. and Ronchetti, E. (1991). Robust M-type testing procedures for linear models. In: *Directions in Robust Statistics and Diagnostics (Part I)*, eds. W. Stahel and S. Weisberg, Springer, New York, 201-220.

Maronna, R.A., Bustos, O.H. and Yohai, V.J. (1979). Bias- and efficiency-robustness of general M-estimators for regression with random carriers. In: *Smoothing Techniques for Curve Estimation*, eds. T. Gasser and M. Rosenblatt, Lecture Notes in Mathematics **757**, Springer, Berlin, 91-116.

Maronna, R.A. and Yohai, V.J. (1981). Asymptotic behavior of general M-estimates for regression and scale with random carriers. *Z. Wahrsch. verw. Gebiete* **58**, 7-20.

Martin, R.J., Eccleston, J.A. and Gleeson, A.C. (1993). Robust linear block designs for a suspected LV model. *J. Statist. Plann. Inference* **34**, 433-450.

Martin, R.D., Yohai, V.J. and Zamar, R.H. (1989). Min-max bias robust regression. *Ann. Statist.* **17**, 1608-1630.

Mathew, T. and Bhaumik, D.K. (1989). The model-robustness and optimality of randomized designs. *J. Statist. Plann. Inference* **23**, 371-379.

McKean, J.W., Sheather, S.J. and Hettmansperger, T.P. (1993). The use and interpretation of residuals based on robust estimation. *J. Amer. Statist. Assoc.* **88**, 1254-1263.

McKean, J.W., Sheather, S.J. and Hettmansperger, T.P. (1994). Robust and high breakdown fits of polynomial models. *Technometrics* **36**, 409-415.

Mili, L. and Coakley, C.W. (1993). Robust estimation in structured linear regression. *Technical Report No. 93-13*, Department of Statistics, Virginia Polytechnic Institute and State University. *To appear in Ann. Statist.* **24**.

Mizera, I. and Müller, Ch.H. (1996). Breakdown points and variation exponents of robust M-estimators in linear models. *Preprint No. A-22-96*, Freie Universität Berlin, Fachbereich Mathematik und Informatik.

Morgenthaler, S. (1991). A note on efficient regression estimators with positive breakdown point. *Statist. Probab. Lett.* **11**, 469-472.

Morgenthaler, S. (1994). Small sample efficiency and exact fit for Cauchy regression models. *Statist. Probab. Lett.* **19**, 381-385.

Müller, Ch.H. (1987). Optimale Versuchspläne für robuste Schätzfunktionen in linearen Modellen. *Ph. D. thesis.* Freie Universität Berlin.

Müller, Ch.H. (1992a). L_1-estimation and testing in conditionally contaminated linear models. In *L_1-Statistical Analysis and Related Methods*, ed. Y. Dodge, North-Holland, Amsterdam, 69-76.

Müller, Ch.H. (1992b). Robust estimation with minimum bias and A-optimal designs. In *PROBASTAT'91, Proceedings of the International Conference on Probability and Mathematical Statistics, Bratislava 1991*, Pázman, A. and J. Volaufová (eds.), 109-115.

Müller, Ch.H. (1992c). Robust testing in conditionally contaminated linear models. *Preprint No. A-92-12*, Freie Universität Berlin, Fachbereich Mathematik und Informatik.

Müller, Ch.H. (1993a). Behaviour of asymptotically optimal designs for robust estimation at finite sample sizes. In *Model-Oriented Data Analysis*, eds. W.G. Müller, H.P. Wynn, A.A. Zhigljavsky, Physica-Verlag, Heidelberg, 53-62.

Müller, Ch.H. (1993b). Robustness against m-dependent errors in linear models. *Preprint No. A-36-93*, Freie Universität Berlin, Fachbereich Mathematik und Informatik.

Müller, Ch.H. (1994a). Optimal designs for robust estimation in conditionally contaminated linear models. *J. Statist. Plann. Inference.* **38**, 125-140.

Müller, Ch.H. (1994b). Asymptotic behaviour of one-step-M-estimators in contaminated non-linear models. In *Asymptotic Statistics*, eds. P. Mandl, M. Hušková, Physica-Verlag, Heidelberg, 395-404.

Müller, Ch.H. (1994c). On the calculation of MSE minimizing robust estimators. In *COMPSTAT'94, Proceedings in Computational Statistics*, eds. R. Dutter, W. Grossmann, Physica-Verlag, Heidelberg, 257-262.

Müller, Ch.H. (1994d). Optimal bias bounds for robust estimation in linear models. In *Proceedings of the International Conference on Linear Statistical Inference LINSTAT'93*, eds. T. Caliński, R. Kala, Kluwer Academic Publishers, Dordrecht, 97-102.

Müller, Ch.H. (1994e). One-step-M-estimators in conditionally contaminated linear models. *Stat. Decis.* **12**, 331-342.

Müller, Ch.H. (1995a). Maximin efficient designs for estimating nonlinear aspects in linear models. *J. Statist. Plann. Inference.* **44**, 117-132.

Müller, Ch.H. (1995b). Breakdown points for designed experiments. *J. Statist. Plann. Inference.* **45**, 413-427.

Müller, Ch.H. (1996a). Optimal breakdown point maximizing designs. In *PROBASTAT'94, Proceedings of the Second International Conference on Mathematical Statistics, Smolenice 1994*, eds. A. Pázman, V. Witkovský. *Tatra Mountains Mathematical Publications* **7**, 79-85.

Müller, Ch.H. (1996b). High breakdown point designs. In *Robust Statistics, Data Analysis, and Computer Intensive Methods - In Honor of Peter Huber's 60th Birthday*, ed. H. Rieder. *Lecture Notes in Statistics* **109**, Springer, New York, 353-360.

Naranjo, J.D. and Hettmansperger, T.P. (1994). Bounded influence rank regression. *J. R. Statist. Soc. B* **56**, 209-220.

Näther, W. and Reinsch, V. (1981). D_s-optimality and Whittle's equivalence theorem. *Math. Operationsforsch. Statist., Ser. Statist.* **12**, 307-316.

Noether, G.E. (1963). Note on the Kolmogorov statistic in the discrete case. *Metrika* **7**, 115-116.

Notz, W.I. (1989). Optimal designs for regression models with possible bias. *J. Statist. Plann. Inference* **22**, 43-54.

Notz, W.I., Whittinghill, D.C. and Zhu, Y. (1994). Robustness to the unavailability of data in block designs. *Metrika* **41**, 263-275.

Pázman, A. (1980). Singular experimental designs (standard and Hilbert-space approaches). *Math. Operationsforsch. Statist., Ser. Statist.* **11**, 137-149.

Pázman, A. (1986). *Foundations of Optimum Experimental Design.* Reidel, Dordrecht.

Pázman, A. (1993). *Nonlinear Statistical Models.* Kluwer, Dordrecht.

Pesotchinsky, L. (1982). Optimal robust designs: Linear regression in R^k. *Ann. Statist.* **10**, 511-525.

Pesotchinsky, L. (1984). Robust designs and optimality of least squares for regression problems. *J. Statist. Plann. Inference* **9**, 103-117.

Polasek, W. and Pötzelberger, K. (1994). Robust Bayesian methods in simple ANOVA models. *J. Statist. Plann. Inference 40*, 295-312.

Pronzato, L. and Walter, E. (1991). Robustness to outliers of bounded-error estimators, consequences on experiment design. In: *Proceedings of 9th IFAC/IFORS Symposium on Identification and System Parameter Estimation, Vol. 1*, Budapest, 821-826.

Pukelsheim, F. (1981). On c-optimal design measures. *Math. Operationsforsch. Statist., Ser. Statistics* **12**, 13-20.

Pukelsheim, F. (1987). Information increasing orderings in experimental design theory. *Int. Statist. Rev.* **55**, 203-219.

Pukelsheim, F. (1993). *Optimal Design of Experiments.* John Wiley, New York.

Pukelsheim, F. and Titterington, D.M. (1983). General differential and Lagrangian theory for optimal design. *Ann. Statist.* **11**, 1060-1068.

Pukelsheim, F. and Torsney, B. (1991). Optimal weights for experimental designs on linearly independent support points. *Ann. Statist.* **19**, 1614-1625.

Rao, C.R. (1973). *Linear Statistical Inference and Its Applications.* Wiley, New York.

Ren, J.-J. (1994). On Hadamard differentiability and its application to R-estimation in linear models. *Stat. Decis.* **12**, 1-22.

Rieder, H. (1978). A robust asymptotic testing model. *Ann. Statist.* **6**, 1080-1094.

Rieder, H. (1980). Estimates derived from robust tests. *Ann. Statist.* **8**, 106-115.

Rieder, H. (1985). Robust estimation of functionals. *Technical Report.* Universität Bayreuth.

Rieder, H. (1987). Robust regression estimators and their least favorable contamination curves. *Stat. Decis.* **5**, 307-336.

Rieder, H. (1989). A finite-sample minimax regression estimator. *Statistics* **20**, 211-221.

Rieder, H. (1994). *Robust Asymptotic Statistics*, Springer, New York.

Ronchetti, E. (1982). Robust testing in linear models: The infinitesimal approach. *Ph. D. thesis.* Swiss Federal Institute of Technology, Zürich.

Ronchetti, E. and Rousseeuw, P.J. (1985). Change-of-variance sensitivities in regression analysis. *Z. Wahrsch. verw. Gebiete* **68**, 503-519.

Rousseeuw, P.J. (1984). Least median of squares regression. *J. Amer. Statist. Assoc.* **79**, 871-880.

Rousseeuw, P.J. (1994). Unconventional features of positive-breakdown estimators. *Statist. Probab. Lett.* **19**, 417-431.

Rousseeuw, P.J. and Croux, C. (1993). Alternatives to the median absolute derivation. *J. Amer. Statist. Assoc.* **88**, 1273-1283.

Rousseeuw, P.J. and Croux, C. (1994). The bias of k-step M-estimators. *Statist. Probab. Lett.* **20**, 411-420.

Rousseeuw, P.J. and Leroy, A.M. (1987). *Robust Regression and Outlier Detection.* John Wiley, New York.

Rousseeuw, P.J. and Yohai, V. (1984). Robust regression by means of S-estimators. In: *Robust and Nonlinear Time Series Analysis. Lecture Notes in Statist. 26,* Springer, New York, 256-272.

Ruppert, D. (1985). On the bounded-influence regression estimator of Krasker and Welsch. *J. Amer. Statist. Assoc.* **80**, 205-208.

Sacks, J. and Ylvisaker, D. (1978). Linear estimation for approximately linear models. *Ann. Statist.* **6**, 1122-1137.

Sacks, J. and Ylvisaker, D. (1984). Some model robust designs in regression. *Ann. Statist.* **12**, 1324-1348.

Sakata, S. and White, H. (1995). An alternative definition of finite-sample breakdown point with applications to regression model estimators. *J. Amer. Statist. Assoc.* **90**, 1099-1106.

Samarov, A.M. (1985). Bounded-influence regression via local minimax mean squared error. *J. Amer. Statist. Assoc.* **80**, 1032-1040.

Sathe, Y.S. and Satam, M.R. (1992). Some more robust block designs against the unavailability of data. *J. Statist. Plann. Inference* **30**, 93-98.

Schmidt, W.H. (1975). Asymptotic normality of least-square estimators in multivariate singular linear models. *Math. Operationsforsch. Statist.* **6**, 285-300.

Schrader, R.M. and Hettmansperger, T.P. (1980). Robust analysis of variance based upon a likelihood ratio criterion. *Biometrika* **67**, 93-101.

Schwabe, R. (1995). Experimental design for linear models with higher order interaction terms. In: *Symposia Gaussiana. Proceedings of the 2nd Gauss Symposium, Conference B: Statistical Sciences, München 1993*, eds. V. Mammitzsch and H. Schneeweiss, De Gruyter, Berlin, 281-288.

Schwabe, R. (1996a). *Optimum Designs for Multi-Factor Models.* Lecture Notes in Statistics **113**, Springer, New York.

Schwabe, R. (1996b). Model robust experimental design in the presence of interactions: The orthogonal case. In *PROBASTAT'94, Proceedings of the Second International Conference on Mathematical Statistics, Smolenice 1994*, eds. A. Pázman, V. Witkovský. *Tatra Mountains Mathematical Publications* **7**, 97-104.

Seber, G.A.F. and Wild, C.J. (1989). *Nonlinear Regression.* Wiley, New York.

Sen, P.K. (1982). On M-tests in linear models. *Biometrika* **69**, 245-248.

Shao, J. and Wu, C.F.J. (1987). Heteroscedasticity-robustness of jackknife variance estimators in linear models. *Ann. Statist.* **15**, 1563-1579.

Silvapulle, M.J. (1992a). Robust Wald-type tests of one-sided hypothesis in the linear model. *J. Amer. Statist. Assoc.* **87**, 156-161.

Silvapulle, M.J. (1992b). Robust tests of inequality constraints and one-sided hypotheses in the linear model. *Biometrika* **79**, 621-630.

Silvey, S.D. (1978). Optimal design measures with singular information matrices. *Biometrika* **65**, 553-559.

Silvey, S.D. (1980). *Optimal Design*. Chapman and Hall, London.

Simpson, D.G., Ruppert, D. and Carroll, R.J. (1992). On one-step GM estimates and stability of inferences in linear regression. *J. Amer. Statist. Assoc.* **87**, 439-450.

Spruill, C. (1985). Model robustness of Hoel-Levine optimal designs. *J. Statist. Plann. Inference* **11**, 217-225.

Staudte, R.G. and Sheather, S.J. (1990). *Robust Estimation and Testing*. Wiley, New York.

Stefanski, L.A. (1991). A note on high-breakdown estimators. *Statist. Probab. Lett.* **11**, 353-358.

Stefanski, L.A., Carroll, R.J. and Ruppert, D. (1986). Optimally bounded score functions for generalized linear models with applications to logistic regression. *Biometrika 73*, 413-424.

Stromberg, A.J. and Ruppert, D. (1992). Breakdown in nonlinear regression. *J. Amer. Statist. Assoc.* **87**, 991-997.

Studden, W.J. (1982). Some robust-type D-optimal designs in polynomial regression. *J. Amer. Statist. Assoc.* **77**, 916-921.

Tiku, M.L., Tan, W.Y. and Balakrishnan, N. (1986). *Robust Inference*. Marcel Dekker, New York.

Van der Vaart, A.W. (1991). On differentiable functionals. *Ann. Statist.* **19**, 178-204.

Van der Vaart, A.W. (1991). Efficiency and Hadamard differentiability. *Scand. J. Statist.* **18**, 63-75.

Vuchkov, I.N. and Boyadjieva, L.N. (1983). The robustness of experimental designs against errors in the factor levels. *J. Statist. Comput. Simulations* **17**, 31-41.

Wald, A. (1943). Tests of statistical hypothesis concerning several parameters when the number of observations is large. *Transactions Amer. Math. Soc.* **3**, 426-482.

Wang, C.Y. and Carroll, R.J. (1993). On robust estimation in logistic case-control studies. *Biometrika* **80**, 237-241.

Wang, C.Y. and Carroll, R.J. (1995). On robust logistic case-control studies with response-dependent weights. *J. Statist. Plann. Inference* **43**, 331-340.

Whittle, P. (1973). Some general points in the theory of optimal experimental design. *J. R. Statist. Soc. B* **35**, 123-130.

Wiens, D.P. (1992). Minimax designs for approximately linear regression. *J. Statist. Plann. Inference* **31**, 353-371.

Wiens, D.P. (1994). Robust designs for approximately linear regression: M-estimated parameters. *J. Statist. Plann. Inference* **40**, 135-160.

Wiens, D.P. (1996). Asymptotics of generalized M-estimation of regression and scale with fixed carriers, in an approximately linear model. *Statist. Probab. Lett.* **30**, 271-285.

Wierich, W. (1985). Optimum designs under experimental constraints for a covariate model and an intra-class regression model. *J. Statist. Plann. Inference* **12**, 27-40.

Wong, W.K. and Cook, R.D. (1993). Heteroscedastic G-optimal designs. *J. R. Statist. Soc. B* **55**, 871-880.

Wu, C.F.J. (1981). Asymptotic theory of non-linear least squares estimation. *Ann. Statist.* **9**, 501-513.

Wu, X. and Luo, Z. (1993). Second-order approach to local influence. *J. R. Statist. Soc. B* **55**, 929-936.

Wynn, H. (1982). Optimum submeasures with applications to finite population sampling. In: *Statistical Decision Theory and Related Topics III*, Vol. 2. Academic Press, New York, 485-495.

Ylvisaker, D. (1977). Test resistance. *J. Amer. Statist. Assoc.* **72**, 551-556.

Yohai, V.J. (1987). High breakdown-point and high efficiency robust estimates for regression. *Ann. Statist.* **15**, 642-656.

Yohai, V.J. and Maronna, R.A. (1979). Asymptotic behavior of M-estimators for the linear model. *Ann. Statist.* **7**, 258-268.

Yohai, V.J. and Zamar, R.H. (1988). High breakdown-point estimates of regression by means of the minimization of an efficient scale. *J. Amer. Statist. Assoc.* **83**, 406-413.

Zamar, R.H. (1992). Bias robust estimation in orthogonal regression. *Ann. Statist.* **20**, 1875-1888.

Zhang, J. (1996). The sample breakdown points of tests. *J. Statist. Plann. Inference* **52**, 161-181.

Zmyślony, R. and Zontek, S. (1994). Optimality of the orthogonal block design for robust estimation under mixed models. In *Proceedings of the International Conference on Linear Statistical Inference LINSTAT'93*, eds. T. Caliński, R. Kala, Kluwer Academic Publishers, Dordrecht, 195–202.

List of Symbols

$a(\cdot)$	"regression" function of linear models
A_{d_N}	$= (a(t_{1N}, ..., a(t_{NN}))'$, design matrix
$A_{\mathcal{D}}$	$= (a(\tau_1), ..., a(\tau_I))'$ if $\mathcal{D} = \{\tau_1, ..., \tau_I\}$
$\arg\min$	set of all solutions of the minimization problem
$\arg\max$	set of all solutions of the maximization problem
$b(\psi, \delta)$	asymptotic bias for shrinking neighbourhoods
$b_0^E(\delta, L)$	minimum asymptotic bias for estimation at δ
$b_0^E(\Delta, L)$	minimum asymptotic bias for estimation in Δ
$b_0^T(\delta, L)$	minimum asymptotic bias for testing at δ
$b_0^T(\Delta, L)$	minimum asymptotic bias for testing in Δ
$b_{\epsilon,c}(\widehat{\zeta}, P_\delta)$	asymptotic bias for a given contamination neighbourhood
$b_{\epsilon,K}(\widehat{\zeta}, P_\delta)$	asymptotic bias for a given Kolmogorov neighbourhood
$b_M(\widehat{\zeta}_N, d_N, y_N)$	bias by substituting M observations
$b_M^+(\widehat{\zeta}_N, d_N, y_N)$	bias by adding M observations
$\tilde{b}_M(\widehat{\zeta}_N, d_N, y_N)$	bias for contaminated experimental conditions
$b_\epsilon^1(\tau_N, R)$	asymptotic bias of the first error of a test
$b_\epsilon^2(\tau_N, R, \gamma)$	asymptotic bias of the second error of a test for alternatives given by γ
$b_\epsilon^1(\tau_N, R_N)$	bias of the first error of a test
$b_\epsilon^2(\tau_N, R_N, \theta)$	bias of the second error of a test at θ
$\mathcal{B}$	subset of $\mathbb{R}^r$
β	vector of unknown parameters
$\widehat{\beta}_N$	estimator of β
$\widehat{\beta}_N^{LS}$	least squares estimator of β
$\widehat{\beta}_{h,p}$	h-trimmed L_p estimator for β
C_N^{LS}	estimator of the covariance matrix of the Gauss-Markov estimator
C_N	generalization of C_N^{LS}
$C(\psi)$	$= \int \psi(y,t)\, \psi(y,t)'\, P_\delta(dy, dt)$, asymptotic covariance matrix of an asymptotically linear estimator

$Cov_\theta(\cdot)$	covariance matrix at θ
$\chi^2(s,\nu)$	chi-squared distribution with s degrees of freedom and noncentrality parameter ν
$\chi^2_{\alpha,s,\nu}$	α-quantile of $\chi^2(s,\nu)$
$\mathcal{X}_{s,\nu}(\cdot)$	distribution function of $\chi^2(s,\nu)$
$d_K(\cdot,\cdot)$	generalized Kolmogorov metric on $\mathcal{P}^* \times \mathcal{P}^*$
$d_P(\cdot,\cdot)$	Prohorov metric on $\mathbb{R}^q \times \mathbb{R}^q$
d_N	$= (t_{1N}, ..., t_{NN})$, deterministic (concrete) design
D_N	$= (T_{1N}, ..., T_{NN})$, random design
$\mathcal{D}$	subset of $\mathcal{T}$
$\det(\cdot)$	determinant of a matrix
$\mathrm{diag}(\cdot)$	diagonal matrix
δ	design measure
δ_A	A-optimal design
$\delta_{\beta,A}$	locally A-optimal design
δ_D	D-optimal design
$\delta_{\beta,D}$	locally D-optimal design
δ_M	maximin efficient design
δ_N	$= \frac{1}{N}\sum_{n=1}^{N} e_{t_{nN}}$ (generalized design)
Δ_0	set of all design (probability) measures on $\mathcal{T}$
$\Delta(\varphi)$	set of design measures for which φ is identifiable
$\Delta_N(\varphi)$	set of designs for sample size N for which φ is identifiable
$\Delta_{\mathcal{D}}$	$= \{\delta \in \Delta_0;\ \mathrm{supp}(\delta) = \mathcal{D}\}$
$\Delta^*_{\mathcal{D}}$	$= \{\delta \in \Delta_0;\ \mathrm{supp}(\delta) \subset \mathcal{D}\}$
e_t	one-point (Dirac) measure on t
E_n	$n \times n$ identity matrix
$E(\delta,\beta)$	relative efficiency of δ at β
$E_\theta(\cdot)$	expectation at θ
ϵ	radius of a neighbourhood of the central model
$\epsilon^*_c(\widehat{\zeta}, P_\delta)$	asymptotic breakdown point for contamination neighbourhoods
$\epsilon^*_K(\widehat{\zeta}, P_\delta)$	asymptotic breakdown point for Kolmogorov neighbourhoods
$\epsilon^*(\widehat{\zeta}_N, d_N, y_N)$	breakdown point by substituting observations
$\epsilon^*(\widehat{\zeta}_N, d_N)$	$= \min\{\epsilon^*(\widehat{\zeta}_N, d_N, y_N);\ y_N \in \mathbb{R}^N\}$
$\epsilon^+(\widehat{\zeta}_N, d_N, y_N)$	breakdown point by adding observations
$\epsilon^+(\widehat{\zeta}_N, d_N)$	$= \min\{\epsilon^+(\widehat{\zeta}_N, d_N, y_N);\ y_N \in \mathbb{R}^N\}$
$\tilde{\epsilon}^*(\zeta_N, d_N, y_N)$	breakdown point for contaminated experimental conditions
$\epsilon^1(\tau_N, R_N)$	breakdown point of the first error of a test
$\epsilon^2(\tau_N, R_N, \theta)$	breakdown point of the second error of a test at θ
η	interesting aspect given by $\eta(\beta) = A_{\mathcal{D}}\beta$
$\widehat{\eta}$	estimator for $\eta(\beta) = A_{\mathcal{D}}\beta$

F, F_t	distribution function of P and P_t, respectively
$F_{N,t}$ $F_{\theta,t}$	distribution function of $P_{N,t}$ and $P_{\theta,t}$, respectively
$g(\cdot)$	$= \int \min\{\lvert z\rvert, \cdot\}^2 \, n_{(0,1)}(dz)$
G, G_t	distribution function of Q and Q_t, respectively
$G_{N,t}$ $G_{\theta,t}$	distribution function of $Q_{N,t}$ and $Q_{\theta,t}$, respectively
$H_\gamma(\cdot)$	transformation: $(y,t) \to (y + a(t)'\gamma, t)$
$IF(\widehat{\zeta}, P_{\theta,\delta}, \cdot)$	influence function of $\widehat{\zeta}$ at $P_{\theta,\delta}$
$\mathcal{I}(\delta)$	$= \int a(t)\, a(t)'\, \delta(dt)$, information matrix of δ
$\mathcal{I}_\beta(\delta)$	$= \int \dot{\mu}(t,\beta)\, \dot{\mu}(t,\beta)'\, \delta(dt)$, information matrix of δ at β
$K_{Q_\xi}(\theta)$	$= \int \tilde{\psi}(y,t,\theta)\, Q_\xi(dy,dt)$
L	matrix providing a linear aspect φ, i.e. $\varphi(\beta) = L\beta$
λ	Lebesgue measure on $\mathbb{R}$
$M(\rho)$	$= \int a(t)\, a(t)'\, \rho(z,t)\, z\, P(dz)\, \delta(dt)$
$M(\beta,\rho)$	$= \int \dot{\mu}(t,\beta)\, \dot{\mu}(t,\beta)' \rho(z,t)\, z\, P(dz)\, \delta(dt)$
$M_\beta(\rho,\delta)$	$= M(\beta,\rho)$
$M(\theta)$	derivative of $K_{P_{\theta,\delta}}(\cdot)$ at θ
$\mathcal{M}_\varphi(\delta)$	maximum mass of δ on a nonidentifying set
$\mu(\cdot,\cdot)$	response function on $\mathcal{T} \times \mathbb{R}^r$
$\dot{\mu}(t,\cdot)$	first derivative of $\mu(t,\cdot)$
$\ddot{\mu}(t,\cdot)$	second derivative of $\mu(t,\cdot)$
$n_{(0,1)}$	measure of the standard normal distribution
N	sample size
$\mathcal{N}_\varphi(d_N)$	maximal number of experimental conditions of d_N in a nonidentifying set
$\mathcal{N}(\mu,\sigma^2)$	normal distribution with mean μ and variance σ^2
$\mathbb{N}$	set of all integer numbers
$\mathbb{N}_0$	set of all integer numbers and 0
$\nu(\cdot)$	selection functional for nonunique M-functionals
q	dimension of $\zeta(\theta)$
Q_δ	$= Q \otimes \delta$, error distribution modelling outliers
Q_t	conditional distribution of Q_δ given $t \in \mathcal{T}$
$Q_{\theta,\delta}$	$= Q_\theta \otimes \delta$, distribution of the observations modelling outliers
$Q_{\theta,t}$	conditional distribution of $Q_{\theta,\delta}$ given $t \in \mathcal{T}$
$(Q_{N,\delta})_{N\in\mathbb{N}}$	sequence of outlier modelling error distributions
$(Q_{N,\theta,\delta})_{N\in\mathbb{N}}$	sequence of outlier modelling distributions of the observations
Q^N	$= \bigotimes_{n=1}^N Q_{N,\delta}$ or $= \bigotimes_{n=1}^N Q_{N,t_{nN}}$
Q_θ^N	$= \bigotimes_{n=1}^N Q_{N,\theta,\delta}$ or $= \bigotimes_{n=1}^N Q_{N,\theta,t_{nN}}$
P_δ	$= P \otimes \delta$, ideal error distribution
P_t	conditional distribution of P_δ given $t \in \mathcal{T}$
$P_{\theta,\delta}$	$= P_\theta \otimes \delta$, ideal distribution of the observations
$P_{\theta,t}$	conditional distribution of $P_{\theta,\delta}$ given $t \in \mathcal{T}$

$P_{\beta,\sigma,\delta}$	$= P_{\theta,\delta}$ if $\theta = (\beta', \sigma)'$
P^N	$= \bigotimes_{n=1}^N P_\delta$ or $= \bigotimes_{n=1}^N P_{t_{nN}}$
P_θ^N	$= \bigotimes_{n=1}^N P_{\theta,\delta}$ or $= \bigotimes_{n=1}^N P_{\theta,t_{nN}}$
P_{y_N,d_N}	empirical distribution of (y_N, d_N)
$\mathcal{P}$	set of all probability measures on $\mathbb{R}$ without λ singular part
$\mathcal{P}^*$	set of all probability measures on $\mathbb{R} \times \mathcal{T}$ without λ singular part
$\mathcal{P}_\delta^*$	$= \{Q_\xi \in \mathcal{P}^*;\ \xi = \delta\}$
$\mathcal{P}^*(\text{supp}(\delta))$	$= \{Q_\xi \in \mathcal{P}^*;\ \text{supp}(\xi) \subset \text{supp}(\delta)\}$
$\Phi(\cdot)$	distribution function of the standard normal distribution
$\psi(\cdot)$	score function, influence function
$\psi_\theta(\cdot)$	influence function at θ
$\psi_{\beta,\sigma}(\cdot)$	influence function at $(\beta', \sigma)'$
$\tilde{\psi}(\cdot)$	score function of a M-functional or M-estimator
$\psi_0(\cdot)$	influence function for minimum bias
$\psi_{01}(\cdot)$	influence function of L_1 estimator
$\psi_{b,\delta}$	influence function at δ for bias bound b
$\psi_{\beta,b,\delta}$	influence function of locally optimal AL-estinator at δ for bias bound b
$\Psi(\delta, L)$	set of all influence functions
$\Psi^*(\delta, L)$	set of all influence functions of the form $\psi(\cdot,\cdot) = M\, a(\cdot)\, \rho(\cdot,\cdot)$
r	dimension of β
$\text{rk}(\cdot)$	rank of a matrix
$\mathbb{R}$	set of all real numbers
$\mathbb{R}^r$	set of all r-dimensional real vectors
$\mathbb{R}^{s\times r}$	set of all $s \times r$ matrices
s	dimension of $\varphi(\beta)$
$s^*(\widehat{\zeta}, P_{\theta,\delta})$	gross-error sensitivity of $\widehat{\zeta}$ at $P_{\theta,\delta}$
$\text{sgn}(\cdot)$	signum function
$\text{supp}(\cdot)$	support of a design measure
σ^2	variance
t	experimental condition = design point
$\mathcal{T}$	set of all experimental conditions = design region
t_{nN}	n'th experimental condition of a design d_N
T_{nN}	n'th experimental condition of a random design D_N
$\text{tr}(\cdot)$	trace of a matrix
τ_N^{LS}	test statistic of the classical F-test
τ_N	generalization of τ_N^{LS}
θ	$= (\beta', \sigma)'$ or $= (\beta', \sigma_1, ..., \sigma_J)'$
$\widehat{\theta}_N$	estimator of θ
$\widehat{\theta}$	functional providing an estimator for θ

$\widehat{\theta}_\nu$	M-functional for θ with selection function ν
Θ	set of all θ
$\mathcal{U}_c(P_\delta,\epsilon,c)$	conditional ϵ-contamination neighbourhood of P_δ with contamination curve c
$\mathcal{U}_c(P_\delta,\epsilon)$	ϵ-contamination neighbourhood of P_δ
$\mathcal{U}_c^s(P_\delta,\epsilon)$	simple ϵ-contamination neighbourhood of P_δ
$\mathcal{U}_{c,\epsilon}(P_\delta)$	shrinking ϵ-contamination neighbourhood of P_δ
$\mathcal{U}_{c,\epsilon}^0(P_\delta)$	restricted shrinking ϵ-contamination neighbourhood of P_δ
$\mathcal{U}_K^0(P_\delta,\epsilon)$	restricted Kolmogorov neighbourhood of P_δ with radius ϵ
$\mathcal{U}_K(P_\delta,\epsilon)$	full Kolmogorov neighbourhood of P_δ with radius ϵ
$\mathcal{U}_{K,\epsilon}(P_\delta)$	shrinking Kolmogorov neighbourhood of P_δ
φ	interesting aspect of β
φ^*	given by $\varphi(\beta)=\varphi^*(A_\mathcal{D}\beta)$
$\widehat{\varphi}_N$	estimator of $\varphi(\beta)$
$\widehat{\varphi}_N^{LS}$	Gauss-Markov estimator of $\varphi(\beta)$
$\widehat{\varphi}_{h,p}$	h-trimmed L_p estimator for $\varphi(\beta)$
$\dot{\varphi}_\beta$	derivative of φ with respect to β
$\dot{\varphi}_\beta^*$	derivative of φ^* with respect to η at $\eta=A_\mathcal{D}\beta$
ξ	alternative design measure
y_N	$=(y_{1N},...,y_{NN})'$, realized observation vector
y_{nN}	realized n'th observation
Y_N	$=(Y_{1N},...,Y_{NN})'$, observation vector
Y_{nN}	nth observation
$\mathcal{Y}_M(y_N)$	set of all $\overline{y}_N$, where M observations of y_N are replaced
$\mathcal{Y}_M^+(y_N)$	set of all $\overline{y}_{N+M}$, where M observations are added to y_N
$\mathcal{Y}_M(y_N,d_N)$	set of all $(\overline{y}_N,\overline{d}_N)$, where M observations and corresponding experimental conditions of (y_N,d_N) are replaced
z_N	$=(z_{1N},...,z_{NN})'$, realized error vector
z_{nN}	realized n'th error
Z_N	$=(Z_{1N},...,Z_{NN})'$, error vector
Z_{nN}	n'th error
ζ	interesting aspect of θ
$\widehat{\zeta}$	functional providing estimators for $\zeta(\theta)$
$\widehat{\zeta}_N$	estimator of $\zeta(\theta)$
$\tilde{\zeta}_\theta(\cdot)$	Fréchet derivative of $\widehat{\zeta}$ at $P_{\theta,\delta}$
$\dot{\zeta}(\theta)$	derivative of $\zeta(\theta)$
$\lvert\cdot\rvert$	euclidean norm in $\mathbb{R}$ or $\mathbb{R}^r$
$\lVert\cdot\rVert_\delta$	essential supremum with respect to $n_{(0,1)}\otimes\delta$
$\lfloor z\rfloor$	$\max\{n\in\mathbb{N};\ n\le z\}$

$\cdot'$	a vector or matrix transposed
$\cdot^-$	generalized inverse of a matrix: $M\,M^-\,M = M$
1_r	r-dimensional vector of ones
0_r	r-dimensional vector of zeros
$0_{r\times s}$	$r \times s$ matrix of zeros
#	number of elements of a set

Index

Lecture Notes in Statistics

For information about Volumes 1 to 49
please contact Springer-Verlag

Vol. 50: O.E. Barndorff-Nielsen, Parametric Statistical Models and Likelihood. vii, 276 pages, 1988.

Vol. 51: J. Hüsler, R.-D. Reiss (Editors). Extreme Value Theory, Proceedings, 1987. x, 279 pages, 1989.

Vol. 52: P.K. Goel, T. Ramalingam, The Matching Methodology: Some Statistical Properties. viii, 152 pages, 1989.

Vol. 53: B.C. Arnold, N. Balakrishnan, Relations, Bounds and Approximations for Order Statistics. ix, 173 pages, 1989.

Vol. 54: K.R. Shah, B.K. Sinha, Theory of Optimal Designs. viii, 171 pages, 1989.

Vol. 55: L. McDonald, B. Manly, J. Lockwood, J. Logan (Editors), Estimation and Analysis of Insect Populations. Proceedings, 1988. xiv, 492 pages, 1989.

Vol. 56: J.K. Lindsey, The Analysis of Categorical Data Using GLIM. v, 168 pages, 1989.

Vol. 57: A. Decarli, B.J. Francis, R. Gilchrist, G.U.H. Seeber (Editors), Statistical Modelling. Proceedings, 1989. ix, 343 pages, 1989.

Vol. 58: O.E. Barndorff-Nielsen, P. Bl¾sild, P.S. Eriksen, Decomposition and Invariance of Measures, and Statistical Transformation Models. v, 147 pages, 1989.

Vol. 59: S. Gupta, R. Mukerjee, A Calculus for Factorial Arrangements. vi, 126 pages, 1989.

Vol. 60: L. Gyorfi, W. Härdle, P. Sarda, Ph. Vieu, Nonparametric Curve Estimation from Time Series. viii, 153 pages, 1989.

Vol. 61: J. Breckling, The Analysis of Directional Time Series: Applications to Wind Speed and Direction. viii, 238 pages, 1989.

Vol. 62: J.C. Akkerboom, Testing Problems with Linear or Angular Inequality Constraints. xii, 291 pages, 1990.

Vol. 63: J. Pfanzagl, Estimation in Semiparametric Models: Some Recent Developments. iii, 112 pages, 1990.

Vol. 64: S. Gabler, Minimax Solutions in Sampling from Finite Populations. v, 132 pages, 1990.

Vol. 65: A. Janssen, D.M. Mason, Non-Standard Rank Tests. vi, 252 pages, 1990.

Vol 66: T. Wright, Exact Confidence Bounds when Sampling from Small Finite Universes. xvi, 431 pages, 1991.

Vol. 67: M.A. Tanner, Tools for Statistical Inference: Observed Data and Data Augmentation Methods. vi, 110 pages, 1991.

Vol. 68: M. Taniguchi, Higher Order Asymptotic Theory for Time Series Analysis. viii, 160 pages, 1991.

Vol. 69: N.J.D. Nagelkerke, Maximum Likelihood Estimation of Functional Relationships. V, 110 pages, 1992.

Vol. 70: K. Iida, Studies on the Optimal Search Plan. viii, 130 pages, 1992.

Vol. 71: E.M.R.A. Engel, A Road to Randomness in Physical Systems. ix, 155 pages, 1992.

Vol. 72: J.K. Lindsey, The Analysis of Stochastic Processes using GLIM. vi, 294 pages, 1992.

Vol. 73: B.C. Arnold, E. Castillo, J.-M. Sarabia, Conditionally Specified Distributions. xiii, 151 pages, 1992.

Vol. 74: P. Barone, A. Frigessi, M. Piccioni, Stochastic Models, Statistical Methods, and Algorithms in Image Analysis. vi, 258 pages, 1992.

Vol. 75: P.K. Goel, N.S. Iyengar (Eds.), Bayesian Analysis in Statistics and Econometrics. xi, 410 pages, 1992.

Vol. 76: L. Bondesson, Generalized Gamma Convolutions and Related Classes of Distributions and Densities. viii, 173 pages, 1992.

Vol. 77: E. Mammen, When Does Bootstrap Work? Asymptotic Results and Simulations. vi, 196 pages, 1992.

Vol. 78: L. Fahrmeir, B. Francis, R. Gilchrist, G. Tutz (Eds.), Advances in GLIM and Statistical Modelling: Proceedings of the GLIM92 Conference and the 7th International Workshop on Statistical Modelling, Munich, 13-17 July 1992. ix, 225 pages, 1992.

Vol. 79: N. Schmitz, Optimal Sequentially Planned Decision Procedures. xii, 209 pages, 1992.

Vol. 80: M. Fligner, J. Verducci (Eds.), Probability Models and Statistical Analyses for Ranking Data. xxii, 306 pages, 1992.

Vol. 81: P. Spirtes, C. Glymour, R. Scheines, Causation, Prediction, and Search. xxiii, 526 pages, 1993.

Vol. 82: A. Korostelev and A. Tsybakov, Minimax Theory of Image Reconstruction. xii, 268 pages, 1993.

Vol. 83: C. Gatsonis, J. Hodges, R. Kass, N. Singpurwalla (Editors), Case Studies in Bayesian Statistics. xii, 437 pages, 1993.

Vol. 84: S. Yamada, Pivotal Measures in Statistical Experiments and Sufficiency. vii, 129 pages, 1994.

Vol. 85: P. Doukhan, Mixing: Properties and Examples. xi, 142 pages, 1994.

Vol. 86: W. Vach, Logistic Regression with Missing Values in the Covariates. xi, 139 pages, 1994.

Vol. 87: J. Müller, Lectures on Random Voronoi Tessellations.vii, 134 pages, 1994.

Vol. 88: J. E. Kolassa, Series Approximation Methods in Statistics.viii, 150 pages, 1994.

Vol. 89: P. Cheeseman, R.W. Oldford (Editors), Selecting Models From Data: AI and Statistics IV. xii, 487 pages, 1994.

Vol. 90: A. Csenki, Dependability for Systems with a Partitioned State Space: Markov and Semi-Markov Theory and Computational Implementation. x, 241 pages, 1994.

Vol. 91: J.D. Malley, Statistical Applications of Jordan Algebras. viii, 101 pages, 1994.

Vol. 92: M. Eerola, Probabilistic Causality in Longitudinal Studies. vii, 133 pages, 1994.

Vol. 93: Bernard Van Cutsem (Editor), Classification and Dissimilarity Analysis. xiv, 238 pages, 1994.

Vol. 94: Jane F. Gentleman and G.A. Whitmore (Editors), Case Studies in Data Analysis. viii, 262 pages, 1994.

Vol. 95: Shelemyahu Zacks, Stochastic Visibility in Random Fields. x, 175 pages, 1994.

Vol. 96: Ibrahim Rahimov, Random Sums and Branching Stochastic Processes. viii, 195 pages, 1995.

Vol. 97: R. Szekli, Stochastic Ordering and Dependence in Applied Probability. viii, 194 pages, 1995.
Vol. 98: Philippe Barbe and Patrice Bertail, The Weighted Bootstrap. viii, 230 pages, 1995.

Vol. 99: C.C. Heyde (Editor), Branching Processes: Proceedings of the First World Congress. viii, 185 pages, 1995.

Vol. 100: Wlodzimierz Bryc, The Normal Distribution: Characterizations with Applications. viii, 139 pages, 1995.

Vol. 101: H.H. Andersen, M.Højbjerre, D. Sørensen, P.S.Eriksen, Linear and Graphical Models: for the Multivariate Complex Normal Distribution. x, 184 pages, 1995.

Vol. 102: A.M. Mathai, Serge B. Provost, Takesi Hayakawa, Bilinear Forms and Zonal Polynomials. x, 378 pages, 1995.

Vol. 103: Anestis Antoniadis and Georges Oppenheim (Editors), Wavelets and Statistics. vi, 411 pages, 1995.

Vol. 104: Gilg U.H. Seeber, Brian J. Francis, Reinhold Hatzinger, Gabriele Steckel-Berger (Editors), Statistical Modelling: 10th International Workshop, Innsbruck, July 10-14th 1995. x, 327 pages, 1995.

Vol. 105: Constantine Gatsonis, James S. Hodges, Robert E. Kass, Nozer D. Singpurwalla(Editors), Case Studies in Bayesian Statistics, Volume II. x, 354 pages, 1995.

Vol. 106: Harald Niederreiter, Peter Jau-Shyong Shiue (Editors), Monte Carlo and Quasi-Monte Carlo Methods in Scientific Computing. xiv, 372 pages, 1995.

Vol. 107: Masafumi Akahira, Kei Takeuchi, Non-Regular Statistical Estimation. vii, 183 pages, 1995.

Vol. 108: Wesley L. Schaible (Editor), Indirect Estimators in U.S. Federal Programs. viii, 195 pages, 1995.

Vol. 109: Helmut Rieder (Editor), Robust Statistics, Data Analysis, and Computer Intensive Methods. xiv, 427 pages, 1996.

Vol. 110: D. Bosq, Nonparametric Statistics for Stochastic Processes. xii, 169 pages, 1996.

Vol. 111: Leon Willenborg, Ton de Waal, Statistical Disclosure Control in Practice. xiv, 152 pages, 1996.

Vol. 112: Doug Fischer, Hans-J. Lenz (Editors), Learning from Data. xii, 450 pages, 1996.

Vol. 113: Rainer Schwabe, Optimum Designs for Multi-Factor Models. viii, 124 pages, 1996.

Vol. 114: C.C. Heyde, Yu. V. Prohorov, R. Pyke, and S. T. Rachev (Editors), Athens Conference on Applied Probability and Time Series Analysis Volume I: Applied Probability In Honor of J.M. Gani. viii, 424 pages, 1996.

Vol. 115: P.M. Robinson, M. Rosenblatt (Editors), Athens Conference on Applied Probability and Time Series Analysis Volume II: Time Series Analysis In Memory of E.J. Hannan. viii, 448 pages, 1996.

Vol. 116: Genshiro Kitagawa and Will Gersch, Smoothness Priors Analysis of Time Series. x, 261 pages, 1996.

Vol. 117: Paul Glasserman, Karl Sigman, David D. Yao (Editors), Stochastic Networks. xii, 298, 1996.

Vol. 118: Radford M. Neal, Bayesian Learning for Neural Networks. xv, 183, 1996.

Vol. 119: Masanao Aoki, Arthur M. Havenner, Applications of Computer Aided Time Series Modeling. ix, 329 pages, 1997.

Vol. 120: Maia Berkane, Latent Variable Modeling and Applications to Causality. vi, 288 pages, 1997.

Vol. 121: Constantine Gatsonis, James S. Hodges, Robert E. Kass, Robert McCulloch, Peter Rossi, Nozer D. Singpurwalla (Editors), Case Studies in Bayesian Statistics, Volume III. xvi, 487 pages, 1997.

Vol. 122: Timothy G. Gregoire, David R. Brillinger, Peter J. Diggle, Estelle Russek-Cohen, William G. Warren, Russell D. Wolfinger (Editors), Modeling Longitudinal and Spatially Correlated Data. x, 402 pages, 1997.

Vol. 123: D. Y. Lin and T. R. Fleming (Editors), Proceedings of the First Seattle Symposium in Biostatistics: Survival Analysis. xiii, 308 pages, 1997.

Vol. 124: Christine H. Müller, Robust Planning and Analysis of Experiments, x, 234 pages, 1997.